T0367967

Fracture Mechanics of Electromagnetic Materials

Nonlinear Field Theory and Applications

FRACTURE MECHANICS OF ELECTROMAGNETIC MATERIALS

Nonlinear Field Theory and Applications

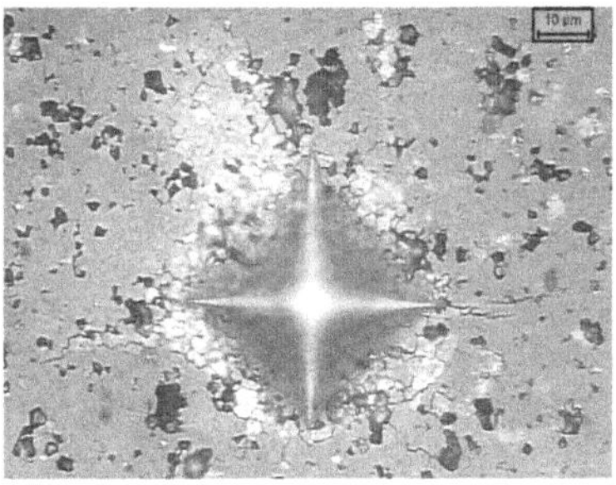

XIAOHONG CHEN
Goodrich Aerostructures, USA

YIU-WING MAI
University of Sydney, Australia

Imperial College Press

ICP

Published by

Imperial College Press
57 Shelton Street
Covent Garden
London WC2H 9HE

Distributed by

World Scientific Publishing Co. Pte. Ltd.
5 Toh Tuck Link, Singapore 596224
USA office: 27 Warren Street, Suite 401-402, Hackensack, NJ 07601
UK office: 57 Shelton Street, Covent Garden, London WC2H 9HE

British Library Cataloguing-in-Publication Data
A catalogue record for this book is available from the British Library.

FRACTURE MECHANICS OF ELECTROMAGNETIC MATERIALS
Nonlinear Field Theory and Applications
Copyright © 2013 by Imperial College Press

All rights reserved. This book, or parts thereof, may not be reproduced in any form or by any means, electronic or mechanical, including photocopying, recording or any information storage and retrieval system now known or to be invented, without written permission from the Publisher.

For photocopying of material in this volume, please pay a copying fee through the Copyright Clearance Center, Inc., 222 Rosewood Drive, Danvers, MA 01923, USA. In this case permission to photocopy is not required from the publisher.

ISBN 978-1-84816-663-9

Printed in Singapore by B & Jo Enterprise Pte Ltd

To my parents for their love, understanding, and support.

Xiaohong Chen

To my mother, Yuet-Yau Tsui, and my wife, Louisa Kit-Ling Mai, whose love and unfailing support have enabled me to pursue my academic dreams for many years!

Yiu-Wing Mai

Foreword

I take pleasure in writing this foreword for the excellent new book entitled *Fracture Mechanics of Electromagnetic Materials: Nonlinear Field Theory and Applications*, written by my former colleagues, Drs. Xiaohong Chen and Yiu-Wing Mai, at the Hong Kong University of Science and Technology.

Novel multifunctional materials have tremendous potential for high-performance structural and functional applications in aeronautical, mechanical, and civil engineering, as well as in microelectronic and biomedical devices, due to their versatile actuating, sensing, healing, and other functional properties. The susceptibility of such advanced materials to cracking in service is of fundamental concern and has become a very popular area of research. Attempts to describe the failure behavior of these advanced materials and structures have clearly shown that linear piezoelectric/piezomagnetic fracture mechanics does not adequately explain the crack growth behavior under combined magnetic, electric, thermal, and mechanical loadings. It appears that coupling and dissipative effects play an important role in the growth and propagation of cracks.

Significant discrepancies still exist between theoretical predictions and experimental observations. Both new and modified theories have been proposed to overcome the discrepancies, with only limited success. These failures may be because there is no comprehensive guide to the theoretical basis and application of nonlinear dynamic fracture mechanics, especially in cases involving multiple coupled fields with dissipation effects.

This book is the first monograph on the subject of nonlinear dynamic field theory for piezoelectric/piezomagnetic materials. It provides an overview of the current state of the art of fracture mechanics and some of the authors' recent research outcomes. In developing the theoretical models for application to fracture characterization of materials and structures in the presence of magneto-electro-thermo-mechanical coupling and dissipative effects, the authors emphasize the physical interpretation of the fundamental concepts of fracture mechanics. One of the book's unique contributions is the development of a nonlinear fracture mechanics theory which rigorously treats the dynamic crack problems involving coupled magnetic, electric, thermal, and mechanical fields. By unifying the coupling of these fields, this book fills a gap in the literature of fracture mechanics involving multifield interactions. This book is a valuable resource which sheds light on the still-developing multidisciplinary subject of multifield fracture mechanics.

The book has an extensive list of references reflecting the most recent developments. It can be used as a textbook for graduate students as well as a reference for researchers and engineers studying and/or applying the concepts of advanced fracture mechanics to design and practical applications in the presence of multifield coupling and dissipative effects.

Pin Tong
San Diego
April, 2012

Preface

This book covers the current status of conventional fracture mechanics methodologies and presents a new formulation of a nonlinear field theory of fracture mechanics for electromagnetic materials. The study of classic fracture mechanics problems is concerned with the mechanical state of a deformable body containing a crack or cracks. Nevertheless, recent advances in multifunctional smart materials have created new research frontiers due to the occurrence of magneto-electro-thermo-mechanical coupling and dissipative effects accompanying crack propagation.

Electromagnetic materials have broad civilian and defense applications such as infrastructure health monitoring, microelectronic packaging, novel antenna designs, and biomedical devices owing to their remarkable multifunctional properties. Fracture of these smart material systems has become the subject of active research because of their susceptibility to cracking in service. A major challenge is how to resolve the fundamental discrepancy between theoretical predictions and experimental observations on the fracture behaviors of piezoelectric and piezomagnetic materials.

A highly important question in the development of a fracture mechanics theory for electromagnetic materials is whether there is any particular thermodynamic quantity of a cracked body that can be interpreted as the "driving force" for crack propagation under combined magneto-electro-thermo-mechanical loadings. The answer to this question has been pursued for decades, but no satisfactory agreement has yet been reached. Thus, the establishment of a physically sound fracture criterion becomes the hallmark of an advanced fracture mechanics treatment for electromagnetic materials.

As the first monograph on the subject of nonlinear field theory of fracture mechanics for deformable electromagnetic materials, this book offers postgraduate students, academic researchers, and engineering specialists who are active in this challenging multidisciplinary area a sketch of the history, an overview of the current status, and a description of some most recent research outcomes based on our own work. It gives first priority to physical interpretation of fundamental concepts, development of theoretical models, and exploration of their applications to fracture characterization in the presence of magneto-electro-thermo-mechanical coupling and dissipative effects. A general formulation of a nonlinear field theory of fracture mechanics and a rigorous treatment of dynamic crack problems involving coupled magnetic, electric, thermal, and mechanical fields fill a gap in the literature.

We would like to express our sincere appreciation and gratitude to those who have provided helpful discussions and support to this book project, especially Professors Pin Tong and Tongyi Zhang (Hong Kong University of Science and Technology), Cun-Fa Gao (Nanjing University of Aeronautics & Astronautics), Baolin Wang (Harbin Institute of Technology), Qinghua Qin (Australian National University), and Meng Lu (CSIRO). XHC is also heartily indebted to the late Professor Ren Wang for his guidance and inspiration during her graduate studies in solid mechanics at Peking University and for his care and encouragement throughout all these years. The Centre for Advanced Materials Technology at the University of Sydney, where XHC previously worked and YWM was Founding Director, has provided an intellectually stimulating environment for advanced fracture mechanics research. Special thanks are due to Lance Sucharov, Tasha D'Cruz, Rajesh Babu, Lindsay Robert Wilson, Gregory Lee, and Romén Reyes-Peschl from Imperial College Press and World Scientific Publishing for their commitments to excellence in publishing this book from proposal review to proofreading. Permissions from professional societies and publishers to use cited materials in the book are also gratefully acknowledged.

Xiaohong Chen & Yiu-Wing Mai
San Diego & Sydney
August, 2012

Contents

List of Tables

List of Figures

Chapter 1

Fundamentals of Fracture Mechanics

Classic fracture mechanics is concerned with the study of the mechanical state of a deformable body containing a crack or cracks by application of analytical mechanics to calculate the driving force for crack propagation and experimental mechanics to characterize the resistance of materials to crack extension. A highly important question in the development of a fracture mechanics theory is whether there is any particular thermodynamic quantity of a cracked body that can be interpreted as the "driving force" for crack propagation.

1.1 Historical Perspective

The establishment of fracture mechanics as an engineering discipline dates back to the early work of Griffith (1921), Orowan (1948) and Irwin (1948, 1956, 1957, 1958). In Griffith's famous paper "The phenomena of rupture and flow in solids" (Griffith, 1921), which quantitatively relates the flaw size to the fracture stress, he proposed an energy balance approach for the fracture of brittle materials with the introduction of the surface energy term by realizing that the relatively low strength and the size dependence of strength were due to the presence of crack-like flaws in the materials.

The Griffith energy balance leads to a critical condition for fracture of an ideal elastic-brittle material:

$$\frac{dW}{dA} - \frac{dU}{dA} = \frac{d\Gamma_s}{dA}, \tag{1.1}$$

where A is the crack area, W is the work done on the cracked body by external forces, U is the strain energy stored in the system, and Γ_s is the surface energy.

For a through-thickness crack with length $2a$ in an infinite plate under remote tensile stress σ (Fig. 1.1), Griffith (1921, 1924) used the solution of Inglis (1913) to show that the fracture stress, σ_f, is given by

$$\sigma_f = \sqrt{\frac{2E'\gamma_s}{\pi a}}, \tag{1.2}$$

where $E' = E$ for plane stress and $E' = E/(1-\nu^2)$ for plane strain, E is Young's modulus, ν is Poisson's ratio, and $\gamma_s[=(1/2)d\Gamma_s/dA]$ is the specific surface energy.

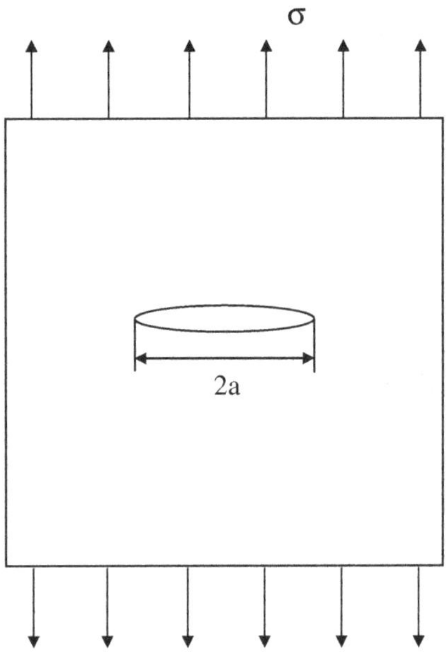

Fig. 1.1. Griffith crack with length $2a$ under remote tensile stress σ.

Although the Griffith energy balance approach provides excellent agreement with experimental data for brittle materials such as glass, the surface energy predicted by Griffith's fracture criterion is usually unrealistically high for ductile materials such as steel. Irwin (1948) and

Orowan (1948) independently modified Griffith's fracture criterion to account for the energy dissipated by local plastic flow. Under small-scale yielding (SSY) conditions, the plastic work required to grow the crack is a material constant that can be added to the surface energy. The modified expression is given by

$$\sigma_f = \sqrt{\frac{2E'(\gamma_s + \gamma_p)}{\pi a}}, \tag{1.3}$$

where γ_p is the plastic work dissipated during the creation of a unit area of crack surface, which is typically orders of magnitude larger than the specific surface energy γ_s.

It is also feasible to extend the modified model to account for any type of energy dissipation, that is,

$$\sigma_f = \sqrt{\frac{E'R}{\pi a}}, \tag{1.4}$$

where R is the crack resistance, including viscoelastic or viscoplastic effects, depending on the material type.

Subsequently, Irwin (1956, 1957, 1958) found a way to relate the global amount of energy available for fracture to the local crack tip parameter called the stress intensity factor. Linear elastic fracture mechanics (LEFM) is also known as the Griffith–Irwin–Orowan theory because of their leading roles in its establishment. When large-scale inelastic deformation or a significant amount of crack growth occurs, nonlinear approaches must be adopted instead. Rice (1968) developed a path-independent line integral called the *J*-integral, which has dominated the development of nonlinear fracture mechanics (NLFM) in the USA. In the meantime, Wells (1961, 1963) advanced an alternative approach by employing the crack opening displacement (COD) as the fracture parameter, which has guided fracture mechanics research under general yielding conditions in the UK and Europe.

1.2 Stress Intensity Factors (SIF)

Irwin (1957, 1958) and Williams (1957) realized that the stresses near a crack tip in a linear elastic solid have an inverse square-root singularity, that is, they are inversely proportional to the square root of the distance from the crack tip. The near-tip fields in plane elasticity problems are associated with three basic modes, shown in Fig. 1.2. Mode I is the opening (tensile) mode where the displacements are normal to the plane of the crack surface, mode II is the sliding (in-plane shear) mode where the displacements are parallel to the plane of the crack surface and normal to the crack front, and mode III is the tearing (out-of-plane shear) mode where the displacements are parallel to the plane of the crack surface and parallel to the crack front.

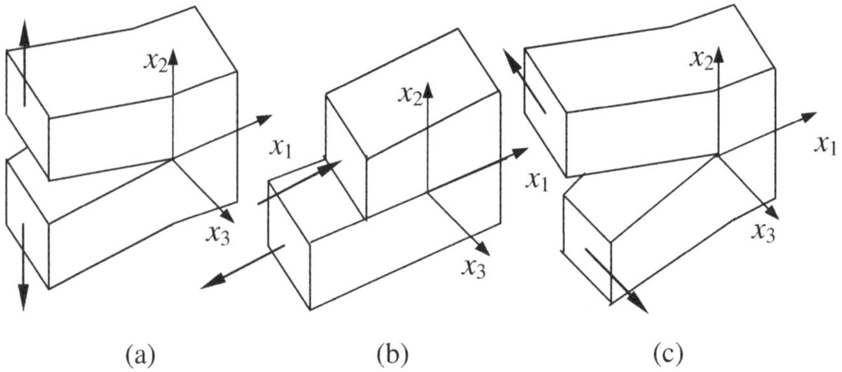

Fig. 1.2. Three fracture modes: (a) mode-I crack – opening mode, (b) mode-II crack – sliding mode, and (c) mode-III crack – tearing mode.

The asymptotic expressions for the near-tip stress fields under mode-I, II, or III fracture are given by

$$\begin{cases} \sigma_{11} = \dfrac{K_I}{\sqrt{2\pi r}} \cos(\theta/2)[1-\sin(\theta/2)\sin(3\theta/2)] \\ \sigma_{22} = \dfrac{K_I}{\sqrt{2\pi r}} \cos(\theta/2)[1+\sin(\theta/2)\sin(3\theta/2)], \\ \sigma_{12} = \dfrac{K_I}{\sqrt{2\pi r}} \cos(\theta/2)\sin(\theta/2)\cos(3\theta/2) \end{cases} \tag{1.5}$$

$$\begin{cases} \sigma_{11} = -\dfrac{K_{II}}{\sqrt{2\pi r}} \sin(\theta/2)[2 + \cos(\theta/2)\cos(3\theta/2)] \\ \sigma_{22} = \dfrac{K_{II}}{\sqrt{2\pi r}} \sin(\theta/2)\cos(\theta/2)\cos(3\theta/2) \\ \sigma_{12} = \dfrac{K_{II}}{\sqrt{2\pi r}} \cos(\theta/2)[1 - \sin(\theta/2)\sin(3\theta/2)] \end{cases}, \tag{1.6}$$

$$\begin{cases} \sigma_{13} = -\dfrac{K_{III}}{\sqrt{2\pi r}} \sin(\theta/2) \\ \sigma_{23} = \dfrac{K_{III}}{\sqrt{2\pi r}} \cos(\theta/2) \end{cases}, \tag{1.7}$$

where r, θ are polar coordinates, K_I, K_{II} and K_{III} are mode-I, II, and III stress intensity factors with units of $\mathrm{MPa}\sqrt{\mathrm{m}}$.

For the classic Griffith crack, the stress intensity factor is given by

$$K_I = \sigma\sqrt{\pi a}\,. \tag{1.8}$$

Since failure at the crack tip in a linear elastic solid is driven solely by the stress intensity factor, a fracture criterion based on the stress intensity factor approach can be expressed as

$$K = K_c\,, \tag{1.9}$$

where K_c is the critical stress intensity factor as a measure of material resistance to fracture, which is called the fracture toughness.

The American Society for Testing and Materials (ASTM) standards E399 and D5045 describe the experimental procedure for measurement of fracture toughness of metallic and plastic specimens, respectively. The size requirement for obtaining a valid measurement of K_{IC} is given by

$$B, a, W - a > 2.5\left(\frac{K_{IC}}{\sigma_y}\right)^2, \tag{1.10}$$

where B is the specimen thickness, a is the crack length, W is the specimen width, and σ_y is the yield strength.

1.3 Energy Release Rate (ERR)

Griffith (1921) was the first to propose the energy approach for fracture of brittle materials, but Irwin (1948, 1956) was primarily responsible for defining the present version of the energy release rate (also referred to as the strain energy release rate), G,

$$G = -\frac{d\Pi}{dA} = -\left(\frac{\partial U}{\partial A}\right)_{\Delta}, \tag{1.11}$$

where Π is the potential energy of the system, U is the strain energy stored in the system, and Δ is the load-point displacement.

Irwin (1957) showed that the energy release rate for a planar crack in a linear elastic body subjected to mixed-mode loading is related to mode-I, II, and III stress intensity factors by performing crack closure analysis:

$$\begin{aligned} G &= \lim_{\delta a \to 0} \frac{1}{2\delta a} \int_a^{a+\delta a} \sigma_{2i}(x_1, 0)\Delta u_i(x_1 - \delta a)dx_1 \\ &= \frac{1}{E'}(K_I^2 + K_{II}^2) + \frac{1+\nu}{E} K_{III}^2, \end{aligned} \tag{1.12}$$

where Δ denotes the jump between the upper and lower surfaces of the crack.

Crack initiation occurs when G reaches a critical value, G_c,

$$G = G_c. \tag{1.13}$$

The energy release rate, also referred to as the crack extension force, provides the thermodynamic driving force for fracture. The onset of crack extension is determined by (1.13), but crack growth may be stable or unstable depending on how the crack driving force and the crack resistance vary with crack extension (Atkins and Mai, 1985; Cotterell and Mai, 1996). In general, the conditions for stable crack growth can be expressed as

$$G = G_R, \tag{1.14}$$

$$\frac{dG}{dA} \le \frac{dG_R}{dA}. \tag{1.15}$$

Unstable crack growth occurs when

$$\frac{dG}{dA} > \frac{dG_R}{dA}. \tag{1.16}$$

A plot of G versus crack extension gives the crack driving force curve, whereas a plot of G_R versus crack extension gives the crack growth resistance curve (Fig. 1.3). The transition from stable to unstable fracture occurs when the crack driving force curve is tangent to the crack growth resistance curve (R curve).

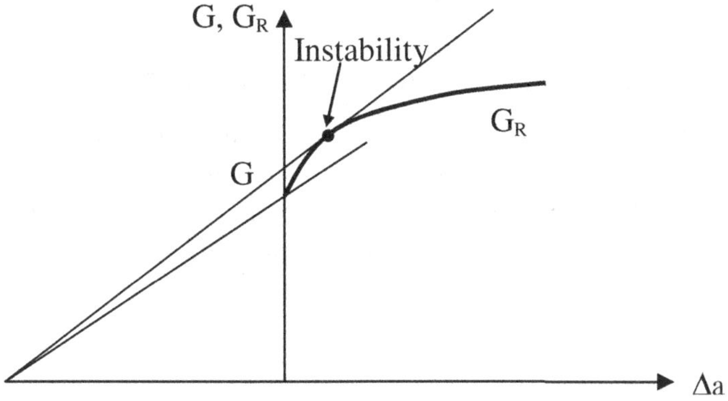

Fig. 1.3. Illustration of crack driving force and resistance curves.

1.4 *J*-Integral

Path-independent integrals have been extensively employed to study bodies with defects or cracks since the pioneering work of Eshelby (1951, 1956, 1970, 1975), Cherepanov (1967, 1968, 1979) and Rice (1968). As indicated by Kannimen and Popelar (1985), the number of path-independent integrals appears to be unlimited. The energy release rate in a nonlinear elastic body containing a crack can be expressed as a contour integral called the J-integral (Rice, 1968). The J-integral has also been related to the crack-tip stress fields in a power-law hardening material (Hutchinson, 1968; Rice and Rosengren, 1968). The path-independent I-integral formulated from the complementary energy density (Bui, 1974) can be taken as the dual of the J-integral. Further discussions on the J-integral and other invariant integrals can be found in the papers by Knowles and Sternberg (1972), Kishimoto *et al.* (1980), Atluri (1982), Atluri *et al.* (1984), Freund and Hutchinson (1985), Li *et al.* (1985), Shih *et al.* (1986), Moran and Shih (1987a–b), and Simo and Honein (1990), among others.

Consider a path Γ in a nonlinear elastic body extended counter clockwise from the lower crack face to the upper crack face, as shown in Fig. 1.4. The *J*-integral is defined as

$$J = \int_\Gamma (w dx_2 - \sigma_{ij} n_j \frac{\partial u_i}{\partial x_1} ds) , \tag{1.17}$$

where $w = \int_0^{\varepsilon_{ij}} \sigma_{ij} d\varepsilon_{ij}$ is the strain energy density, σ_{ij} are the components of the Cauchy stress tensor, ε_{ij} are the components of the infinitesimal strain tensor, n_j are the components of the unit outer normal vector, u_i are the components of the displacement vector, *ds* is the length increment along the path Γ, and the x_2-direction is perpendicular to the crack line.

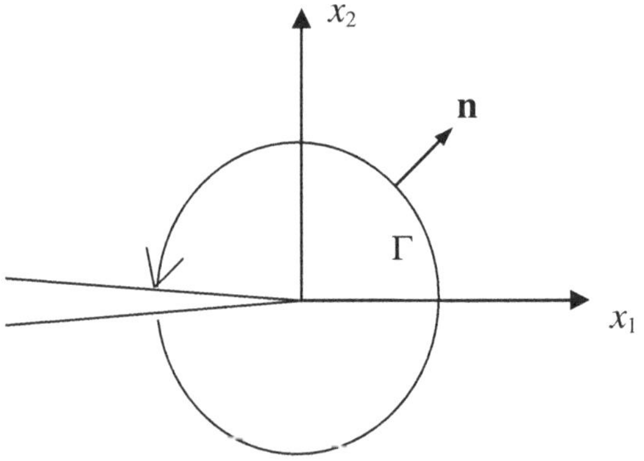

Fig. 1.4. Contour for the *J*-integral.

Most importantly, Rice (1968) showed that, for deformation plasticity (i.e., nonlinear elasticity), the value of the *J*-integral in a two-dimensional cracked body free of body forces is independent of the path around the crack, provided it is taken outside the fracture process zone. The *J*-integral criterion for crack initiation gives

$$J = J_c . \tag{1.18}$$

The dimensionless tearing moduli (Paris *et al.*, 1979; Atkins and Mai, 1985; Anderson, 2005) may be introduced by

$$T_{app} = \frac{E}{\sigma_0^2}\frac{dJ}{da}, \tag{1.19}$$

$$T_R = \frac{E}{\sigma_0^2}\frac{dJ_R}{da}, \tag{1.20}$$

where σ_0 is an appropriate flow stress.

The conditions for stable crack growth can be expressed as

$$J = J_R, \tag{1.21}$$

$$T_{app} \leq T_R. \tag{1.22}$$

Unstable crack growth occurs when

$$T_{app} > T_R. \tag{1.23}$$

The *J*-integral method has great success in nonlinear fracture mechanics (NLFM). Nevertheless, the *J*-integral might lose its path independence when there is a combination of significant plasticity and crack growth (Kanninen and Popelar, 1985; Anderson, 2005).

1.5 Dynamic Fracture

A comprehensive overview on the development of dynamic fracture mechanics, which is concerned with fracture problems in which either the applied load or the crack size changes rapidly, can be found in the monographs by Freund (1990) and Ravi-Chandar (2004). Linear elastodynamic fracture mechanics is the dynamic version of LEFM, incorporating inertia effects but neglecting nonlinear material behavior. A review paper by Cox *et al.* (2005) focuses on modern topics and challenges in dynamic fracture. For example, hyperelasticity may play an important role in the dynamics of fracture where the linear elastic theory is incapable of fully capturing all failure phenomena (Geubelle and Knauss, 1995; Geubelle, 1995; Gao, 1996; Abraham *et al.*, 1997; Buehler *et al.*, 2003; Tarantino, 2005).

The first important dynamic crack propagation analysis was conducted by Yoffe (1951) for the problem of a moving Griffith crack of fixed length gliding through an otherwise unbounded solid at constant speed with the crack opening at the leading edge and closing at the trailing edge. This is referred to as the Yoffe problem. The next important dynamic crack propagation solutions were contributed by Craggs (1960) and Broberg (1960). Craggs (1960) considered the extension of a semi-infinite crack at constant speed with the crack face loading moving with the same speed as the crack tip. The Broberg problem is that of self-similar expansion of a crack from zero initial length at constant speed under uniform remote tension. While these models were not physically realistic, they provided an indication of the influence of the crack speed on the stress state near the moving crack tip.

In a series of papers, Freund (1972b–c, 1973, 1974a–b) provided results for dynamic crack growth in an elastic solid subjected to general loading. The asymptotic stress fields near a moving crack tip in linear elastic materials still have the inverse square-root singularity and are generally expressed as

$$\sigma_{ij}(t) = \frac{\tilde{K}_I(t)}{\sqrt{2\pi r}}\Sigma_{ij}^{I}(\theta, V_C) + \frac{\tilde{K}_{II}(t)}{\sqrt{2\pi r}}\Sigma_{ij}^{II}(\theta, V_C) + \frac{\tilde{K}_{III}(t)}{\sqrt{2\pi r}}\Sigma_{ij}^{III}(\theta, V_C), \quad (1.24)$$

where the functions $\Sigma_{ij}^{I}(\theta, V_C)$, $\Sigma_{ij}^{II}(\theta, V_C)$, and $\Sigma_{ij}^{III}(\theta, V_C)$ describe angular variation for any value of crack tip velocity V_C for the cases of mode-I, II, and III crack growth (Freund, 1990).

The mode-I and II dynamic stress intensity factors, $\tilde{K}_I$ and $\tilde{K}_{II}$, tend to zero as the crack velocity approaches the Rayleigh wave speed, whereas the mode-III dynamic stress intensity factor, $\tilde{K}_{III}$, tends to zero as the crack velocity approaches the shear wave speed. The dependence of dynamic propagation toughness on crack speed, loading rate and temperature can be measured by means of photoelasticity and caustics.

Mott (1947) realized that the inertia effect on crack advance could become significant at high crack speed and did the first energetic balance analysis of a dynamic crack. An extension of the Griffith energy balance approach to dynamic fracture problems with inclusion of the kinetic energy, E^k, over the cracked body leads to the following expression for the dynamic energy release rate:

$$\tilde{G} = \frac{dW}{dA} - \frac{dU}{dA} - \frac{dE^k}{dA}. \tag{1.25}$$

Consider a two-dimensional body $\tilde{A}$ that contains an extending crack (Fig. 1.5). A contour $\tilde{\Gamma}$ enclosing the crack tip translates with the crack tip moving at instantaneous speed $\mathbf{V}_C$. A dynamic contour integral (Atkinson and Eshelby, 1968; Freund, 1972a, 1990) is given by

$$\tilde{J}_{\tilde{\Gamma}} = \frac{1}{V_C} \int_{\tilde{\Gamma}} [\sigma_{ij} n_j \dot{u}_i + (w + \rho \hat{k}) V_C n_1] d\tilde{\Gamma}, \tag{1.26}$$

where $w = \int_{-\infty}^{t} \sigma_{ij} \dot{u}_{i,j} dt'$ is the stress work density, $\rho \hat{k} = \rho \dot{u}_i \dot{u}_i / 2$ is the kinetic energy density, and $V_C = |\mathbf{V}_C|$ is the magnitude of the crack speed.

The dynamic contour integral given by Eq. (1.26) is generally not path independent in elastodynamics. For quasi-static crack problems, the dynamic contour integral is reduced to the conventional *J*-integral.

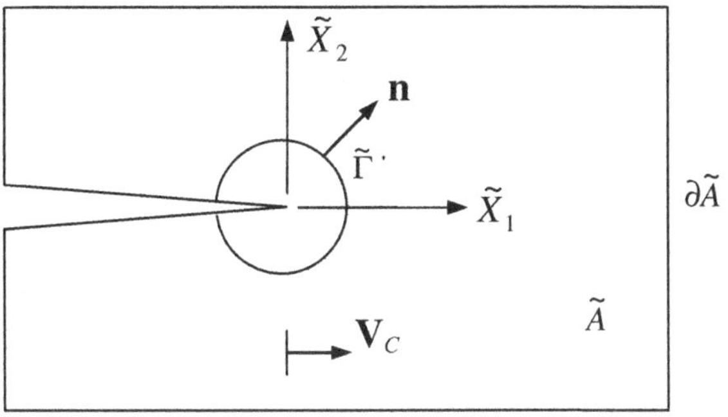

Fig. 1.5. A two-dimensional body $\tilde{A}$ containing an extending crack with contour $\tilde{\Gamma}$, translating with the crack tip moving at instantaneous speed $\mathbf{V}_C$.

The dynamic energy release rate is the rate of energy flow out of the body and into the crack tip per unit crack advance, that is,

$$\tilde{G} = \tilde{J}_0 = \lim_{\tilde{\Gamma} \to 0} \left\{ \frac{1}{V_C} \int_{\tilde{\Gamma}} [\sigma_{ij} n_j \dot{u}_i + (w + \rho \hat{k}) V_C n_1] d\tilde{\Gamma} \right\}, \tag{1.27}$$

where the limit implies that $\tilde{\Gamma}$ is shrunk onto the crack tip.

For the special case of steady-state crack growth, the displacement field $u_i(X_1, X_2, t)$ is invariant in the reference frame affixed to the crack tip moving at uniform speed $\mathbf{V}_C$, that is, $u_i(X_1, X_2, t) = u_i(\tilde{X}_1, \tilde{X}_2)$, where $\tilde{X}_1 = X_1 - V_C t$ and $\tilde{X}_2 = X_2$. Thus, the dynamic contour integral takes the special form

$$\tilde{J}_{\tilde{\Gamma}} = \int_{\tilde{\Gamma}} [(w + \rho\hat{k}) d\tilde{X}_2 - \sigma_{ij} n_j \frac{\partial u_i}{\partial \tilde{X}_1} d\tilde{\Gamma}]. \tag{1.28}$$

A particular choice of the contour $\tilde{\Gamma}$ (see Fig. 1.6) enables the generalization of the Irwin relationship (1.12) to the dynamic case. If the contour $\tilde{\Gamma}$ is shrunk onto the crack tip by first letting $\delta_2 \to 0$ and then $\delta_1 \to 0$, there is no contribution to the dynamic energy release rate from the segments parallel to the $\tilde{X}_2$-axis. Consequently, the dynamic energy release rate can be computed by evaluating only the first term on the right-hand side of Eq. (1.27) along the segments parallel to the $\tilde{X}_1$-axis, that is,

$$\tilde{G} = \tilde{J}_0 = \frac{2}{V_C} \lim_{\delta_1 \to 0} \lim_{\delta_2 \to 0} \int_{-\delta_1}^{\delta_1} \sigma_{2j}(\tilde{X}_1, \delta_2, t) \dot{u}_j(\tilde{X}_1, \delta_2, t) d\tilde{X}_1, \tag{1.29}$$

where the factor 2 is introduced to account for the sides of the rectangle at $\tilde{X}_2 = -\delta_2$, by symmetry.

Thus, the dynamic energy release rate can be related to the dynamic stress intensity factors by

$$\tilde{G} = \tilde{J}_0 = \frac{1}{E'} [A_I(V_C) \tilde{K}_I^2 + A_{II}(V_C) \tilde{K}_{II}^2] + \frac{1+\nu}{E} A_{III}(V_C) \tilde{K}_{III}^2, \tag{1.30}$$

where A_I, A_{II}, and A_{III} are universal functions of crack speed and material properties (Freund, 1990).

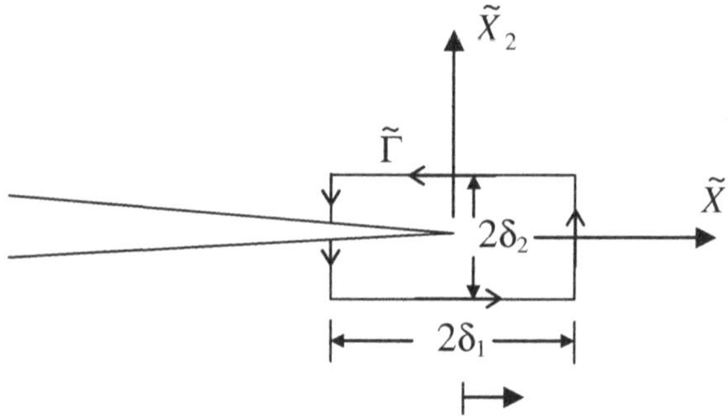

Fig. 1.6. A convenient selection of the contour $\tilde{\Gamma}$.

1.6 Viscoelastic Fracture

Increasing interest in engineering applications of polymeric materials has consequently stimulated the study of viscoelastic fracture mechanics which incorporates a time-dependent response. Willis (1967) first considered anti-plane, steady-state crack propagation in an infinite viscoelastic medium. Later, Atkinson and Popelar (1979) and Popelar and Atkinson (1980) investigated dynamic crack propagation in a viscoelastic strip subjected to mode-I and III loadings. Knauss (1970, 1973, 1974), McCartney (1977), and Christensen (1979, 1982) explored the possibilities of generalizing the Griffith energy balance approach to viscoelastic fracture problems. Discussions on viscoelastic fracture criteria occurred between Christensen and McCartney (Christensen, 1980, 1981; McCartney, 1980, 1981; Christensen and McCartney, 1983), leading to further studies on this subject (Nikitin, 1984). Meanwhile, Schapery (1975a–c) developed a crack-tip model for viscoelastic materials based on an assumption of the material behavior in the fracture process zone. Schapery (1984) also developed correspondence principles and a generalized J-integral for large deformation and fracture analysis of viscoelastic media. An overview of the constitutive equations, fracture and strength models for nonlinear viscoelastic solids can be found in Schapery (2000).

Fracture characterization of polymers, polymer blends, and composites in hygrothermal environments is essential for safety assessment and life prediction for many practical applications. The major challenge lies in the coupling among thermal, mechanical and other physicochemical effects involved in time-dependent fracture. For example, experimental and analytical studies have shown evidence of a large temperature rise in the vicinity of a fast moving crack tip in metals and polymers (Weichert and Schonert, 1978; Maugin, 1992; Kotousov, 2002). This temperature rise may participate in governing the fast-running crack through influencing the energy release rate. Diffusion adds another complexity to time-dependent fracture problems. High stress and temperature gradients associated with the crack tip lead to thermally- and mechanically-enhanced fluid transport, which, in turn, affect the energy release rate. For polymeric materials subjected to combined mechanical loading and hygrothermal exposure, Chen (2007) developed a consistent thermodynamic formulation of a coupled hygro-thermo-viscoelastic fracture theory from the global energy balance equation and the non-negative global dissipation requirement based on fundamental principles of thermodynamics.

An extension of the Griffith–Irwin–Orowan theory to hygro-thermo-viscoelastic fracture problems, with incorporation of the coupling and dissipative effects, leads to the expression

$$\begin{aligned}\frac{dW}{dt}-\frac{dH}{dt}-\frac{dE^k}{dt}=\tilde{G}\dot{A}+\int_V \rho\hat{\Lambda}dV+\int_V \rho\hat{s}\dot{T}dV\\ -\int_V \rho(\hat{\mu}^{(f)}-\hat{\mu}^{(s)})\dot{c}^{(f)}dV\\ +\int_V \mathbf{j}_m^{(f)}\cdot(\hat{\mathbf{f}}^{(f)}-\hat{\mathbf{f}}^{(s)}-\dot{\mathbf{v}}^{(f)}+\dot{\mathbf{v}}^{(s)})dV,\end{aligned} \tag{1.31}$$

where $\tilde{G}$ is the generalized energy release rate serving as the thermodynamic driving force conjugate to the crack variable A, H is the Helmholtz free energy over the cracked body, $\rho=\rho^{(f)}+\rho^{(s)}$ is the total mass density, $\rho^{(f)}$ and $\rho^{(s)}$ are the densities of the fluid and solid phases, $\hat{\Lambda}$ is the rate of viscous dissipation per unit mass, T is the absolute temperature, $\hat{s}$ is the entropy per unit mass, $\hat{\mu}^{(f)}$ and $\hat{\mu}^{(s)}$ are the chemical potentials of the fluid and solid phases, $c^{(f)}=\rho^{(f)}/\rho$ is the mass fraction of the fluid phase, $\mathbf{j}_m^{(f)}$ is the mass flux of the fluid phase, $\hat{\mathbf{f}}^{(f)}$ and $\hat{\mathbf{f}}^{(s)}$ are body forces acting on the fluid and solid phases, and $\dot{\mathbf{v}}^{(f)}$ and $\dot{\mathbf{v}}^{(s)}$ are the accelerations of the fluid and solid phases.

Under isothermal conditions in the absence of fluid diffusion, Eq. (1.31) is reduced to the global energy balance approach for viscoelastic fracture (Knauss, 1970; McCartney, 1977; Christensen, 1979, 1982):

$$\frac{dW}{dt}-\frac{dH}{dt}-\frac{dE^k}{dt}=\tilde{G}\dot{A}+\int_V \rho\hat{\Lambda}dV\,. \tag{1.32}$$

For a two-dimensional crack problem, the generalized contour integral is related to the energy flux through the contour $\tilde{\Gamma}$ affixed to the crack tip moving at instantaneous speed $\mathbf{V}_C$ by

$$\tilde{J}_{\tilde{\Gamma}}=\frac{1}{V_C}\int_{\tilde{\Gamma}}[\sigma_{ij}\dot{u}_i+\tilde{\rho}(\hat{h}+\hat{k})V_C\delta_{1j}]n_j d\tilde{\Gamma}\,, \tag{1.33}$$

where $\hat{h}$ is the Helmholtz free energy per unit mass and $\hat{k}$ is the kinetic energy per unit mass.

The difference between the generalized contour integrals along the paths $\tilde{\Gamma}_1$ and $\tilde{\Gamma}_2$ is caused by unsteady, viscous, thermal and hygroscopic effects as well as the total body force, that is,

$$\begin{aligned}\tilde{J}_{\tilde{\Gamma}_2} - \tilde{J}_{\tilde{\Gamma}_1} = \frac{1}{V_C}[\int_{\tilde{A}_{12}} \frac{\tilde{\partial}}{\tilde{\partial} t}(\tilde{\rho}\hat{h} + \tilde{\rho}\hat{k})d\tilde{A} - \int_{\tilde{A}_{12}} \tilde{\rho}\hat{\mathbf{f}} \cdot \mathbf{v} d\tilde{A} + \int_{\tilde{A}_{12}} \tilde{\rho}\hat{\Lambda} d\tilde{A} \\ + \int_{\tilde{A}_{12}} \tilde{\rho}\hat{s}\dot{T} d\tilde{A} - \int_{\tilde{A}_{12}} \tilde{\rho}(\hat{\mu}^{(f)} - \hat{\mu}^{(s)})\dot{c}^{(f)} d\tilde{A}],\end{aligned} \tag{1.34}$$

where $\tilde{A}_{12}$ is the difference in the areas enclosed by the contours $\tilde{\Gamma}_1$ and $\tilde{\Gamma}_2$ including the crack faces, and $\hat{\mathbf{f}} = (\rho^{(f)}\hat{\mathbf{f}}^{(f)} + \rho^{(s)}\hat{\mathbf{f}}^{(s)}) / \rho$ is the total body force per unit mass.

The generalized energy release rate is the rate of energy flow out of the body and into the crack tip per unit crack advance, that is,

$$\tilde{G} = \tilde{J}_0 = \lim_{\tilde{\Gamma}\to 0}\left\{\frac{1}{V_C}\int_{\tilde{\Gamma}}[\sigma_{ij}\dot{u}_i + \tilde{\rho}(\hat{h} + \hat{k})V_C\delta_{1j}]n_j d\tilde{\Gamma}\right\}. \tag{1.35}$$

For quasi-static and dynamic fracture characterization of elastic materials, Eq. (1.33) is reduced to the conventional J-integral and dynamic contour integral, respectively. Without accounting for fluid diffusion, Schapery's crack-tip model (Schapery, 1975) relies on a special form of Eq. (1.35).

The generalized energy release rate method and the generalized contour integral method should give consistent results, independent of material systems, loading and environmental conditions. An experimental study by Frassine and Pavan (1990) has verified that the observed behavior of an elastomeric epoxy resin is in qualitative agreement with the theoretical predictions by the global and local approaches for viscoelastic fracture, which are the special cases of the generalized energy release rate method and the generalized contour integral method presented here. Another experimental and numerical investigation on crack propagation in carbon/epoxy composite (Gamby and Delaumenie, 1993; Gamby and Chaoufi, 1999) has also demonstrated the agreement between Christensen's model and Schapery's model.

1.7 Essential Work of Fracture (EWF)

Fracture characterization for new ductile materials, such as polymeric thin films, toughened polymers and polymer blends, has greatly stimulated the development of fracture mechanics, which, in turn, plays an important role in design and safety evaluation with an optimum combination of stiffness, strength and toughness. Energy release rate and stress intensity factor in LEFM are widely used to characterize fracture toughness of glassy polymeric materials under brittle fracture (Atkins and Mai, 1985). If plastic flow occurs, the energy approach becomes more complicated. The J-integral (Rice, 1968) based upon deformation plasticity is used as an alternative.

Nevertheless, crack advance in an elastoplastic material involves elastic unloading and nonproportional loading around the crack tip, neither of which can be adequately accommodated by deformation plasticity. Hence, the J-integral theory might break down for a combination of significant plasticity and crack growth. In addition, it is difficult and cumbersome to use for the evaluation of impact fracture toughness. Similarly, the J-integral testing procedure for fracture toughness characterization of polymeric thin films is cumbersome. Accordingly, a simple yet elegant method, i.e., the essential work of fracture (EWF) method, was developed by Cotterell, Mai and co-workers (Cotterell and Reddel, 1977; Mai and Cotterell, 1980, 1986; Mai *et al.*, 2000) from the unified theory of fracture (Broberg, 1971, 1975). It has been adopted by many research groups for the experimental measurement of fracture toughness for thin metal sheets, polymeric thin films, toughened plastics, and blends. A European Structural Integrity Society (ESIS) Test Protocol for Essential Work of Fracture has also been established (1997). The advantage of this technique lies in its experimental simplicity and ease of test data analysis.

The general concept of the EWF Method for toughness measurement is demonstrated in Fig. 1.7. There exists an inner autonomous zone, which is crucial to the fracture process, called the fracture process zone (FPZ). As crack growth is accompanied by permanent deformation of the surrounding material, plastic dissipation in the outer region is not directly associated with the crucial fracture process. The total work of fracture,

W_f, can be partitioned into the essential work imported into the fracture process zone (a material property) and the nonessential work absorbed by the outer plastic zone (geometry-dependent), that is,

$$W_f = W_e + W_p. \tag{1.36}$$

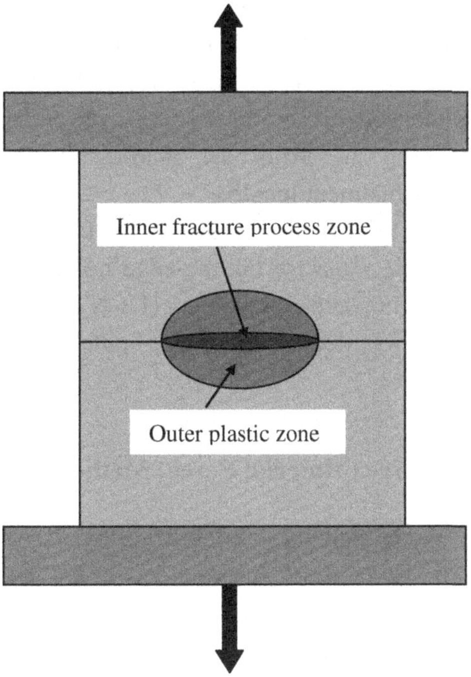

Fig. 1.7. A schematic of a double-edge notched tensile specimen, showing the inner fracture process zone and the outer plastic zone.

The specific essential work of fracture can be conveniently determined using deep-crack specimens, where the height of the outer plastic zone may be proportional to the ligament length. Hence, the essential work of fracture is proportional to the ligament length and the nonessential work of fracture is proportional to the square of the ligament length, leading to the expression

$$w_f = w_e + \beta w_p l, \tag{1.37}$$

where $w_f = (W_f / Bl)$ is the specific total work of fracture, w_e is the specific essential work of fracture, w_p is the specific nonessential work of fracture, and β is a geometry-dependent plastic-zone shape factor.

On the assumption that w_e is a material property and w_p and β are independent of l in all tested specimens, there should be a linear relation when w_f is plotted against l according to Eq. (1.37). By extrapolation of this line to zero ligament length, the intercept at the Y-axis and the slope of the line give w_e and βw_p, respectively. Therefore, Eq. (1.37) provides a sound theoretical basis for a simple yet elegant experimental method of determining w_e from the load-displacement curves for specimens of different ligament lengths.

Mai and Cotterell (1986) also showed the following equivalence: $w_e = J_c$, $\beta w_p = (1/4) dJ_R / da$ for double-edge notched tension (DENT) and deep center notched tension (DCNT) specimens, and $\beta w_p = (1/2) dJ_R / da$ for deep single-edge notched tension (DSEN) specimens.

1.8 Configuration Force (Material Force) Method

The notion of the Newtonian force is clarified by its role in describing the motion of a body. By contrast, the concepts of the energy-momentum tensor (also referred to as the Eshelby stress tensor) and the configuration force (also referred to as the material force) are introduced in the interpretation of the evolution of material microstructures such as defects (Eshelby, 1951, 1956, 1970). The nature of the configuration force and its application to fracture have been discussed by Stumpf and Le (1990), Maugin and Trimarco (1992), Maugin and Berezovski (1999), Gurtin (2000), Steinmann (2000), Steinmann *et al.* (2001), and Nguyen *et al.* (2005), among others.

Eshelby (1970) recognized the use of the energy-momentum tensor in the J-integral, that is,

$$J = \int_\Gamma \mathbf{n} \cdot \mathbf{b} d\Gamma \cdot \overline{\mathbf{e}}_1 = \mathbf{J} \cdot \overline{\mathbf{e}}_1 , \tag{1.38}$$

where $\mathbf{b} = w\mathbf{I} - \boldsymbol{\sigma} \cdot \mathbf{u}\nabla$ is the energy-momentum tensor, $\mathbf{J} = \int_\Gamma \mathbf{n} \cdot \mathbf{b} d\Gamma$ is the configuration force (also referred to as the J_k-integral vector), and $\overline{\mathbf{e}}_1$ is the unit vector along the crack advance direction.

For steady-state crack growth, the dynamic contour integral can be rewritten as

$$\tilde{J} = \int_{\tilde{\Gamma}} \mathbf{n} \cdot \tilde{\mathbf{b}} d\tilde{\Gamma} \cdot \overline{\mathbf{e}}_1 = \tilde{\mathbf{J}} \cdot \overline{\mathbf{e}}_1 , \tag{1.39}$$

where $\tilde{\mathbf{b}} = (w+k)\mathbf{I} - \boldsymbol{\sigma} \cdot \mathbf{u}\nabla$ is the dynamic energy-momentum tensor, and $\tilde{\mathbf{J}} = \int_{\Gamma} \mathbf{n} \cdot \tilde{\mathbf{b}} d\Gamma$ is the dynamic configuration force.

Hence, the classic J-integral or its dynamic counterpart, the $\tilde{J}$-integral, are the projection of the configuration force on the crack advance direction. The material force (configuration force) method affords a uniform treatment of complex material behaviors in inelastic fracture, as demonstrated by Nguyen *et al.* (2005) for the quasi-static case. The formulation requires only that the constitutive relations are derived from a free energy density and that the evolution of inelastic strain conforms to a dissipation potential.

For a simple illustration, the presentation is restricted to quasi-static small-strain problems. An internal variable description of associative elastoplasticity assumes the existence of a free energy density function $\Psi(\boldsymbol{\varepsilon}^e, \boldsymbol{\alpha})$ for an additive decomposition of the strain tensor into elastic and plastic parts, that is,

$$\boldsymbol{\varepsilon} = \boldsymbol{\varepsilon}^e + \boldsymbol{\varepsilon}^p , \tag{1.40}$$

where $\boldsymbol{\alpha}$ is an internal strain governing the hardening behavior.

The energy-momentum tensor is defined as

$$\mathbf{b} \equiv \Psi(\boldsymbol{\varepsilon}^e, \boldsymbol{\alpha})\mathbf{I} - \boldsymbol{\sigma} \cdot \mathbf{u}\nabla . \tag{1.41}$$

Under quasi-static conditions, the resulting local balance of energy-momentum relates the divergence of the energy-momentum tensor to two material body force terms, each of which is the product of the gradient of one internal strain and its thermodynamic conjugate stress, i.e.,

$$\nabla \cdot \mathbf{b} + \boldsymbol{\sigma} : \nabla \boldsymbol{\varepsilon}^p + \mathbf{q} : \nabla \boldsymbol{\alpha} = 0 . \tag{1.42}$$

Consider a crack in an otherwise homogeneous body as shown in Fig. 1.8. In the material force framework, crack growth is treated formally as a change in the material configuration. A contour Γ with outward unit normal $\mathbf{N}$ is defined to trace the external boundary of the body Ω and the crack surfaces. It is joined to a similarly defined contour Γ_δ surrounding the infinitesimal volume Ω_δ containing the crack tip.

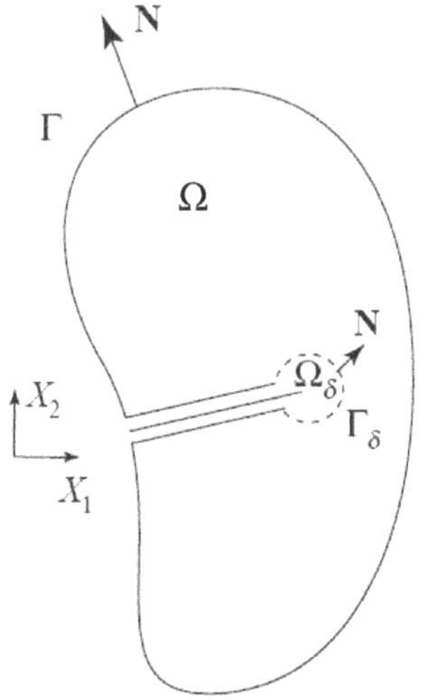

Fig. 1.8. A crack in an otherwise homogeneous body. (From Nguyen *et al.*, 2005, with permission from Elsevier).

The global energy-momentum balance is obtained as

$$\int_\Gamma \mathbf{N}\cdot\mathbf{b}dS - \lim_{\delta\to 0}\int_{\Omega-\Omega_\delta}[-(\boldsymbol{\sigma}:\nabla\boldsymbol{\varepsilon}^p + \mathbf{q}:\nabla\boldsymbol{\alpha})]dV = \lim_{\delta\to 0}\int_{\Gamma_\delta}\mathbf{N}\cdot\mathbf{b}dS\,, \quad (1.43)$$

where the global material and dissipation forces are defined by

$$\mathbf{F}^{\text{mat}} = \int_\Gamma \mathbf{N}\cdot\mathbf{b}dS - \mathbf{F}^{\text{dissip}}\,, \quad (1.44)$$

$$\mathbf{F}^{\text{dissip}} = \lim_{\delta\to 0}\int_{\Omega-\Omega_\delta} -(\boldsymbol{\sigma}:\nabla\boldsymbol{\varepsilon}^p + \mathbf{q}:\nabla\boldsymbol{\alpha})dV\,. \quad (1.45)$$

The global material force corresponds to the path-domain integral developed by Moran and Shih (1987a–b) and Simo and Honein (1990) for elastoplasticity. For elastic problems, the global dissipation force vanishes so that the global material force becomes path independent.

If the elastic strain energy density, U^e, instead of the free energy density, Ψ, is used to define the energy-momentum tensor in Eq. (1.41), the result, in the absence of strain hardening, becomes the $\hat{J}_k$-integral given by Kishimoto *et al.* (1980):

$$\hat{J}_k = \int_\Gamma (U^e n_k - n_i\sigma_{ij}u_{j,k})dS + \lim_{\delta\to 0}\int_{\Omega-\Omega_\delta}\sigma_{ij}\varepsilon^*_{ij,k}dV\,, \quad (1.46)$$

where $\varepsilon_{ij}^{*} = \varepsilon_{ij} - \varepsilon_{ij}^{e}$ are the components of eigenstrains (thermal strain, plastic strain, etc.) in the formulation of Kishimoto and co-workers.

1.9 Cohesive Zone and Virtual Internal Bond Models

The cohesive zone or yielded strip models have been developed for examining the crack-tip behavior of materials (e.g., metal, polymer, and concrete) which may exhibit nonlinearity and viscosity by many researchers, including Barenblatt (1959a–c, 1962), Dugdale (1960), Irwin (1961), Knauss (1974), Schapery (1975), and Hillerborg *et al.* (1976). In order to describe the inelastic behavior in the fracture process zone ahead of a crack tip, it is assumed that the material along the crack path obeys a specified traction-separation function in the cohesive surface model (e.g., Tvergaard and Hutchinson, 1992; Xu and Needleman, 1994).

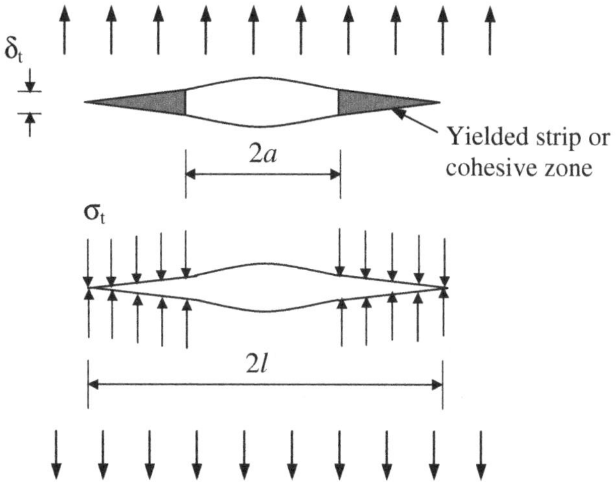

Fig. 1.9. Dugdale–Barenblatt model.

As shown in Fig. 1.9, Dugdale (1960) applied this concept to the problem of a Griffith crack in an elastic-plastic material on the assumption that the opening of prospective crack surfaces ahead of the crack tip be opposed by a closing stress equal to the yield strength of the material. The elastoplastic problem was thus turned into a simple elastic one, which is similar to the analysis of the cohesive forces at a crack tip by Barenblatt (1959a–c, 1962). The condition that the total stress intensity factor for the Dugdale model must be zero for nonsingular stresses yields

$$K_I = -(\sigma_y - \sigma)\sqrt{\pi l} + 2\sigma_y \sqrt{\frac{l}{\pi}} \sin^{-1}\left(\frac{a}{l}\right) = 0 . \tag{1.47}$$

The extent of the cohesive zone is thus given by

$$d_p = (l - a) = a(\sec\beta - 1) , \tag{1.48}$$

where

$$\beta = \frac{\pi}{2}\frac{\sigma}{\sigma_y} . \tag{1.49}$$

If σ/σ_y is small, Eq. (1.48) can be approximated as

$$d_p = \frac{\pi}{8}\left(\frac{K_I}{\sigma_y}\right)^2 . \tag{1.50}$$

This expression can be compared with the estimation of the plastic zone size by the Irwin approach (1961):

$$r_p = 2r_y = \frac{1}{\pi}\left(\frac{K_I}{\sigma_y}\right)^2 . \tag{1.51}$$

Wells (1961, 1963) suggested that fracture in metals occurs when the crack-tip opening displacement (CTOD) reaches a critical value. The CTOD can be calculated from the elastic field (Goodier and Field, 1963) as

$$\delta_t = \left(\frac{8\sigma_y a}{\pi E}\right) \ln(\sec\beta) . \tag{1.52}$$

Generally speaking, if the extent of the cohesive zone is small enough compared to characteristic dimensions, regardless of the force-separation law, the *J*-integral, the energy release rate, the stress intensity factor, and the crack-tip opening displacement are all equivalent fracture mechanics parameters under small-scale yielding conditions, that is,

$$J = G = K_I^2 \big/ E^{'} = \int_0^{\delta_t} \sigma(\delta) d\delta . \tag{1.53}$$

For limited amounts of crack extension, the tearing modulus approach generally gives a more accurate assessment than that based only on the critical J-integral at crack initiation. The crack-tip opening angle (CTOA) appears to be an attractive alternative parameter in elastic-plastic analysis of extended amounts of stable crack growth accompanied by significant elastic unloading (Kannimen and Popelar, 1985).

Recent advances in nanotechnology have provided a strong impetus for understanding the fracture behavior of nanoscale materials. Many classical fracture mechanics concepts will no longer be applicable as the characteristic dimension of a structure becomes comparable to or smaller than the size of the cohesive zone near a crack tip. For example, no well-defined crack front was observed in molecular dynamics simulations of nanowire fracture (Walsh *et al.*, 2001). Gao and Klein (1998) and Klein and Gao (1998) developed a virtual internal bond (VIB) method with direct incorporation of a cohesive interactive law into the constitutive model so that crack initiation and growth become natural consequence of the method without a presumed fracture criterion. Gao and Ji (2003) applied the VIB method to study fracture in nanomaterials with a focus on the features that are unique at nanoscale. They investigated the transition of the fracture mechanism from the classical Griffith fracture to one of homogeneous failure near the theoretical strength of the solid with no stress concentration at the crack tip. Ji and Gao (2004) also studied fracture mechanisms in biological nanocomposites via the VIB method with a focus on the effects of protein and protein-mineral interfaces. Volokh and Gao (2005) further proposed a modified VIB (MVIB) formulation, which allows for two independent linear elastic constants.

The VIB method was developed based on an extension of the so-called Cauchy–Born rule for establishing continuum constitutive equations with the use of atomic-like bond potentials. This is a multi-scale assumption that relates the motion of atoms to continuum deformation measures. Under this assumption, atoms in a crystal move according to a single mapping from the undeformed to the deformed configurations. As the mapping is taken to be the deformation gradient **F**, a link between the discrete microstructural description and the continuum constitutive model is attained by equating the strain energy density to the potential energy stored in a virtual network of internal cohesive bonds, that is,

$$w(\mathrm{E}_{IJ}) = \langle U(l) \rangle = \left\langle U(l_0\sqrt{1+2\xi_I \mathrm{E}_{IJ}\xi_J}) \right\rangle, \tag{1.54}$$

where $\mathrm{E}_{IJ} = (C_{IJ} - \delta_{IJ})/2$ is the Lagrange strain tensor, also called the Green–Lagrange strain tensor, $C_{IJ} = F_{kI}F_{kJ}$ is the right Cauchy–Green deformation tensor, $U(l)$ is the bond potential, ξ_I denotes the bond orientation, l is the bond length, l_0 is the length of the unstretched bond, and $\langle \cdots \rangle$ is a weighted average with respect to a bond density function.

Thus, the symmetric (second) Piola–Kirchhoff stress can be obtained from the derivative of the strain energy density function as

$$\Sigma_{IJ} = \frac{\partial w}{\partial \mathrm{E}_{IJ}} = \left\langle \frac{l_0^2 U'(l)}{l} \xi_I \xi_J \right\rangle. \tag{1.55}$$

The finite-deformation form of the J_K-integral is given by

$$J_K = \int_\Gamma (w\delta_{KJ} - F_{iK}T_{iJ})N_J d\Gamma, \tag{1.56}$$

where T_{iJ} is the asymmetric (first) Piola–Kirchhoff stress tensor satisfying the relation

$$T_{iJ}F_{jJ} = F_{iI}\Sigma_{IJ}F_{jJ} = j\sigma_{ij}. \tag{1.57}$$

The onset of fracture predicted by the VIB model is not only determined by the choice of the bond potential but also by the state of deformation in the fracture process zone (Klein and Gao, 1998; Gao and Ji, 2003; Ji and Gao, 2004). The size of the fracture localization zone h can be evaluated via J-integral analysis by selecting a contour that lies along the upper and lower edges of the localization zone (see Fig. 1.10):

$$h = -\frac{J_c}{\pi D_0 U(l_0^*)}. \tag{1.58}$$

The size of the fracture localization zone is correlated with the fracture energy and the virtual bond potential of the VIB model. The cohesive surface models only apply the traction-separation law to the crack plane rather than to the bulk of the material, whereas the VIB method does not assume pre-existing weak paths and directly incorporates the cohesive interactive law into the constitutive model on the continuum level. A VIB-based finite element method (VIB-FEM) is typically used to simulate the fracture process with crack nucleation and growth represented by separation of two adjacent nodes near the crack tip, resulting in localization of strain within one overstretched sheet of mesh. An important difference between VIB-FEM and conventional FEM lies in the specific physical meaning of the mesh size in VIB-FEM, which is no longer a purely numerical concept as in conventional FEM.

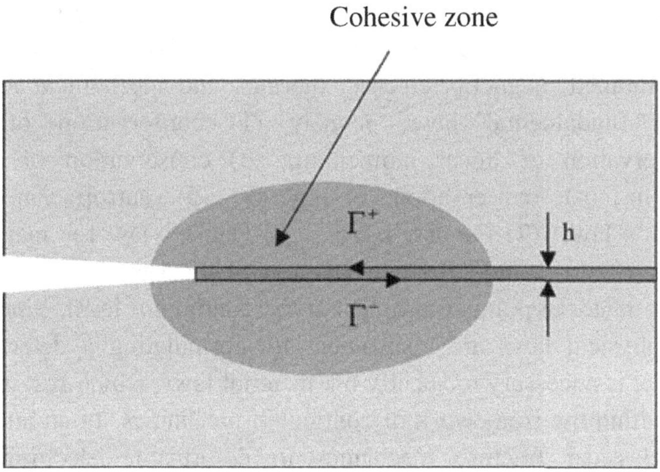

Fig. 1.10. Illustration of fracture localization zone and *J*-integral contour. (From Gao and Ji, 2003, with permission from Elsevier).

Chapter 2

Elements of Electrodynamics of Continua

Electrodynamics is a branch of physics which studies electric charges in motion, whereas mechanics is the science of force and motion of matter (Fung and Tong, 2001). Electrodynamics of continua or continuum mechanics of electromagnetic materials is concerned with the behavior of deformable electromagnetic materials modeled as continuous media under combined magnetic, electric, thermal, and mechanical loadings. The ten fundamental laws, namely (1) conservation of mass, (2) conservation of linear momentum, (3) conservation of angular momentum, (4) conservation of energy, (5) entropy inequality, (6) Gauss's law, (7) Faraday's law, (8) Gauss's law for magnetism, (9) Ampere's law, and (10) conservation of electric charges, may be applied to material points (particles) at the continuum level. Since these general physical laws are insufficient for formulating a deterministic problem, it is necessary to specify the material laws, which rest upon the axioms within the framework of continuum mechanics. In an attempt to extend classical fracture mechanics to deformable electromagnetic materials, we shall make a summary of the elements of electrodynamics of continua in this chapter, including conventional terms and notations, Maxwell equations, balance equations of mass, linear momentum, angular momentum and energy, constitutive relations and transport laws from the general nonlinear formulation to the simple linearized theory. The reader who desires more information may refer to the literature, from classical treatises (e.g., Landau and Lifshitz, 1960; Eringen, 1980; Maugin, 1988) to recent papers (e.g., Dorfmann and Ogden, 2003–2006; McMeeking and Landis, 2005; McMeeking *et al.*, 2007; Vu and Steinmann, 2007; Suo *et al.*, 2008; Kuang, 2008; Bustamante *et al.*, 2009; Trimarco, 2009).

2.1 Notations

In this section, physical terms and notational conventions are presented for the statement of basic field equations and they will be used throughout the book. Modeling a body as a continuum is a mathematical approximation that is highly accurate when the characteristic length in macroscopic phenomena is much larger than the atomic size. Under this hypothesis, the atomistic structure of the body is ignored and neighboring points remain as neighbors under any loading condition.

2.1.1 *Eulerian and Lagrangian descriptions*

The Eulerian (spatial) description, in terms of the spatial coordinates and time, focuses on what is occurring at a fixed point in space as time progresses, whereas the Lagrangian (material) description, in terms of the material or referential coordinates and time, gives attention to individual particles as they move through space and time. The choice of two distinct coordinate systems (see Fig. 2.1), one for the reference configuration at time t_0 and the other for the current configuration at time t, has many advantages in describing the motion and deformation of a continuous body.

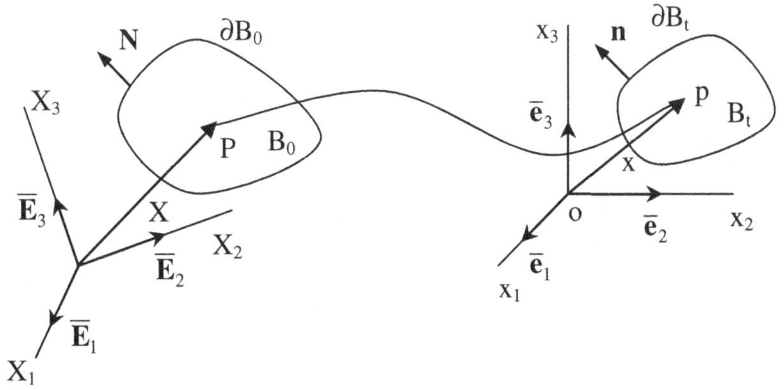

Fig. 2.1. Two coordinate systems for reference and current configurations.

A position vector is used to characterize a particle within a deformable body in the current configuration as

$$\mathbf{x} = \sum_{k=1}^{3} x_k \overline{\mathbf{e}}_k = x_k \overline{\mathbf{e}}_k \,, \tag{2.1}$$

where x_k are the spatial coordinates of the particle and $\overline{\mathbf{e}}_k$ are mutually orthogonal unit base vectors in a Cartesian coordinate system for the current configuration. The summation over the repeated index implied by the last entry in the continued equality is adopted as a convention. An index that is summed over is called the dummy index, and one that is not summed over is called the free index.

This vector can be expressed as a function of the particle position in some reference configuration; for example, the configuration at initial time, that is,

$$\mathbf{X} = \sum_{K=1}^{3} X_K \overline{\mathbf{E}}_K = X_K \overline{\mathbf{E}}_K \,, \tag{2.2}$$

$$\mathbf{x} = \boldsymbol{\chi}(\mathbf{X}, t) \,, \tag{2.3}$$

with inverse

$$\mathbf{X} = \boldsymbol{\chi}^{-1}(\mathbf{x}, t) \,, \tag{2.4}$$

where X_K are the material coordinates of the particle and $\overline{\mathbf{E}}_K$ are mutually orthogonal unit base vectors in a Cartesian coordinate system for the reference configuration.

The vector joining the positions of a particle in the initial and current configurations is called the displacement vector, that is,

$$\mathbf{u} = \mathbf{x} - \mathbf{X} \,. \tag{2.5}$$

Vectors and tensors are represented by boldface and their components are represented by means of subscripts throughout the book. The rectangular components of a vector **n** and a second-order tensor **A** are

$$n_k = \mathbf{n} \cdot \overline{\mathbf{e}}_k \,,\ n_K = \mathbf{n} \cdot \overline{\mathbf{E}}_K \,, \tag{2.6}$$

$$A_{ij} = \overline{\mathbf{e}}_i \cdot \mathbf{A} \cdot \overline{\mathbf{e}}_j \,,\ A_{IJ} = \overline{\mathbf{E}}_I \cdot \mathbf{A} \cdot \overline{\mathbf{E}}_J \,, \tag{2.7}$$

where the dot denotes the inner product operator.

The inner product or the dot product of two vectors **m** and **n** is a scalar with value

$$\mathbf{m}\cdot\mathbf{n}=\mathbf{n}\cdot\mathbf{m}=\delta_{ij}m_i n_j=m_i n_i\,, \tag{2.8}$$

where δ_{ij} is the Kronecker delta symbol

$$\delta_{ij}=\overline{\mathbf{e}}_i\cdot\overline{\mathbf{e}}_j=\begin{cases}1 & \text{for } i=j\\ 0 & \text{for } i\neq j\end{cases}. \tag{2.9}$$

The right or left dot product of a second-order tensor **A** and a vector **n** is a vector with components

$$(\mathbf{A}\cdot\mathbf{n})_i=A_{ij}n_j\,,\quad (\mathbf{n}\cdot\mathbf{A})_i=A_{ji}n_j\,. \tag{2.10}$$

The double dot product of two second-order tensors is a scalar with value

$$\mathbf{A}:\mathbf{B}=\mathbf{B}:\mathbf{A}=A_{ij}B_{ji}=B_{ij}A_{ji}\,. \tag{2.11}$$

The cross product of two vectors **m** and **n** is a vector **l** with components

$$l_k=(\mathbf{m}\times\mathbf{n})_k=-(\mathbf{n}\times\mathbf{m})_k=\varepsilon_{ijk}m_i n_j\,, \tag{2.12}$$

where ε_{ijk} is the permutation symbol

$$\varepsilon_{ijk}=\begin{cases}+1 & \text{if } ijk \text{ is an even permution of } 123\\ -1 & \text{if } ijk \text{ is an odd permution of } 123\\ 0 & \text{otherwise}\end{cases}. \tag{2.13}$$

The nabla notations in the Eulerian and Lagrangian descriptions, respectively, are given by

$$\nabla=\overline{\mathbf{e}}_k\frac{\partial}{\partial x_k}\,, \tag{2.14}$$

$$\nabla_R=\overline{\mathbf{E}}_K\frac{\partial}{\partial X_K}\,. \tag{2.15}$$

The deformation gradient is defined as

$$\mathbf{F}=\frac{\partial\mathbf{x}}{\partial\mathbf{X}}=\frac{\partial x_i}{\partial X_K}\overline{\mathbf{e}}_i\otimes\overline{\mathbf{E}}_K=x_{i,K}\overline{\mathbf{e}}_i\otimes\overline{\mathbf{E}}_K\,, \tag{2.16}$$

where the symbol $\otimes$ is the dyadic product as an outer product of two vectors and a comma is used to denote partial differentiation when Cartesian coordinates are used. The index K after the comma denotes the partial derivative with respect to X_K, whereas the index k after the comma denotes the partial derivative with respect to x_k.

The Jacobian determinant is given by

$$j = \det(\mathbf{F}) = \begin{vmatrix} \dfrac{\partial x_1}{\partial X_1} & \dfrac{\partial x_1}{\partial X_2} & \dfrac{\partial x_1}{\partial X_3} \\ \dfrac{\partial x_2}{\partial X_1} & \dfrac{\partial x_2}{\partial X_2} & \dfrac{\partial x_2}{\partial X_3} \\ \dfrac{\partial x_3}{\partial X_1} & \dfrac{\partial x_3}{\partial X_2} & \dfrac{\partial x_3}{\partial X_3} \end{vmatrix} . \tag{2.17}$$

The right and left Cauchy–Green deformation tensors are defined as

$$\mathbf{C} = \mathbf{F}^T \cdot \mathbf{F} , \tag{2.18}$$

$$\mathbf{b} = \mathbf{F} \cdot \mathbf{F}^T . \tag{2.19}$$

The Lagrange strain tensor is defined as

$$\mathbf{E} = \frac{1}{2}(\mathbf{C} - \mathbf{I}) = \frac{1}{2}(\mathbf{F}^T \cdot \mathbf{F} - \mathbf{I}) . \tag{2.20}$$

The material time derivative of a tensor **A**, denoted either by the symbol d/dt or a superimposed dot, is defined as

$$\frac{d\mathbf{A}}{dt} = \dot{\mathbf{A}} = \left.\frac{\partial \mathbf{A}}{\partial t}\right|_{\mathbf{X}} , \tag{2.21}$$

where **X** is kept constant during material time differentiation.

The velocity vector **v** is the material time derivative of the position vector or the displacement vector of a particle, that is,

$$\mathbf{v} = \frac{d\mathbf{u}}{dt} = \dot{\mathbf{u}} = \left.\frac{\partial \mathbf{u}}{\partial t}\right|_{\mathbf{X}} . \tag{2.22}$$

The deformation rate tensor **d** is the symmetric part of the velocity gradient, that is,

$$\mathbf{d} = \frac{1}{2}(\mathbf{v}\nabla + \nabla\mathbf{v}) . \tag{2.23}$$

The vorticity vector $\boldsymbol{\omega}$ is defined as

$$\boldsymbol{\omega} = \frac{1}{2}\nabla \times \mathbf{v}\,. \tag{2.24}$$

The convective time derivative of a vector **m**, denoted by a superimposed asterisk, is defined as

$$\overset{*}{\mathbf{m}} = \dot{\mathbf{m}} - (\mathbf{m} \cdot \nabla)\mathbf{v} + \mathbf{m}(\nabla \cdot \mathbf{v})\,. \tag{2.25}$$

If a vector field $\hat{\mathbf{m}}$ in the reference configuration is associated with a vector field $\mathbf{m}$ in the current configuration by

$$\hat{\mathbf{m}} = j\mathbf{F}^{-1} \cdot \mathbf{m}, \tag{2.26}$$

then the material time derivative of the vector field $\hat{\mathbf{m}}$ is related to the convective time derivative of the vector field $\mathbf{m}$ by

$$\dot{\hat{\mathbf{m}}} = j\mathbf{F}^{-1} \cdot \overset{*}{\mathbf{m}}\,. \tag{2.27}$$

2.1.2 *Electromagnetic field*

The electromagnetic field can be viewed as a combination of the electric field and the magnetic field, which are mathematically represented as vectors. The electric charge in a body may be positive or negative. The motion of charged particles in a given direction is known as the electric current. The Maxwell equations and the Lorentz force law have been used to describe the way that charges and currents interact with the electromagnetic field. It is known that the Maxwell equations are form-invariant with respect to the Lorentz transformations. Since a high velocity close to light speed is not easily achievable in solid material media, to which this book is devoted, the Galilean approximation is adopted hereafter instead of a relativistic treatment.

The electromagnetic field quantities in a fixed Galilean frame R_G, also referred to as the laboratory frame, are denoted by **P**, **E**, **D**, **M**, **B**, **H**, $\mathbf{j}_e$, where **P** is polarization, **E** is electric field, $\mathbf{D} = \varepsilon_0\mathbf{E} + \mathbf{P}$ is electric displacement, **M** is magnetization, **B** is magnetic induction, $\mathbf{H} = \mathbf{B}/\mu_0 - \mathbf{M}$ is magnetic field, $\mathbf{j}_e$ is total electric current, ε_0 is vacuum permittivity, and μ_0 is vacuum permeability.

The field quantities in the co-moving frame R_C are introduced by

$$\boldsymbol{E} = \mathbf{E} + \mathbf{v} \times \mathbf{B}, \tag{2.28}$$

$$\boldsymbol{M} = \mathbf{M} + \mathbf{v} \times \mathbf{P}, \tag{2.29}$$

$$\boldsymbol{H} = \mathbf{H} - \mathbf{v} \times \mathbf{D}, \tag{2.30}$$

$$\boldsymbol{j}_e = \mathbf{j}_e - q_f \mathbf{v}, \tag{2.31}$$

where $\boldsymbol{E}$ is electromotive intensity, $\boldsymbol{j}_e$ is conduction current, and q_f is the free electric charge density.

2.1.3 *Electromagnetic body force and couple*

The force that the electromagnetic field exerts on electrically charged particles is called the electromagnetic force, which is one of the fundamental forces in nature. The other fundamental forces are the strong interaction, the weak interaction, and the gravitational force. All other forces are ultimately derived from these four fundamental forces. The electromagnetic field contains electromagnetic energy with a density proportional to the square of the field intensities. The Lorentz force law that describes the force acting on a point charge due to the electromagnetic field has been used to construct the expressions for the electromagnetic body force and couple in continuous media.

The Lorentz force acting on a point charge δq^α in a microscopic volume element (see Fig. 2.2) is

$$\delta \mathbf{f}^\alpha = \delta q^\alpha \left[\mathbf{e}(\mathbf{x}^\alpha) + \mathbf{v}^\alpha \times \mathbf{b}(\mathbf{x}^\alpha) \right], \tag{2.32}$$

where $\mathbf{e}(\mathbf{x}^\alpha)$ and $\mathbf{b}(\mathbf{x}^\alpha)$ are the microscopic electric field and the microscopic magnetic induction at $\mathbf{x}^\alpha$, respectively.

The electromagnetic force and the electromagnetic couple acting on the microscopic volume element are given by, respectively,

$$_{em}\mathbf{f}\Delta v = \sum_\alpha \delta q^\alpha \mathbf{e}(\mathbf{x} + \boldsymbol{\xi}^\alpha) + \sum_\alpha \delta q^\alpha (\mathbf{v} + \dot{\boldsymbol{\xi}}^\alpha + \dot{\hat{\boldsymbol{\xi}}}^\alpha) \times \mathbf{b}(\mathbf{x} + \boldsymbol{\xi}^\alpha), \tag{2.33}$$

$$\begin{aligned} _{em}\mathbf{l}\Delta v &= \sum_\alpha \delta q^\alpha (\mathbf{x} + \boldsymbol{\xi}^\alpha) \times \mathbf{e}(\mathbf{x} + \boldsymbol{\xi}^\alpha) \\ &\quad + \sum_\alpha \delta q^\alpha (\mathbf{x} + \boldsymbol{\xi}^\alpha) \times [(\mathbf{v} + \dot{\boldsymbol{\xi}}^\alpha + \dot{\hat{\boldsymbol{\xi}}}^\alpha) \times \mathbf{b}(\mathbf{x} + \boldsymbol{\xi}^\alpha)], \end{aligned} \tag{2.34}$$

where $\mathbf{x}+\boldsymbol{\xi}^{\alpha}$ is the average position of the point charge δq^{α} and $\dot{\hat{\boldsymbol{\xi}}}^{\alpha}$ is the fluctuation velocity.

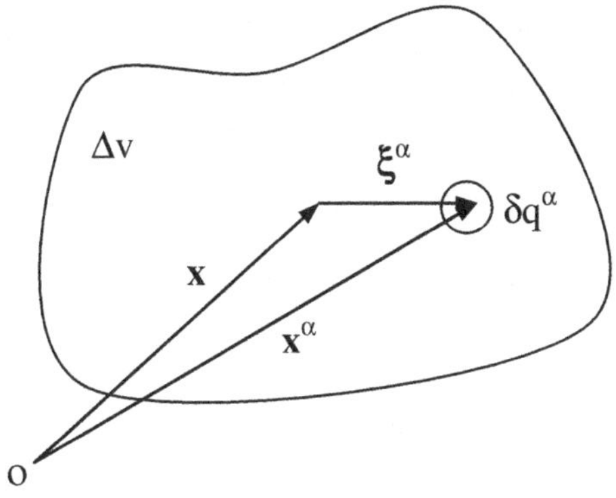

Fig. 2.2. Microscopic volume element.

The macroscopic electromagnetic field quantities are introduced by

$$\mathbf{E}(\mathbf{x})=\mathbf{e}(\mathbf{x}), \tag{2.35}$$

$$\mathbf{B}(\mathbf{x})=\mathbf{b}(\mathbf{x}), \tag{2.36}$$

$$q_f\Delta v=\sum_{\alpha}\delta q^{\alpha}, \tag{2.37}$$

$$\mathbf{P}\Delta v=\sum_{\alpha}\delta q^{\alpha}\boldsymbol{\xi}^{\alpha}, \tag{2.38}$$

$$\boldsymbol{M}\Delta v=\frac{1}{2}\sum_{\alpha}\delta q^{\alpha}\boldsymbol{\xi}^{\alpha}\times\dot{\boldsymbol{\xi}}^{\alpha}, \tag{2.39}$$

$$\boldsymbol{j}_e\Delta v=\sum_{\alpha}\delta q^{\alpha}\dot{\hat{\boldsymbol{\xi}}}^{\alpha}. \tag{2.40}$$

Thus, the expressions for the electromagnetic body force ${}_{em}\mathbf{f}$ and the electromagnetic body couple ${}_{em}\mathbf{c}$ are

$${}_{em}\mathbf{f}=q_f\boldsymbol{E}+(\boldsymbol{j}_e+\overset{*}{\mathbf{P}})\times\mathbf{B}+(\mathbf{P}\cdot\nabla)\boldsymbol{E}+(\nabla\mathbf{B})\cdot\boldsymbol{M}, \tag{2.41}$$

$${}_{em}\mathbf{c}={}_{em}\mathbf{l}-\mathbf{x}\times{}_{em}\mathbf{f}=\mathbf{P}\times\boldsymbol{E}+\boldsymbol{M}\times\mathbf{B}. \tag{2.42}$$

Hence, polarization is related to the electric dipole moment, magnetization is related to the magnetic moment, and conduction current is related to the fluctuation velocity of charges (Mazur and Nijboer, 1953). A material is electrically polarized if **P** is nonzero, whereas a material is magnetized if **M** is nonzero. Materials can also be classified as conducting, semiconducting, or insulating, based on their ability to conduct electric current.

2.1.4 *Electromagnetic stress tensor and momentum vector*

There exist an electromagnetic stress tensor ${}_{em}\sigma_{ij}$ and an electromagnetic momentum vector G_k (Eringen, 1980; Maugin, 1988), such that

$${}_{em}f_k = {}_{em}\sigma_{ik,i} - \frac{\partial G_k}{\partial t}, \tag{2.43}$$

$${}_{em}c_k = \varepsilon_{kij}\,{}_{em}\sigma_{ij}. \tag{2.44}$$

One solution is

$$\begin{aligned}{}_{em}\boldsymbol{\sigma} &= \mathbf{P}\otimes \boldsymbol{E} - \mathbf{B}\otimes \boldsymbol{M} + (\boldsymbol{M}\cdot\mathbf{B})\mathbf{I} + \varepsilon_0\mathbf{E}\otimes\mathbf{E} \\ &+ \mathbf{B}\otimes\mathbf{B}/\mu_0 - {}_{em}u^f\,\mathbf{I},\end{aligned} \tag{2.45}$$

$$\mathbf{G} = \varepsilon_0\mathbf{E}\times\mathbf{B}, \tag{2.46}$$

where ${}_{em}u^f = \varepsilon_0\mathbf{E}\cdot\mathbf{E}/2 + \mathbf{B}\cdot\mathbf{B}/(2\mu_0)$ is the energy density of the free electromagnetic field, and **I** is the second-order unit tensor.

By introducing the Maxwell stress tensor involving only the free electromagnetic field

$${}_{F}\boldsymbol{\sigma} = \varepsilon_0\mathbf{E}\otimes\mathbf{E} + \mathbf{B}\otimes\mathbf{B}/\mu_0 - {}_{em}u^f\,\mathbf{I}, \tag{2.47}$$

the electromagnetic stress tensor ${}_{em}\boldsymbol{\sigma}$ can be decomposed as

$${}_{em}\boldsymbol{\sigma} = {}_{F}\boldsymbol{\sigma} + {}_{em}\bar{\boldsymbol{\sigma}}, \tag{2.48}$$

with

$${}_{em}\bar{\boldsymbol{\sigma}} = \mathbf{P}\otimes\boldsymbol{E} - \mathbf{B}\otimes\boldsymbol{M} + (\boldsymbol{M}\cdot\mathbf{B})\mathbf{I}. \tag{2.49}$$

Therefore, the electromagnetic body force and couple can also be expressed as

$$ {}_{em}f_k = {}_L\bar{f}_k + {}_{em}\bar{\sigma}_{ik,i}\,, \tag{2.50} $$

$$ {}_{em}c_k = \varepsilon_{kij}\,{}_{em}\bar{\sigma}_{ij}\,, \tag{2.51} $$

where ${}_L\bar{\mathbf{f}} = q^{eff}\boldsymbol{E} + \boldsymbol{j}_e^{eff} \times \mathbf{B} = \nabla\cdot_F \boldsymbol{\sigma} - \partial\mathbf{G}/\partial t$ is the effective Lorentz force, $q^{eff} = q_f - \nabla\cdot\mathbf{P}$ is the effective charge density, and $\boldsymbol{j}_e^{eff} = \boldsymbol{j}_e + \overset{*}{\mathbf{P}} + \nabla\times\boldsymbol{M}$ is the effective current.

Hence, the electromagnetic stress tensor can be taken as an extension of the Maxwell stress tensor in classical electrodynamics, which also involves the contribution due to polarization and magnetization defined by Eq. (2.49). For the special case of electrostatics or magnetostatics, we have the electric stress ${}_e\boldsymbol{\sigma} = \mathbf{D}\otimes\mathbf{E} - {}_e u^f\mathbf{I}$ or the magnetic stress ${}_m\boldsymbol{\sigma} = \mathbf{B}\otimes\mathbf{H} + (\mathbf{M}\cdot\mathbf{B})\mathbf{I} - {}_m u^f\mathbf{I}$ with the corresponding energy density denoted by ${}_e u^f = \varepsilon_0\mathbf{E}\cdot\mathbf{E}/2$ or ${}_m u^f = \mathbf{B}\cdot\mathbf{B}/(2\mu_0)$.

2.1.5 *Electromagnetic power*

The electromagnetic power is the rate of work done by the electromagnetic forces, that is,

$$ {}_{em}w\Delta v = \sum_{\alpha} \delta q^{\alpha}(\mathbf{v} + \dot{\boldsymbol{\xi}}^{\alpha} + \dot{\hat{\boldsymbol{\xi}}}^{\alpha})\cdot\mathbf{e}(\mathbf{x} + \boldsymbol{\xi}^{\alpha})\,. \tag{2.52} $$

The useful equivalent expressions for the electromagnetic power density in terms of different time derivatives are listed as follows:

$$ {}_{em}w = \mathbf{E}\cdot\frac{\partial\mathbf{P}}{\partial t} - \mathbf{M}\cdot\frac{\partial\mathbf{B}}{\partial t} + \nabla\cdot[\mathbf{v}(\mathbf{E}\cdot\mathbf{P})] + \mathbf{j}_e\cdot\mathbf{E}\,, \tag{2.53} $$

$$ {}_{em}w = {}_{em}\mathbf{f}\cdot\mathbf{v} + \rho\boldsymbol{E}\cdot\dot{\boldsymbol{\pi}} - \boldsymbol{M}\cdot\dot{\mathbf{B}} + \boldsymbol{j}_e\cdot\boldsymbol{E}\,, \tag{2.54} $$

$$ {}_{em}w = {}_{em}\mathbf{f}\cdot\mathbf{v} + {}_{em}\mathbf{c}\cdot\boldsymbol{\omega} + {}_{em}\bar{\boldsymbol{\sigma}}:\mathbf{d} + \boldsymbol{E}\cdot\overset{*}{\mathbf{P}} - \boldsymbol{M}\cdot\overset{*}{\mathbf{B}} + \boldsymbol{j}_e\cdot\boldsymbol{E}\,, \tag{2.55} $$

where $\boldsymbol{\pi} = \mathbf{P}/\rho$ is the polarization per unit mass and ρ is the mass density.

2.1.6 *Poynting theorem*

The Poynting vector, which represents the flux of the electromagnetic energy, is denoted by $\mathbf{S}=\mathbf{E}\times\mathbf{H}$ in the laboratory frame R_G and by $\boldsymbol{S}=\boldsymbol{E}\times\boldsymbol{H}$ in the co-moving frame R_C.

The Poynting theorem in R_G gives the identity

$$\mathbf{H}\cdot\frac{\partial\mathbf{B}}{\partial t}+\mathbf{E}\cdot\frac{\partial\mathbf{D}}{\partial t}=-\mathbf{j}_e\cdot\mathbf{E}-\nabla\cdot\mathbf{S}. \tag{2.56}$$

With use of this identity, the electromagnetic power density can be expressed in a new form

$$_{em}w=-\frac{\partial\,_{em}u^f}{\partial t}-\nabla\cdot[\mathbf{S}-\mathbf{v}(\mathbf{E}\cdot\mathbf{P})]. \tag{2.57}$$

The Poynting theorem in R_C gives the identity

$$\boldsymbol{H}\cdot\overset{*}{\mathbf{B}}+\boldsymbol{E}\cdot\overset{*}{\mathbf{D}}=-\boldsymbol{j}_e\cdot\boldsymbol{E}-\nabla\cdot\boldsymbol{S}. \tag{2.58}$$

Using this identity, the electromagnetic power density can be expressed in another form

$$_{em}w=-\rho\frac{d}{dt}\left(\frac{_{em}u^f}{\rho}\right)+\nabla\cdot[(_{em}\boldsymbol{\sigma}+\mathbf{v}\otimes\mathbf{G})\cdot\mathbf{v}-\boldsymbol{S}]. \tag{2.59}$$

2.2 Maxwell Equations

With use of the notations in Section 2.1, the set of equations named after James Clerk Maxwell are expressed in a fixed Galilean frame R_G as

$$\nabla\cdot\mathbf{D}=q_f, \tag{2.60}$$

$$\nabla\times\mathbf{E}+\frac{\partial\mathbf{B}}{\partial t}=0, \tag{2.61}$$

$$\nabla\cdot\mathbf{B}=0, \tag{2.62}$$

$$\nabla\times\mathbf{H}-\frac{\partial\mathbf{D}}{\partial t}=\mathbf{j}_e. \tag{2.63}$$

Equation (2.60) is Gauss's law that relates the divergence of the electrical displacement to free charge density, Eq. (2.61) is Faraday's law that describes how a changing magnetic field is related to the induced electric field, Eq. (2.62) is Gauss's law for magnetism which shows that the magnetic induction has zero divergence, and Eq. (2.63) is Ampere's law which states that magnetic field can be generated either by an electrical current or by a changing electric field. From Eqs. (2.60) and (2.63), the electric charge balance equation is obtained as

$$-\frac{\partial q_f}{\partial t} = \nabla \cdot \mathbf{j}_e \,. \tag{2.64}$$

In the co-moving frame R_C, the Maxwell equations are rewritten as

$$\nabla \cdot \mathbf{D} = q_f \,, \tag{2.65}$$

$$\nabla \times \boldsymbol{E} + \overset{*}{\mathbf{B}} = 0 \,, \tag{2.66}$$

$$\nabla \cdot \mathbf{B} = 0 \,, \tag{2.67}$$

$$\nabla \times \boldsymbol{H} - \overset{*}{\mathbf{D}} = \boldsymbol{j}_e \,. \tag{2.68}$$

With the introduction of the corresponding Lagrangian fields $Q_f = jq_f$, $\hat{\mathbf{D}} = j\mathbf{F}^{-1} \cdot \mathbf{D}$, $\hat{\mathbf{B}} = j\mathbf{F}^{-1} \cdot \mathbf{B}$, $\hat{\boldsymbol{E}} = \boldsymbol{E} \cdot \mathbf{F}$, $\hat{\boldsymbol{H}} = \boldsymbol{H} \cdot \mathbf{F}$, and $\boldsymbol{J}_e = j\mathbf{F}^{-1} \cdot \boldsymbol{j}_e$, the material formulations of the Maxwell equations (Lax and Nelson, 1976; Trimarco, 2002, 2009) are given by

$$\nabla_R \cdot \hat{\mathbf{D}} = Q_f \,, \tag{2.69}$$

$$\nabla_R \times \hat{\boldsymbol{E}} + \dot{\hat{\mathbf{B}}} = 0 \,, \tag{2.70}$$

$$\nabla_R \cdot \hat{\mathbf{B}} = 0 \,, \tag{2.71}$$

$$\nabla_R \times \hat{\boldsymbol{H}} - \dot{\hat{\mathbf{D}}} = \boldsymbol{J}_e \,, \tag{2.72}$$

where $\dot{\hat{\mathbf{B}}} = j\mathbf{F}^{-1} \overset{*}{\mathbf{B}}$ and $\dot{\hat{\mathbf{D}}} = j\mathbf{F}^{-1} \overset{*}{\mathbf{D}}$ are used following the relation (2.27).

The material formulation of the electric charge balance equation is obtained from Eqs. (2.69) and (2.72) as

$$-\dot{Q}_f = \nabla_R \cdot \boldsymbol{J}_e \,. \tag{2.73}$$

If field quantities are discontinuous across a surface moving at a speed v_s, the Maxwell equations and the charge balance equation are replaced by the following jump conditions in the Eulerian description (Eringen, 1980):

$$\mathbf{n}\cdot[[\mathbf{D}]]=\varpi_f\,, \tag{2.74}$$

$$\mathbf{n}\times[[\boldsymbol{E}+\mathbf{B}\times(\mathbf{v}-\boldsymbol{v}_s)]]=0\,, \tag{2.75}$$

$$\mathbf{n}\cdot[[\mathbf{B}]]=0, \tag{2.76}$$

$$\mathbf{n}\times[[\boldsymbol{H}-\mathbf{D}\times(\mathbf{v}-\boldsymbol{v}_s)]]=0\,, \tag{2.77}$$

$$\mathbf{n}\cdot[[\boldsymbol{j}_e+q_f(\mathbf{v}-\boldsymbol{v}_s)]]+\frac{\bar{\delta}\varpi_f}{\bar{\delta}t}=0\,, \tag{2.78}$$

where $[[\cdots]]$ represents the jump of the field quantity inside the double square brackets across the moving surface, $\bar{\delta}/\bar{\delta}t$ denotes the convective time derivative operator following the motion of the surface along its normal, $\mathbf{n}$ is the unit normal to the surface, and ϖ_f is the free surface charge density.

Interface or boundary conditions for electromagnetic field quantities in the Eulerian description can be obtained by considering a special surface of discontinuity moving at the speed $\boldsymbol{v}_s=\mathbf{v}$, that is,

$$\mathbf{n}\cdot[[\mathbf{D}]]=\varpi_f\,, \tag{2.79}$$

$$\mathbf{n}\times[[\boldsymbol{E}]]=0, \tag{2.80}$$

$$\mathbf{n}\cdot[[\mathbf{B}]]=0, \tag{2.81}$$

$$\mathbf{n}\times[[\boldsymbol{H}]]=0\,, \tag{2.82}$$

$$\mathbf{n}\cdot[[\boldsymbol{j}_e]]+\frac{\bar{\delta}\varpi_f}{\bar{\delta}t}=0\,. \tag{2.83}$$

In the Lagrangian description, the jump conditions across a surface of discontinuity moving at a speed $\hat{\boldsymbol{V}}_s$ through the material can be rewritten as

$$\mathbf{N}\cdot[[\hat{\mathbf{D}}]]=\hat{\varpi}_f\,, \tag{2.84}$$

$$\mathbf{N}\times[[\hat{\boldsymbol{E}}-\hat{\mathbf{B}}\times\hat{\boldsymbol{V}}_s]]=0\,, \tag{2.85}$$

$$\mathbf{N}\cdot[[\hat{\mathbf{B}}]]=0\,, \tag{2.86}$$

$$\mathbf{N}\times[[\hat{\boldsymbol{H}}+\hat{\mathbf{D}}\times\hat{\boldsymbol{V}}_s]]=0\,, \tag{2.87}$$

$$\mathbf{N}\cdot[[\boldsymbol{J}_e-Q_f\hat{\boldsymbol{V}}_s]]+\frac{\bar{\delta}\,\hat{\varpi}_f}{\bar{\delta}t}=0, \tag{2.88}$$

where $V_s = \mathbf{F}^{-1} \cdot v_s$, $\mathbf{V} = \mathbf{F}^{-1} \cdot \mathbf{v}$, $\hat{\varpi}_f = \varpi_f da / dA$, $\mathbf{N} = j^{-1}\mathbf{n} \cdot \mathbf{F} da/dA$ is the unit normal of the moving surface in the initial configuration, and $da / dA = j\sqrt{N_K C_{KL} N_L}$ is the ratio of the area elements in the current configuration to those in the reference configuration.

Interface or boundary conditions for electromagnetic field quantities in the Lagrangian description can be obtained as the surface of discontinuity coincides with the interface or boundary considered, that is,

$$\mathbf{N} \cdot [[\hat{\mathbf{D}}]] = \hat{\varpi}_f , \tag{2.89}$$

$$\mathbf{N} \times [[\hat{\boldsymbol{E}}]] = 0 , \tag{2.90}$$

$$\mathbf{N} \cdot [[\hat{\mathbf{B}}]] = 0 , \tag{2.91}$$

$$\mathbf{N} \times [[\hat{\boldsymbol{H}}]] = 0 , \tag{2.92}$$

$$\mathbf{N} \cdot [[\boldsymbol{J}_e]] + \frac{\bar{\delta} \hat{\varpi}_f}{\bar{\delta} t} = 0 . \tag{2.93}$$

2.3 Balance Equations of Mass, Momentum, Moment of Momentum, and Energy

Following the notations in Section 2.1 we use conservation laws in order to derive the local field equations for balance of mass, momentum (linear momentum), moment of momentum (angular momentum), and energy, in addition to the Maxwell equations. These balance equations are valid irrespective of material constitutive laws.

Based on the law of conservation of mass, the local mass balance equation is given by

$$\frac{d\rho}{dt} = -\rho \nabla \cdot \mathbf{v} , \tag{2.94}$$

where ρ is the mass density in the current configuration.

In the material formulation, the local mass balance equation can be rewritten as

$$\rho_0 = j\rho , \tag{2.95}$$

where ρ_0 is the mass density in the initial configuration.

Based on the law of conservation of momentum, the local linear momentum balance equation is given by

$$\rho\frac{d\mathbf{v}}{dt}=\nabla\cdot\boldsymbol{\sigma}+\rho\hat{\mathbf{f}}+{}_{em}\mathbf{f}\,, \tag{2.96}$$

where $\boldsymbol{\sigma}$ is the Cauchy stress tensor and $\hat{\mathbf{f}}$ is the mechanical body force per unit mass.

Based on the law of conservation of angular momentum, the local angular momentum balance leads to

$$\varepsilon_{kij}\sigma_{ij}+{}_{em}c_k=0\,. \tag{2.97}$$

Using the electromagnetic stress tensor ${}_{em}\boldsymbol{\sigma}$ and the electromagnetic momentum vector $\mathbf{G}$ defined in Eqs. (2.45) and (2.46), Eqs. (2.96) and (2.97) become

$$\rho\frac{d\mathbf{v}}{dt}=\nabla\cdot(\boldsymbol{\sigma}+{}_{em}\boldsymbol{\sigma})+\rho\hat{\mathbf{f}}-\frac{\partial\mathbf{G}}{\partial t}\,, \tag{2.98}$$

$$\varepsilon_{kij}(\sigma_{ij}+{}_{em}\sigma_{ij})=0\,. \tag{2.99}$$

Thus, the total stress tensor ${}_t\boldsymbol{\sigma}=\boldsymbol{\sigma}+{}_{em}\boldsymbol{\sigma}$ is symmetric, that is, ${}_t\sigma_{ij}={}_t\sigma_{ji}$, although the Cauchy stress tensor or the electromagnetic stress tensor may not be symmetric.

In the material formulation, the local linear and angular momentum balance equations can be rewritten as

$$\rho_0(\dot{\mathbf{v}}+\dot{\hat{\mathbf{g}}})=\nabla_R\cdot({}_t\mathbf{T}+\rho_0\mathbf{V}\otimes\hat{\mathbf{g}})+\rho_0\hat{\mathbf{f}}\,, \tag{2.100}$$

$$\varepsilon_{kij}F_{iL}\,{}_tT_{Lj}=0\,, \tag{2.101}$$

where $\hat{\mathbf{g}}=\mathbf{G}/\rho$ is the electromagnetic momentum per unit mass and ${}_t\mathbf{T}=j\mathbf{F}^{-1}\cdot{}_t\boldsymbol{\sigma}$ is the first Piola–Kirchhoff total stress tensor.

Based on the conservation law of energy, the local energy balance equation is given by

$$\rho\frac{d\hat{e}}{dt}=-\nabla\cdot\mathbf{j}_q+\boldsymbol{\sigma}:\mathbf{v}\nabla+\rho\boldsymbol{E}\cdot\dot{\boldsymbol{\pi}}-\boldsymbol{M}\cdot\dot{\mathbf{B}}+\boldsymbol{E}\cdot\boldsymbol{j}_e\,, \tag{2.102}$$

where $\hat{e}$ is the internal energy per unit mass and $\mathbf{j}_q$ is the heat flux.

Applying Poynting's theorem as discussed in Section 2.1.6, the local energy balance equation (2.102) becomes

$$\rho\frac{d}{dt}(\hat{e}+\hat{k}+{}_{em}\hat{u}^{f}) = -\nabla\cdot\mathbf{j}_q + \nabla\cdot[({}_t\boldsymbol{\sigma}+\mathbf{v}\otimes\mathbf{G})\cdot\mathbf{v} - S] + \rho\hat{\mathbf{f}}\cdot\mathbf{v}, \tag{2.103}$$

where ${}_{em}\hat{u}^{f} = {}_{em}u^{f}/\rho$ is the energy of the free electromagnetic field per unit mass.

In the material formulation, the local energy balance equation can be rewritten as

$$\rho_0\dot{\hat{e}} + \rho_0\dot{\hat{k}} + \rho_{0\,em}\dot{\hat{u}}^{f} = -\nabla_R\cdot\mathbf{J}_q + \nabla_R\cdot[({}_t\mathbf{T}+\mathbf{V}\otimes\hat{\mathbf{g}})\cdot\mathbf{v} - \hat{S}] + \rho_0\hat{\mathbf{f}}\cdot\mathbf{v}, \tag{2.104}$$

where $\mathbf{J}_q = j\mathbf{F}^{-1}\cdot\mathbf{j}_q$ and $\hat{S} = j\mathbf{F}^{-1}\cdot S$.

If field quantities are discontinuous across a surface moving at a speed $\boldsymbol{v}_s$, Eqs. (2.94), (2.98), and (2.103) are replaced by the following jump conditions:

$$\mathbf{n}\cdot[[\rho(\mathbf{v}-\boldsymbol{v}_s)]] = 0, \tag{2.105}$$

$$\mathbf{n}\cdot[[(\mathbf{v}-\boldsymbol{v}_s)\otimes\rho\mathbf{v} - ({}_t\boldsymbol{\sigma}+\boldsymbol{v}_s\otimes\mathbf{G})]] = 0, \tag{2.106}$$

$$\mathbf{n}\cdot[[(\mathbf{v}-\boldsymbol{v}_s)(\rho\hat{e}+\rho\hat{k}+{}_{em}u^{f}) + \mathbf{j}_q - ({}_t\boldsymbol{\sigma}+\mathbf{v}\otimes\mathbf{G})\cdot\mathbf{v} + S]] = 0. \tag{2.107}$$

Interface or boundary conditions for total traction and heat flux in the Eulerian description can be obtained by considering a special surface of discontinuity moving at speed $\boldsymbol{v}_s = \mathbf{v}$, that is,

$$\mathbf{n}\cdot[[{}_t\boldsymbol{\sigma}+\mathbf{v}\otimes\mathbf{G}]] = 0, \tag{2.108}$$

$$\mathbf{n}\cdot[[\mathbf{j}_q - ({}_t\boldsymbol{\sigma}+\mathbf{v}\otimes\mathbf{G})\cdot\mathbf{v} + S]] = 0. \tag{2.109}$$

In the Lagrangian description, the jump conditions across a surface of discontinuity moving at a speed $\hat{\boldsymbol{V}}_s$ through the material can be rewritten as

$$\mathbf{N}\cdot[[\rho_0(-\hat{\boldsymbol{V}}_s)]] = 0, \tag{2.110}$$

$$\mathbf{N}\cdot[[(-\hat{\boldsymbol{V}}_s)\otimes\rho_0(\mathbf{v}+\hat{\mathbf{g}}) - ({}_t\mathbf{T}+\rho_0\mathbf{V}\otimes\hat{\mathbf{g}})]] = 0, \tag{2.111}$$

$$\mathbf{N}\cdot[[(-\hat{\boldsymbol{V}}_s)(\rho_0\hat{e}+\rho_0\hat{k}+\rho_{0\,em}\hat{u}^{f}) + \mathbf{J}_q - ({}_t\mathbf{T}+\mathbf{V}\otimes\hat{\mathbf{g}})\cdot\mathbf{v} + \hat{S}]] = 0. \tag{2.112}$$

Interface or boundary conditions for total traction and heat flux in the Lagrangian description can be obtained as the surface of discontinuity coincides with the interface or boundary considered, that is,

$$\mathbf{N}\cdot[[{}_t\mathbf{T}+\rho_0\mathbf{V}\otimes\hat{\mathbf{g}}]]=0\,, \tag{2.113}$$

$$\mathbf{N}\cdot[[\mathbf{J}_q-({}_t\mathbf{T}+\mathbf{V}\otimes\hat{\mathbf{g}})\cdot\mathbf{v}+\hat{S}]]=0. \tag{2.114}$$

2.4 Constitutive Relations

The Helmholtz free energy per unit mass is defined as

$$\hat{h}\equiv\hat{e}-T\hat{s}\,, \tag{2.115}$$

where T is the absolute temperature and $\hat{s}$ is the entropy per unit mass.

Substituting (2.115) into (2.102) yields

$$\begin{aligned}\frac{d\hat{s}}{dt}&=-\frac{1}{\rho T}\nabla\cdot\mathbf{j}_q+\frac{1}{\rho T}\boldsymbol{\sigma}:\mathbf{v}\nabla+\frac{1}{T}\boldsymbol{E}\cdot\dot{\boldsymbol{\pi}}-\frac{1}{\rho T}\boldsymbol{M}\cdot\dot{\mathbf{B}}\\&+\frac{1}{\rho T}\boldsymbol{E}\cdot\boldsymbol{j}_e-\frac{1}{T}\hat{s}\dot{T}-\frac{1}{T}\frac{d\hat{h}}{dt}.\end{aligned} \tag{2.116}$$

The entropy production inequality is

$$\frac{d_i\hat{s}}{dt}\equiv\frac{d\hat{s}}{dt}+\frac{1}{\rho}\nabla\cdot\mathbf{j}_s\geq 0\,, \tag{2.117}$$

where $\mathbf{j}_s$ is the entropy flux.

In the reference configuration, the entropy production inequality can be rewritten as

$$\frac{d_i\hat{s}}{dt}=\frac{d\hat{s}}{dt}+\frac{1}{\rho_0}\nabla_R\cdot\mathbf{J}_s\geq 0\,, \tag{2.118}$$

where $\mathbf{J}_s=j\mathbf{F}^{-1}\cdot\mathbf{j}_s$.

Substituting (2.116) into (2.118) gives

$$\begin{aligned}\frac{d_i\hat{s}}{dt}&=\frac{1}{\rho_0}\nabla_R\cdot(\mathbf{J}_s-\frac{\mathbf{J}_q}{T})+\frac{1}{\rho_0}\mathbf{J}_q\cdot\nabla_R\frac{1}{T}+\frac{1}{\rho_0 T}\hat{\boldsymbol{E}}\cdot\boldsymbol{J}_e\\&+\frac{1}{2\rho_0 T}{}_E\boldsymbol{\Sigma}:\dot{\mathbf{C}}+\frac{1}{\rho_0 T}\hat{\boldsymbol{E}}\cdot\dot{\hat{\boldsymbol{\Pi}}}-\frac{1}{\rho_0 T}\hat{\boldsymbol{M}}\cdot\dot{\hat{\mathbf{B}}}-\frac{1}{T}\hat{s}\dot{T}-\frac{1}{T}\frac{d\hat{h}}{dt}\geq 0,\end{aligned} \tag{2.119}$$

where ${}_E\mathbf{\Sigma} = j\mathbf{F}^{-1}\cdot{}_E\boldsymbol{\sigma}\cdot(\mathbf{F}^{-1})^T$ is the second Piola–Kirchhoff stress tensor conjugate to the right Cauchy–Green deformation tensor $\mathbf{C}$, ${}_E\boldsymbol{\sigma} = \boldsymbol{\sigma} + {}_{em}\boldsymbol{\sigma} - {}_F\boldsymbol{\sigma} = {}_t\boldsymbol{\sigma} - {}_F\boldsymbol{\sigma}$ is symmetric, $\hat{\boldsymbol{E}} = \boldsymbol{E}\cdot\mathbf{F}$, $\hat{\mathbf{\Pi}} = j\mathbf{F}^{-1}\cdot\mathbf{P}$, $\hat{\boldsymbol{M}} = \boldsymbol{M}\cdot\mathbf{F}$, $\hat{\mathbf{B}} = j\mathbf{F}^{-1}\cdot\mathbf{B}$, $\mathbf{J}_q = j\mathbf{F}^{-1}\cdot\mathbf{j}_q$, and $\boldsymbol{J}_e = j\mathbf{F}^{-1}\cdot\boldsymbol{j}_e$.

Material laws should satisfy the restrictions imposed by the fundamental principles of thermodynamics. As an illustration, we focus on the behavior of a typical magneto-electro-thermo-elastic solid for which the Helmholtz free energy $\hat{h}$ is taken to be a function of deformation, temperature, temperature gradient, polarization, and magnetic induction in the reference configuration V_R with respect to which the deformation gradient $\mathbf{F}$ is measured, that is,

$$\hat{h} = \hat{h}(\mathbf{C}, T, \nabla_R T, \hat{\mathbf{\Pi}}, \hat{\mathbf{B}}; \mathbf{X}) . \tag{2.120}$$

Since the entropy production inequality (2.119) should always be valid, it is necessary and sufficient that the state equations fulfill the following conditions:

$$\frac{\partial \hat{h}}{\partial T_{,K}} = 0 , \tag{2.121}$$

$${}_E\Sigma_{KL} = 2\rho_0 \frac{\partial \hat{h}}{\partial C_{KL}} , \tag{2.122}$$

$$\hat{s} = -\frac{\partial \hat{h}}{\partial T} , \tag{2.123}$$

$$\hat{E}_K = \rho_0 \frac{\partial \hat{h}}{\partial \hat{\Pi}_K} , \tag{2.124}$$

$$\hat{M}_K = -\rho_0 \frac{\partial \hat{h}}{\partial \hat{B}_K} , \tag{2.125}$$

$$\mathbf{J}_s = \frac{1}{T}\mathbf{J}_q , \tag{2.126}$$

$$\frac{d_i \hat{s}}{dt} = \frac{1}{\rho_0}\mathbf{J}_q \cdot \nabla_R \frac{1}{T} + \frac{1}{\rho_0 T}\hat{\boldsymbol{E}}\cdot\boldsymbol{J}_e \geq 0 . \tag{2.127}$$

From Eq. (2.121), the Helmholtz free energy does not depend on the temperature gradient. Since the inequality (2.127) should always be satisfied, transport laws for coupled heat conduction and electricity conduction can be determined accordingly.

It is proposed that the thermodynamic fluxes for heat conduction and electricity conduction depend linearly on the corresponding thermodynamic forces with the Onsager reciprocity relations, that is,

$$\mathbf{J}_q = \hat{\mathbf{L}}^{qq} \cdot \nabla_R \frac{1}{T} + \frac{1}{T} \hat{\mathbf{L}}^{qe} \cdot \hat{\boldsymbol{E}}, \tag{2.128}$$

$$\boldsymbol{J}_e = \hat{\mathbf{L}}^{eq} \cdot \nabla_R \frac{1}{T} + \frac{1}{T} \hat{\mathbf{L}}^{ee} \cdot \hat{\boldsymbol{E}}, \tag{2.129}$$

where the coefficient matrix

$$\begin{bmatrix} \hat{\mathbf{L}}^{qq} & \hat{\mathbf{L}}^{qe} \\ \hat{\mathbf{L}}^{eq} & \hat{\mathbf{L}}^{ee} \end{bmatrix}^T = \begin{bmatrix} \hat{\mathbf{L}}^{qq} & \hat{\mathbf{L}}^{qe} \\ \hat{\mathbf{L}}^{eq} & \hat{\mathbf{L}}^{ee} \end{bmatrix}, \tag{2.130}$$

is positive definite. It can be seen that the generalized transport laws (2.128) and (2.129) contain Fourier's law and Ohm's law as well as the Peltier–Seebeck effect.

Thus, the coupled heat transfer equation is obtained from Eq. (2.116) with the use of the state equations as

$$\frac{d}{dt}\left(-\frac{\partial \hat{h}}{\partial T}\right) = -\frac{1}{\rho T} \nabla \cdot \mathbf{j}_q + \frac{1}{\rho T} \boldsymbol{E} \cdot \boldsymbol{j}_e. \tag{2.131}$$

2.5 Linearized Theory

Nonlinear constitutive equations incorporating magneto-electro-thermo-mechanical coupling effects can be formulated when the free energy is expanded in its arguments. In the reference configuration V_R, expansion of the Helmholtz free energy with respect to strain, temperature, polarization, and magnetization for an anisotropic magneto-electro-thermo-elastic solid to the second-order terms gives

$$\begin{aligned}\rho_0\hat{h} = \rho_0\hat{h}_0 - \rho_0 s_0\theta + {}_E\Sigma^0_{KL}\mathrm{E}_{KL} + \hat{E}^0_K\hat{\Pi}_K - \hat{M}^0_K\hat{B}_K \\ + \frac{1}{2}c_{KLMN}\mathrm{E}_{KL}\mathrm{E}_{MN} - f^P_{MKL}\mathrm{E}_{KL}\hat{\Pi}_M - f^B_{MKL}\mathrm{E}_{KL}\hat{B}_M \\ + \frac{1}{2}\zeta_{KL}\hat{\Pi}_K\hat{\Pi}_L - \frac{1}{2}\chi^B_{KL}\hat{B}_K\hat{B}_L - \lambda_{KL}\hat{\Pi}_K\hat{B}_L \\ - \frac{1}{2T_0}C_H\theta^2 - \beta_{KL}\mathrm{E}_{KL}\theta - \gamma^P_K\hat{\Pi}_K\theta - \gamma^B_K\hat{B}_K\theta,\end{aligned} \tag{2.132}$$

where the Lagrange strain tensor $\mathbf{E} = (\mathbf{C} - \mathbf{I})/2$ and the temperature change $\theta = T - T_0$ are used, c_{KLMN}, f^P_{MKL}, f^B_{MKL}, ζ_{KL}, λ_{KL}, χ^B_{KL}, C_H, β_{KL}, γ^P_K, γ^B_K are material properties, $c_{KLMN} = c_{MNKL}$, $\zeta_{KL} = \zeta_{LK}$, $\chi^B_{KL} = \chi^B_{LK}$.

Substitution of Eq. (2.132) into Eqs. (2.122)–(2.125) yields anisotropic constitutive equations in finite deformation:

$${}_E\Sigma_{KL} = {}_E\Sigma^0_{KL} + c_{KLMN}\mathrm{E}_{MN} - f^P_{MKL}\hat{\Pi}_M - f^B_{MKL}\hat{B}_M - \beta_{KL}\theta, \tag{2.133}$$

$$\hat{E}_K = \hat{E}^0_K - f^P_{KLM}\mathrm{E}_{LM} + \zeta_{KL}\hat{\Pi}_L - \lambda_{KL}\hat{B}_L - \gamma^P_K\theta, \tag{2.134}$$

$$\hat{M}_K = \hat{M}^0_K + f^B_{KLM}\mathrm{E}_{LM} + \lambda_{LK}\hat{\Pi}_L + \chi^B_{KL}\hat{B}_L + \gamma^B_K\theta, \tag{2.135}$$

$$\rho_0\hat{s} = \rho_0\hat{s}_0 + \beta_{KL}\mathrm{E}_{KL} + \gamma^P_K\hat{\Pi}_K + \gamma^B_K\hat{B}_K + \frac{1}{T_0}C_H\theta. \tag{2.136}$$

The first terms on the right-hand sides of Eqs. (2.133)–(2.136) stand for the values of ${}_E\Sigma_{KL}$, $\hat{E}_K$, $\hat{M}_K$, and $\rho_0\hat{s}$ in the reference state, the second terms for mechanical contribution, the third terms for electric contribution, the fourth terms for magnetic contribution, and the last terms for thermal contribution.

By linearization about the initial configuration, Eqs. (2.133)–(2.136) become

$${}_E\sigma_{kl} = {}_E\sigma^0_{kl} + c_{klmn}\varepsilon_{mn} - f^P_{mkl}P_m - f^B_{mkl}B_m - \beta_{kl}\theta, \tag{2.137}$$

$$E_k = E^0_k - f^P_{klm}\varepsilon_{lm} + \zeta_{kl}P_l - \lambda_{kl}B_l - \gamma^P_k\theta, \tag{2.138}$$

$$M_k = M^0_k + f^B_{klm}\varepsilon_{lm} + \lambda_{lk}P_l + \chi^B_{kl}B_l + \gamma^B_k\theta, \tag{2.139}$$

$$\rho_0\hat{s} = \rho_0\hat{s}_0 + \beta_{kl}\varepsilon_{kl} + \gamma^P_k P_k + \gamma^B_k B_k + \frac{1}{T_0}C_H\theta, \tag{2.140}$$

where $\varepsilon_{mn} = (u_{m,n} + u_{n,m})/2$ is the infinitesimal strain.

With the use of $\mathbf{P} = \mathbf{D} - \varepsilon_0 \mathbf{E}$ and $\mathbf{M} = \mathbf{B}/\mu_0 - \mathbf{H}$, we can rewrite Eqs. (2.137)–(2.140) with the strain, electric field, magnetic field, and temperature change as independent variables:

$$ {}_E\sigma_{kl} = {}_E\sigma_{kl}^0 + c_{klmn}\varepsilon_{mn} - e_{mkl}E_m - h_{mkl}H_m - \beta_{kl}\theta , \tag{2.141}$$

$$ D_k = D_k^0 + e_{klm}\varepsilon_{lm} + \kappa_{kl}E_l + g_{kl}H_l + \omega_k\theta , \tag{2.142}$$

$$ B_k = B_k^0 + h_{klm}\varepsilon_{lm} + g_{lk}E_l + \mu_{kl}H_l + \mu_0\gamma_k\theta , \tag{2.143}$$

$$ \rho_0\hat{s} = \rho_0\hat{s}_0 + \beta_{kl}\varepsilon_{kl} + \omega_k E_k + \mu_0\gamma_k H_k + \frac{1}{T_0}C_v\theta . \tag{2.144}$$

By transformation, we can also rewrite Eqs. (2.141)–(2.144) with the stress, electric field, magnetic field and temperature change as independent variables:

$$ \varepsilon_{kl} = \varepsilon_{kl}^0 + s_{klmn}\,{}_E\sigma_{mn} + d_{mkl}^E E_m + d_{mkl}^H H_m + \alpha_{kl}\theta , \tag{2.145}$$

$$ D_k = D_k^0 + d_{klm}^E\,{}_E\sigma_{lm} + \kappa_{kl}^\sigma E_l + g_{kl}^\sigma H_l + \omega_k^\sigma\theta , \tag{2.146}$$

$$ B_k = B_k^0 + d_{klm}^H\,{}_E\sigma_{lm} + g_{lk}^\sigma E_l + \mu_{kl}^\sigma H_l + \mu_0\gamma_k^\sigma\theta , \tag{2.147}$$

$$ \rho_0\hat{s} = \rho_0\hat{s}_0 + \alpha_{kl}\,{}_E\sigma_{kl} + \omega_k^\sigma E_k + \mu_0\gamma_k^\sigma H_k + \frac{1}{T_0}C_p\theta . \tag{2.148}$$

It is obvious that the material constants in the equivalent constitutive representations are related, which means that material constants in one constitutive representation can be transformed to those in another constitutive representation. Material constants are subjected to constraints imposed by the thermodynamic requirement for stable materials (Alshits *et al.*, 1992). The physical meanings for commonly used material constants are listed in Table 2.1.

Table 2.1 Physical meanings of material constants

Symbol	Physical Meaning
$c_{klmn} = c_{mnkl} = c_{lkmn} = c_{klnm}$	elastic moduli
$\kappa_{kl} = \kappa_{lk}$	dielectric permittivity
$\mu_{kl} = \mu_{lk}$	magnetic permeability
$e_{mkl} = e_{mlk}$	piezoelectric coefficients
$h_{mkl} = h_{mlk}$	piezomagnetic coefficients
g_{kl}	magnetoelectric coefficients
ω_k	pyroelectric coefficients
γ_k	pyromagnetic coefficients
β_{kl}	thermal moduli
C_v	specific heat

Materials with nonzero e_{mkl} exhibit piezoelectricity, that is, mechanical load can produce electric polarization or electric field (direct effect) and, vice versa, electric load can produce deformation or stress (inverse effect). The direct piezoelectric effect was first discovered by Pierre Curie and Jacques Curie (1880) in some crystals such as tourmaline and quartz. A year later, the inverse piezoelectric effect was theoretically predicted by Lippmann (1881) and subsequently confirmed experimentally by the Curies (1884). By analogy, materials with nonzero h_{mkl} exhibit piezomagnetism, that is, mechanical load can produce magnetization or magnetic field (direct effect) and, vice versa, magnetic load can produce deformation or stress (inverse effect). The first experimental observation of piezomagnetism was made by Borovik-Romanov (1960) in the fluorides of cobalt and manganese.

Materials with nonzero ω_k exhibit pyroelectricity, that is, temperature change can produce electric polarization or electric field (direct effect) and, vice versa, electric load can produce temperature change (inverse effect). Correspondingly, materials with nonzero γ_k exhibit pyromagnetism, that is, temperature change can produce magnetization or magnetic field (direct effect) and, vice versa, magnetic load can produce temperature change (inverse effect).

The cross terms due to nonzero g_{kl} represent the magnetoelectric coupling effect. The co-existence of piezoelectric, piezomagnetic and magnetoelectric coupling is called the magneto-electro-mechanical coupling. Advances in state-of-the-art technology have facilitated the formation of new monolithic materials and the synthesis of composite materials, which, remarkably, breaks down the performance barriers encountered with conventional materials. We refer the readers to the *Handbook of Electromagnetic Materials: Monolithic and Composite Versions and their Applications* (Neelakanta, 1995) for detailed classification of different classes of electromagnetic materials and characterization of various material properties.

In summary, the full set of dynamic field equations of coupled magneto-electro-thermo-elasticity are listed as follows:

Maxwell equations:

$$\nabla \cdot \mathbf{D} = q_f , \tag{2.149}$$

$$\nabla \times \mathbf{E} + \frac{\partial \mathbf{B}}{\partial t} = 0 , \tag{2.150}$$

$$\nabla \cdot \mathbf{B} = 0 , \tag{2.151}$$

$$\nabla \times \mathbf{H} - \frac{\partial \mathbf{D}}{\partial t} = \mathbf{j}_e . \tag{2.152}$$

Equation of continuity:

$$\frac{d\rho}{dt} = -\rho \nabla \cdot \mathbf{v} . \tag{2.153}$$

Equation of motion:

$$\rho \frac{d\mathbf{v}}{dt} = \nabla \cdot_l \boldsymbol{\sigma} + \rho \hat{\mathbf{f}} - \frac{\partial \mathbf{G}}{\partial t} . \tag{2.154}$$

Heat transfer equation:

$$\rho T \frac{d\hat{s}}{dt} = -\nabla \cdot \mathbf{j}_q + \boldsymbol{E} \cdot \boldsymbol{j}_e . \tag{2.155}$$

Infinitesimal strain-displacement relation:

$$\boldsymbol{\varepsilon} = (\nabla \mathbf{u} + \mathbf{u} \nabla)/2 . \tag{2.156}$$

Constitutive relations (linearized theory):

$${}_E\sigma_{kl} = c_{klmn}\varepsilon_{mn} - e_{mkl}E_m - h_{mkl}H_m - \beta_{kl}\theta , \tag{2.157}$$

$$D_k = e_{klm}\varepsilon_{lm} + \kappa_{kl}E_l + g_{kl}H_l + \omega_k\theta , \tag{2.158}$$

$$B_k = h_{klm}\varepsilon_{lm} + g_{lk}E_l + \mu_{kl}H_l + \mu_0\gamma_k\theta , \tag{2.159}$$

$$\rho_0 \hat{s} = \beta_{kl}\varepsilon_{kl} + \omega_k E_k + \mu_0 \gamma_k H_k + \frac{1}{T_0} C_v \theta , \tag{2.160}$$

$$\mathbf{j}_q = -\mathbf{k}^{qq} \cdot \nabla\theta + T_0 \mathbf{k}^{qe} \cdot \boldsymbol{E} , \tag{2.161}$$

$$\boldsymbol{j}_e = -\mathbf{k}^{eq} \cdot \nabla\theta + \mathbf{k}^{ee} \cdot \boldsymbol{E} . \tag{2.162}$$

Jump conditions:

$$\mathbf{n} \cdot [[\mathbf{D}]] = \varpi_f , \tag{2.163}$$

$$\mathbf{n} \times [[\boldsymbol{E}]] = 0 , \tag{2.164}$$

$$\mathbf{n} \cdot [[\mathbf{B}]] = 0 , \tag{2.165}$$

$$\mathbf{n} \times [[\boldsymbol{H}]] = 0 , \tag{2.166}$$

$$\mathbf{n} \cdot [[\boldsymbol{j}_e + q_f(\mathbf{v} - \boldsymbol{v}_s)]] + \frac{\bar{\delta}\varpi_f}{\bar{\delta}t} = 0 , \tag{2.167}$$

$$\mathbf{n} \cdot [[\rho(\mathbf{v} - \boldsymbol{v}_s)]] = 0 , \tag{2.168}$$

$$\mathbf{n} \cdot [[(\mathbf{v} - \boldsymbol{v}_s) \otimes \rho\mathbf{v} - ({}_t\boldsymbol{\sigma} + \boldsymbol{v}_s \otimes \mathbf{G})]] = 0 , \tag{2.169}$$

$$\mathbf{n} \cdot [[(\mathbf{v} - \boldsymbol{v}_s)(\rho\hat{e} + \rho\hat{k} + {}_{em}u^f) + \mathbf{j}_q - ({}_t\boldsymbol{\sigma} + \mathbf{v} \otimes \mathbf{G}) \cdot \mathbf{v} + \boldsymbol{S}]] = 0. \tag{2.170}$$

Initial conditions:

$$\mathbf{u}\big|_{t=t_0} = \mathbf{u}_0 , \tag{2.171}$$

$$\dot{\mathbf{u}}\big|_{t=t_0} = \mathbf{v}_0 , \tag{2.172}$$

$$T\big|_{t=t_0} = T_0 , \tag{2.173}$$

$$\mathbf{E}\big|_{t=t_0} = \mathbf{E}_0 , \tag{2.174}$$

$$\mathbf{H}\big|_{t=t_0} = \mathbf{H}_0 . \tag{2.175}$$

Boundary conditions may be obtained by letting $\boldsymbol{v}_s = \mathbf{v}$ in the jump conditions (2.163)–(2.170). Prescribed boundary and initial conditions must be compatible with uniqueness of solution.

In the classical linear theory of piezoelectricity (Voigt, 1910) and its extension, all fields are small so that stress and momentum due to the electromagnetic effects become second-order and can be omitted in the formulation, i.e., ${}_{em}\boldsymbol{\sigma} \approx 0$, ${}_{F}\boldsymbol{\sigma} \approx 0$, $\mathbf{G} \approx 0$.

In practice, it is sometimes convenient to introduce the so-called Voigt's notation with the relations between the indices 11 → 1, 22 → 2, 33 → 3, 23 → 4, 31 → 5, and 12 → 6 so that constitutive equations (2.157)–(2.162) may be expressed in matrix form as

$$\{\tau_p\} = [c_{pq}]\{\varepsilon_q\} - [e_{mp}]^T\{E_m\} - [h_{mp}]^T\{H_m\} - \{\beta_p\}\theta, \quad (2.176)$$

$$\{D_k\} = [e_{kq}]\{\varepsilon_q\} + [\kappa_{kl}]\{E_l\} + [g_{kl}]\{H_l\} + \{\omega_k\}\theta, \quad (2.177)$$

$$\{B_k\} = [h_{kq}]\{\varepsilon_q\} + [g_{lk}]^T\{E_l\} + [\mu_{kl}]\{H_l\} + \mu_0\{\gamma_k\}\theta, \quad (2.178)$$

$$\rho_0\hat{s} = \{\beta_q\}^T\{\varepsilon_q\} + \{\omega_k\}^T\{E_k\} + \mu_0\{\gamma_k\}^T\{H_k\} + \frac{1}{T_0}C_v\theta, \quad (2.179)$$

$$\{j_k^q\} = -[k_{kl}^{qq}]\cdot\{\theta_{,l}\} + T_0[k_{kl}^{qe}]\{E_l\}, \quad (2.180)$$

$$\{j_k^e\} = -[k_{kl}^{eq}]\cdot\{\theta_{,l}\} + [k_{kl}^{ee}]\{E_l\}, \quad (2.181)$$

where

$$\{\tau_p\} = \{{}_E\sigma_{11}, {}_E\sigma_{22}, {}_E\sigma_{33}, {}_E\sigma_{23}, {}_E\sigma_{31}, {}_E\sigma_{12}\}^T, \quad (2.182)$$

$$\{\varepsilon_p\} = \{\varepsilon_{11}, \varepsilon_{22}, \varepsilon_{33}, 2\varepsilon_{23}, 2\varepsilon_{31}, 2\varepsilon_{12}\}^T, \quad (2.183)$$

$$[c_{pq}] = \begin{bmatrix} c_{11} & c_{12} & c_{13} & c_{14} & c_{15} & c_{16} \\ c_{12} & c_{22} & c_{23} & c_{24} & c_{25} & c_{26} \\ c_{13} & c_{23} & c_{33} & c_{34} & c_{35} & c_{36} \\ c_{14} & c_{24} & c_{34} & c_{44} & c_{45} & c_{46} \\ c_{15} & c_{25} & c_{35} & c_{45} & c_{55} & c_{56} \\ c_{16} & c_{26} & c_{36} & c_{46} & c_{56} & c_{66} \end{bmatrix}, \quad (2.184)$$

$$[\kappa_{kl}] = \begin{bmatrix} \kappa_{11} & \kappa_{12} & \kappa_{13} \\ \kappa_{12} & \kappa_{22} & \kappa_{23} \\ \kappa_{13} & \kappa_{23} & \kappa_{33} \end{bmatrix}, \quad (2.185)$$

$$[\mu_{kl}] = \begin{bmatrix} \mu_{11} & \mu_{12} & \mu_{13} \\ \mu_{12} & \mu_{22} & \mu_{23} \\ \mu_{13} & \mu_{23} & \mu_{33} \end{bmatrix}, \quad (2.186)$$

$$[e_{kq}] = \begin{bmatrix} e_{11} & e_{12} & e_{13} & e_{14} & e_{15} & e_{16} \\ e_{21} & e_{22} & e_{23} & e_{24} & e_{25} & e_{26} \\ e_{31} & e_{32} & e_{33} & e_{34} & e_{35} & e_{36} \end{bmatrix}, \tag{2.187}$$

$$[h_{kq}] = \begin{bmatrix} h_{11} & h_{12} & h_{13} & h_{14} & h_{15} & h_{16} \\ h_{21} & h_{22} & h_{23} & h_{24} & h_{25} & h_{26} \\ h_{31} & h_{32} & h_{33} & h_{34} & h_{35} & h_{36} \end{bmatrix}, \tag{2.188}$$

$$[g_{kl}] = \begin{bmatrix} g_{11} & g_{12} & g_{13} \\ g_{21} & g_{22} & g_{23} \\ g_{31} & g_{32} & g_{33} \end{bmatrix}, \tag{2.189}$$

$$\{\omega_k\}^T = \{\omega_1 \quad \omega_2 \quad \omega_3\}, \tag{2.190}$$

$$\{\gamma_k\}^T = \{\gamma_1 \quad \gamma_2 \quad \gamma_3\}, \tag{2.191}$$

$$\{\beta_q\}^T = \{\beta_1 \quad \beta_2 \quad \beta_3 \quad \beta_4 \quad \beta_5 \quad \beta_6\}, \tag{2.192}$$

$$[k_{kl}^{qq}] = \begin{bmatrix} k_{11}^{qq} & k_{12}^{qq} & k_{13}^{qq} \\ k_{12}^{qq} & k_{22}^{qq} & k_{23}^{qq} \\ k_{13}^{qq} & k_{23}^{qq} & k_{33}^{qq} \end{bmatrix}, \tag{2.193}$$

$$[k_{kl}^{ee}] = \begin{bmatrix} k_{11}^{ee} & k_{12}^{ee} & k_{13}^{ee} \\ k_{12}^{ee} & k_{22}^{ee} & k_{23}^{ee} \\ k_{13}^{ee} & k_{23}^{ee} & k_{33}^{ee} \end{bmatrix}, \tag{2.194}$$

$$[k_{kl}^{eq}] = [k_{kl}^{qe}]^T = \begin{bmatrix} k_{11}^{eq} & k_{12}^{eq} & k_{13}^{eq} \\ k_{21}^{eq} & k_{22}^{eq} & k_{23}^{eq} \\ k_{31}^{eq} & k_{32}^{eq} & k_{33}^{eq} \end{bmatrix}. \tag{2.195}$$

The number of material properties required for coupled multifield analysis depends on the material type. For a general anisotropic material, there are a total of 21 (elastic stiffness) + 6 (dielectric permittivity) + 6 (magnetic permeability) + 18 (piezoelectric coefficients) + 18 (piezomagnetic coefficients) + 9 (magnetoelectric coefficients) + 3 (pyroelectric coefficients) + 3 (pyromagnetic coefficients) + 6 (thermoelastic coefficients) + 1 (specific heat) + 6 (thermal conductivity) +

6 (electric conductivity) + 9 (thermoelectric coefficients) = 112 independent material constants.

The material constant matrices for the special case of transverse isotropy with the x_3-axis in the poling direction become

$$[c_{pq}]=\begin{bmatrix} c_{11} & c_{12} & c_{13} & 0 & 0 & 0 \\ c_{12} & c_{11} & c_{13} & 0 & 0 & 0 \\ c_{13} & c_{13} & c_{33} & 0 & 0 & 0 \\ 0 & 0 & 0 & c_{44} & 0 & 0 \\ 0 & 0 & 0 & 0 & c_{44} & 0 \\ 0 & 0 & 0 & 0 & 0 & \frac{1}{2}(c_{11}-c_{12}) \end{bmatrix}, \tag{2.196}$$

$$[\kappa_{kl}]=\begin{bmatrix} \kappa_{11} & 0 & 0 \\ 0 & \kappa_{11} & 0 \\ 0 & 0 & \kappa_{33} \end{bmatrix}, \tag{2.197}$$

$$[\mu_{kl}]=\begin{bmatrix} \mu_{11} & 0 & 0 \\ 0 & \mu_{11} & 0 \\ 0 & 0 & \mu_{33} \end{bmatrix}, \tag{2.198}$$

$$[e_{kq}]=\begin{bmatrix} 0 & 0 & 0 & 0 & e_{15} & 0 \\ 0 & 0 & 0 & e_{15} & 0 & 0 \\ e_{31} & e_{31} & e_{33} & 0 & 0 & 0 \end{bmatrix}, \tag{2.199}$$

$$[h_{kq}]=\begin{bmatrix} 0 & 0 & 0 & 0 & h_{15} & 0 \\ 0 & 0 & 0 & h_{15} & 0 & 0 \\ h_{31} & h_{31} & h_{33} & 0 & 0 & 0 \end{bmatrix}, \tag{2.200}$$

$$[g_{kl}]=\begin{bmatrix} g_{11} & 0 & 0 \\ 0 & g_{11} & 0 \\ 0 & 0 & g_{33} \end{bmatrix}, \tag{2.201}$$

$$\{\omega_k\}^T=\{0 \quad 0 \quad \omega_3\}, \tag{2.202}$$

$$\{\gamma_k\}^T=\{0 \quad 0 \quad \gamma_3\}, \tag{2.203}$$

$$\{\beta_q\}^T=\{\beta_1 \quad \beta_1 \quad \beta_3 \quad 0 \quad 0 \quad 0\}, \tag{2.204}$$

$$[k_{kl}^{qq}] = \begin{bmatrix} k_{11}^{qq} & 0 & 0 \\ 0 & k_{11}^{qq} & 0 \\ 0 & 0 & k_{33}^{qq} \end{bmatrix}, \tag{2.205}$$

$$[k_{kl}^{ee}] = \begin{bmatrix} k_{11}^{ee} & 0 & 0 \\ 0 & k_{11}^{ee} & 0 \\ 0 & 0 & k_{33}^{ee} \end{bmatrix}, \tag{2.206}$$

$$[k_{kl}^{eq}] = [k_{kl}^{qe}]^T = \begin{bmatrix} k_{11}^{eq} & 0 & 0 \\ 0 & k_{11}^{eq} & 0 \\ 0 & 0 & k_{33}^{eq} \end{bmatrix}. \tag{2.207}$$

It can be seen that the transversely isotropic material symmetry reduces the number of independent material constants from 112 to 28, comprising 5 (elastic stiffness) + 2 (dielectric permittivity) + 2 (magnetic permeability) + 3 (piezoelectric coefficients) + 3 (piezomagnetic coefficients) + 2 (magnetoelectric coefficients) + 1 (pyroelectric coefficient) + 1 (pyromagnetic coefficient) + 2 (thermoelastic coefficients) + 1 (specific heat) + 2 (thermal conductivity) + 2 (electric conductivity) + 2 (thermoelectric coefficients).

In particular, for piezoelectric materials with hexagonal symmetry (class $C_{6v} = 6\text{mm}$) in the absence of electricity conduction, the number of independent material constants may be further reduced to 16, consisting of 5 (elastic stiffness) + 2 (dielectric permittivity) + 3 (piezoelectric coefficients) + 1 (pyroelectric coefficient) + 2 (thermoelastic coefficients) + 1 (specific heat) + 2 (thermal conductivity).

In crystallography, a point group, also called a crystal class, is a set of symmetry operations like rotations or reflections. There are 32 possible combinations of symmetry operations, resulting in 32 point groups. 20 of the 32 crystal classes exhibit piezoelectricity, which is the property of nearly all non-centrosymmetric crystals. Only 10 of the 20 piezoelectric crystal classes exhibit pyroelectricity, which is the property of all polar crystals. Crystal classes are commonly represented in the Schoenflies notation and the Hermann–Mauguin notation (Ikeda, 1990; Hahn, 2005). C_{6v} is in the Schoenflies notation, where the character "C" is for cyclic, the subscript "6" for six-fold rotation axis and "v" for vertical mirror planes containing the axis of rotation. 6mm is in the Hermann–Mauguin notation, where the first character refers to the primary symmetry direction (six-fold rotation axis) and the second and third characters refer to the secondary and tertiary symmetry directions (mirror planes).

The quasi-electrostatic approximation indicates that there is almost no change in the magnetic field with time. Thus, Faraday's law (2.150) can be simplified to

$$\nabla \times \mathbf{E} = 0 . \qquad (2.208)$$

As a result, the electric field **E** is related to a scalar function called the electric potential ϕ through

$$\mathbf{E} = -\nabla \phi . \qquad (2.209)$$

The quasi-magnetostatic approximation indicates that there is almost no change in the electric displacement with time. Thus, Ampere's law (2.152) can be simplified to

$$\nabla \times \mathbf{H} = \mathbf{j}_e . \qquad (2.210)$$

If the electric current can be ignored, the magnetic field **H** is related to a scalar function called the magnetic potential ψ through

$$\mathbf{H} = -\nabla \psi . \qquad (2.211)$$

For some applications, both quasi-electrostatic and quasi-magnetostatic approximations may be made without loss of accuracy. Since the basic field equations for anisotropic magneto-electro-thermo-elastic problems have a similar mathematical structure to those for anisotropic elastic and thermoelastic problems, the existing solution procedures in anisotropy elasticity as reviewed by Ting (1996, 2000) can be readily extended to multifield analysis. Among the powerful techniques for solving two-dimensional problems in anisotropic elastic materials, the Lekhnitskii formalism (Lekhnitskii, 1950) starts with the stress functions and then the compatibility equations, whereas the Stroh formalism (Stroh, 1958) starts with the displacements and then the equilibrium equations. The solution techniques for crack problems in electromagnetic materials will be discussed in Chapters 4–8.

Chapter 3

Introduction to Thermoviscoelasticity

Hysteresis effects pose new challenges for the modeling of deformation and fracture processes in time-dependent materials. In this chapter, we will outline the basic equations of thermoelasticity, viscoelasticity and thermoviscoelasticity as prerequisites for understanding the subject matter in later chapters. Further information may be found in the books by Williams (1973), Eringen (1980), Ferry (1980), Christensen (1982), Ward (1983) and Fung and Tong (2001). The reader who is familiar with these theories may skip this chapter.

3.1 Thermoelasticity

Consider an anisotropic thermoelastic body subjected to external forces and heating. The reference state is taken as the initial stress-free state at the reference absolute temperature, T_0. The temperature change from the reference state is

$$\theta = T - T_0, \tag{3.1}$$

where T is the instantaneous absolute temperature.

A transient coupled thermoelastic problem is mathematically formulated with basic equations and appropriate boundary and initial conditions as follows.

Equation of continuity:

$$\dot{\rho} = -\rho v_{i,i}. \tag{3.2}$$

Equations of motion:

$$\rho \dot{v}_i = \sigma_{ij,j} + \rho \hat{f}_i. \tag{3.3}$$

Infinitesimal strain-displacement relation:

$$\varepsilon_{ij} = \frac{1}{2}(u_{i,j} + u_{j,i}). \tag{3.4}$$

Duhamel–Neumann relation:

$$\sigma_{ij} = c_{ijkl}\varepsilon_{kl} - \beta_{ij}\theta. \tag{3.5}$$

Fourier's law for heat conduction:

$$j_i^q = -k_{ij}\theta_{,j}. \tag{3.6}$$

Heat transfer equation:

$$\rho T_0 \beta_{kl}\dot{\varepsilon}_{kl} + \rho C_v \dot{\theta} = -j_{i,i}^q. \tag{3.7}$$

Initial conditions:

$$\mathbf{u}\big|_{t=t_0} = \mathbf{u}_0, \tag{3.8}$$

$$\dot{\mathbf{u}}\big|_{t=t_0} = \mathbf{v}_0, \tag{3.9}$$

$$\theta\big|_{t=t_0} = 0. \tag{3.10}$$

Boundary conditions:

$$\mathbf{u} = \mathbf{u}_B \text{ on } S_u, \tag{3.11}$$

$$\mathbf{n}\cdot\boldsymbol{\sigma} = \mathbf{t}_B \text{ on } S_\sigma, \tag{3.12}$$

$$\theta = \theta_B \text{ on } S_\theta, \tag{3.13}$$

$$\mathbf{n}\cdot\mathbf{j}_q = q_B \text{ on } S_q, \tag{3.14}$$

where $S_{[\,]}$ refers to a certain part of the boundary: displacement is prescribed on S_u, traction on S_σ (the complement of S_u), temperature on S_θ, and heat flux on S_q (the complement of S_θ). Therefore, we have $S_u \cup S_\sigma = S$ and $S_\theta \cup S_q = S$. Other mixed boundary conditions may also be possible which satisfy the existence and uniqueness theorem (Eringen, 1980; Fung and Tong, 2001).

It is usually rather difficult to solve boundary-initial value problems involving coupled effects under transient conditions. As discussed by Fung and Tong (2001), uncoupled, quasi-static approximations may be made in most engineering applications by omitting the coupling term in the heat transfer equation (3.7) and the inertia term in the equations of motion (3.3).

3.2 Viscoelasticity

Viscoelastic materials such as polymers display the characteristics of both elastic solids and viscous fluids. Accelerated test methods have been developed based on the time-temperature superposition principle (TTSP) under the approximation that at higher temperatures and shorter time periods a polymeric material will behave the same as at lower temperatures and longer time periods. The material behavior can be modeled by various combinations of springs and dashpots to represent elastic and viscous components. The simplest models of linear viscoelasticity are the Maxwell model and the Kelvin–Voigt model (Ward, 1983; Fung and Tong, 2001), which comprise an elastic spring and a viscous dashpot in series or parallel (Fig. 3.1).

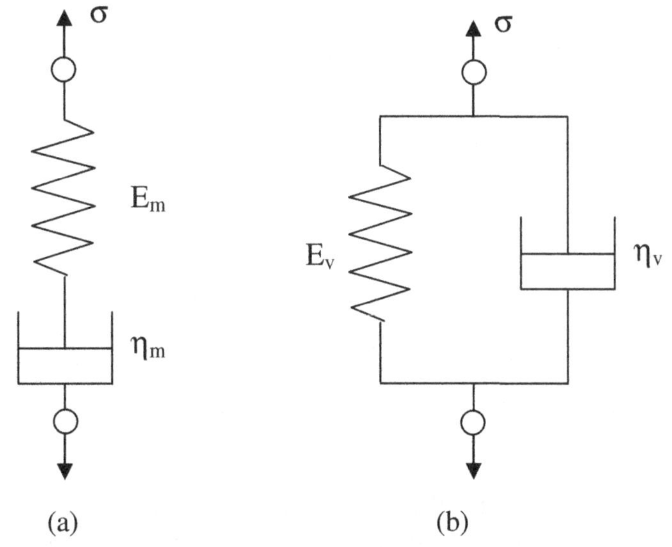

Fig. 3.1. Simplest models of linear viscoelasticity: (a) Maxwell model and (b) Kelvin–Voigt model.

For the Maxwell model, the spring stress is equal to the dashpot stress, and the total strain is a sum of the spring strain and the dashpot strain. Thus, we have

$$\frac{d\varepsilon}{dt} = \frac{1}{E_m}\frac{d\sigma}{dt} + \frac{\sigma}{\eta_m}. \tag{3.15}$$

By integration with a specified step function for strain $\varepsilon = \varepsilon_0 h(t)$, Eq. (3.15) becomes

$$\sigma(t) = E_m \varepsilon_0 \exp\left(-\frac{t}{\tau}\right), \tag{3.16}$$

where $\tau = \eta_m / E_m$ is called the relaxation time.

The generalized Maxwell model is represented by multiple Maxwell elements arranged in parallel, that is,

$$\sigma(t) = \sum_n E_n \int_{-\infty}^{t} \exp\left(-\frac{t-t^{'}}{\tau_n}\right)\frac{d\varepsilon(t^{'})}{dt^{'}}dt^{'}, \tag{3.17}$$

where E_n and τ_n are the spring constant and the relaxation time of the nth Maxwell element.

The above summation can be written as a convolution integral

$$\sigma(t) = \int_{-\infty}^{t} G(t-t^{'})\frac{d\varepsilon(t^{'})}{dt^{'}}dt^{'}, \tag{3.18}$$

where $G(t) = G_r + \int_0^\infty H(\tau)\exp(-t/\tau)d\tau$ is called the relaxation function, G_r is the relaxed modulus, and $H(\tau)$ is the relaxation time spectrum.

Let $\sigma_{ij}(t)$ and $\varepsilon_{ij}(t)$ be the stress and strain tensors defined in the time interval $-\infty < t < \infty$. Equation (3.18) can be extended to a three-dimensional, anisotropic constitutive law of the relaxation type:

$$\sigma_{ij}(t) = \int_{-\infty}^{t} G_{ijkl}(t-t^{'})\frac{d\varepsilon_{kl}(t^{'})}{dt^{'}}dt^{'}, \tag{3.19}$$

where $G_{ijkl}(t)$ is called the tensorial relaxation function.

By contrast, for the Kelvin–Voigt model, the spring strain is equal to the dashpot strain, and the total stress is a sum of the spring stress and the dashpot stress. Thus, we have

$$\sigma = E_v \varepsilon + \eta_v \frac{d\varepsilon}{dt}. \tag{3.20}$$

By integration with a specified step function for stress, $\sigma = \sigma_0 h(t)$, Eq. (3.20) becomes

$$\varepsilon = \frac{\sigma_0}{E_v}\left[1 - \exp\left(-\frac{t}{\tau}\right)\right], \tag{3.21}$$

where $\tau = \eta_v / E_v$ is called the retardation time.

The generalized Kelvin–Voigt model is represented by multiple Kelvin–Voigt elements arranged in series, that is,

$$\varepsilon(t) = \sum_n \frac{1}{E_n}\int_{-\infty}^{t}\left[1 - \exp\left(-\frac{t-t'}{\tau_n}\right)\right]\frac{d\sigma(t')}{dt'}dt', \tag{3.22}$$

where E_n and τ_n are the spring constant and the retardation time of the nth Kelvin–Voigt element, respectively.

The above summation can be written as a convolution integral

$$\varepsilon(t) = \int_{-\infty}^{t} J(t-t')\frac{d\sigma(t')}{dt'}dt', \tag{3.23}$$

where $J(t) = J_u + \int_0^{\infty} L(\tau)[1 - \exp(-t/\tau)]d\tau$ is called the creep function, J_u is the instantaneous compliance, and $L(\tau)$ is the retardation time spectrum.

Similarly, Eq. (3.23) can be extended to a three-dimensional, anisotropic constitutive law of the creep type:

$$\varepsilon_{ij}(t) = \int_{-\infty}^{t} J_{ijkl}(t-t')\frac{d\sigma_{kl}(t')}{dt'}dt', \tag{3.24}$$

where $J_{ijkl}(t)$ is called the tensorial creep function.

It has been shown that the inverse of Eq. (3.19) exists and can be written as Eq. (3.24) if $G_{ijkl}(t)$ is twice differentiable and the initial value of $G_{ijkl}(t)$ at $t = 0$ is nonzero (Gurtin and Sternberg, 1962). Discussions on the thermodynamic restrictions on these functions and applications of the Laplace transform to solving linear viscoelastic problems can be found in the books by Christensen (1982) and Fung and Tong (2001).

3.3 Coupled Theory of Thermoviscoelasticity

The fundamental principles of thermodynamics have been applied to physical, chemical, mechanical, and biological processes in several ways (e.g., Groot and Mazur, 1962; Eringen, 1980; Truesdell, 1984; Muller and Ruggeri, 1993; Fung and Tong, 2001; Truesdell and Noll, 2004). As pointed out by Schapery (2000) in an overview of constitutive, fracture, and strength models for nonlinear viscoelastic solids, nonequilibrium thermodynamic approaches are, essentially, of two types. In state-variable thermodynamics, the free energy is expressed as a function of current strain (stress), temperature, and other variables, including so-called internal state variables (e.g., Coleman and Gurtin, 1967; Rice, 1971; Schapery, 1969, 1994, 1997, 1999; Horstemeyer and Bammann, 2010). In functional thermodynamics, the free energy is expressed as a functional of the histories of strain (stress), temperature, etc. (e.g., Coleman, 1964; Crochet and Naghdi, 1969; Cost, 1973; Eringen, 1980; Christensen, 1982; Lustig *et al.*, 1996; Caruthers *et al.*, 2004; Chen and Wang, 2006). The use of "functional" as a mathematical term originates in the calculus of variations, which is concerned with the minimization of a functional with its arguments as functions. Here, we will summarize the thermodynamic formulation of a coupled theory of thermoviscoelasticity at finite deformation.

3.3.1 *Fundamental principles of thermodynamics*

Based on the first principle of thermodynamics, the local balance equations of mass, momentum, moment of momentum, and energy are given by

$$\frac{d\rho}{dt} = -\rho\nabla\cdot\mathbf{v}\,, \tag{3.25}$$

$$\rho\frac{d\mathbf{v}}{dt} = \nabla\cdot\boldsymbol{\sigma} + \rho\hat{\mathbf{f}}\,, \tag{3.26}$$

$$\boldsymbol{\sigma}^T = \boldsymbol{\sigma}\,, \tag{3.27}$$

$$\rho\frac{d\hat{e}}{dt} = -\nabla\cdot\mathbf{j}_q + \boldsymbol{\sigma}:\mathbf{v}\nabla\,, \tag{3.28}$$

where ρ is mass density, $\mathbf{v} = \dot{\mathbf{u}}$ is velocity, $\mathbf{u}$ is displacement, $\hat{\mathbf{f}}$ is mechanical body force per unit mass, $\boldsymbol{\sigma}$ is the Cauchy stress tensor, $\hat{e}$ is internal energy per unit mass, and $\mathbf{j}_q$ is heat flux.

Based on the second principle of thermodynamics, the entropy production inequality is given by

$$\frac{d_i \hat{s}}{dt} \equiv \frac{d\hat{s}}{dt} + \frac{1}{\rho} \nabla \cdot \mathbf{j}_s \geq 0, \tag{3.29}$$

where $\hat{s}$ is the entropy per unit mass and $\mathbf{j}_s$ is the entropy flux.

In the reference configuration V_R, the entropy production inequality can be rewritten as

$$\frac{d_i \hat{s}}{dt} = \frac{d\hat{s}}{dt} + \frac{1}{\rho_0} \nabla_R \cdot \mathbf{J}_s \geq 0, \tag{3.30}$$

where ρ_0 is the mass density in the reference configuration, $\mathbf{J}_s = j\mathbf{F}^{-1} \cdot \mathbf{j}_s$ is the entropy flux in the reference configuration, $j = \rho_0 / \rho = \det(\mathbf{F})$ is the Jacobian determinant, $\mathbf{F} = \partial \mathbf{x} / \partial \mathbf{X}$ is the deformation gradient, $\mathbf{X}$ is the position in the reference configuration, and $\mathbf{x} = \boldsymbol{\chi}(\mathbf{X}, t)$ is the position in the current configuration.

3.3.2 *Formulation based on Helmholtz free energy functional*

With the use of the Helmholtz free energy $\hat{h} = \hat{e} - T\hat{s}$, the local energy balance equation (3.28) can be rewritten in the reference configuration V_R as

$$\frac{d\hat{s}}{dt} = -\frac{1}{\rho_0} \nabla_R \cdot \frac{\mathbf{J}_q}{T} + \frac{1}{\rho_0} \mathbf{J}_q \cdot \nabla_R \frac{1}{T} + \frac{1}{2\rho_0 T} \boldsymbol{\Sigma} : \dot{\mathbf{C}} - \frac{1}{T} \hat{s}\dot{T} - \frac{1}{T} \frac{d\hat{h}}{dt}, \tag{3.31}$$

where T is absolute temperature, $\mathbf{J}_q = j\mathbf{F}^{-1} \cdot \mathbf{j}_q$ is heat flux in the reference configuration, $\boldsymbol{\Sigma} = j\mathbf{F}^{-1} \cdot \boldsymbol{\sigma} \cdot \mathbf{F}^{-T}$ is the second Piola–Kirchhoff stress tensor, and $\mathbf{C} = \mathbf{F}^T \cdot \mathbf{F}$ is the right Cauchy–Green deformation tensor.

The Lagrange strain measure and the temperature deviation are given by

$$\mathbf{E} = \frac{1}{2}(\mathbf{C} - \mathbf{I}), \tag{3.32}$$

$$\theta = T - T_0 \ . \tag{3.33}$$

The Helmholtz free energy is taken to be a functional of the histories of strain and temperature. In thermodynamics of materials with memory, the fading memory hypothesis states that the influence of long past events should be weaker than that of recent ones in determining the material response (Coleman and Noll, 1960; Coleman, 1964; Truesdell and Noll, 2004). Expansion of the Helmholtz free energy functional for materials with fading memory up to the second order yields

$$\begin{aligned}\rho_0 \hat{h} &= \rho_0 \hat{h}_0 + \int_{-\infty}^{t} L_{IJ}(\mathbf{X}, t-\tau) \frac{\partial \mathrm{E}_{IJ}(\mathbf{X},\tau)}{\partial \tau} d\tau \\ &- \int_{-\infty}^{t} M(\mathbf{X}, t-\tau) \frac{\partial \theta(\mathbf{X},\tau)}{\partial \tau} d\tau \\ &+ \frac{1}{2} \int_{-\infty}^{t} \int_{-\infty}^{t} G_{IJKL}(\mathbf{X}, t-\tau, t-\zeta) \frac{\partial \mathrm{E}_{IJ}(\mathbf{X},\tau)}{\partial \tau} \frac{\partial \mathrm{E}_{KL}(\mathbf{X},\zeta)}{\partial \zeta} d\tau d\zeta \\ &- \int_{-\infty}^{t} \int_{-\infty}^{t} \beta_{IJ}(\mathbf{X}, t-\tau, t-\zeta) \frac{\partial \mathrm{E}_{IJ}(\mathbf{X},\tau)}{\partial \tau} \frac{\partial \theta(\mathbf{X},\zeta)}{\partial \zeta} d\tau d\zeta \\ &- \frac{1}{2T_0} \int_{-\infty}^{t} \int_{-\infty}^{t} C_H(\mathbf{X}, t-\tau, t-\zeta) \frac{\partial \theta(\mathbf{X},\tau)}{\partial \tau} \frac{\partial \theta(\mathbf{X},\zeta)}{\partial \zeta} d\tau d\zeta,\end{aligned} \tag{3.34}$$

where $\hat{h}_0$ is the value of the Helmholtz free energy in the reference state (i.e., $\mathbf{E} = 0$, $T = T_0$), $G_{IJKL}(\mathbf{X}, t-\tau, t-\zeta) = G_{KLIJ}(\mathbf{X}, t-\zeta, t-\tau)$, and $C_H(\mathbf{X}, t-\tau, t-\zeta) = C_H(\mathbf{X}, t-\zeta, t-\tau)$.

Substituting Eqs. (3.31) and (3.34) into (3.30) and performing differentiation with respect to time using the Leibnitz rule, we obtain

$$\begin{aligned}\frac{d_i \hat{s}}{dt} &= \frac{1}{\rho_0 T} [\Sigma_{IJ} - L_{IJ}^0 - \int_{-\infty}^{t} G_{IJKL}(\mathbf{X}, 0, t-\zeta) \frac{\partial \mathrm{E}_{KL}(\mathbf{X},\zeta)}{\partial \zeta} d\zeta \\ &+ \int_{-\infty}^{t} \beta_{IJ}(\mathbf{X}, 0, t-\zeta) \frac{\partial \theta(\mathbf{X},\zeta)}{\partial \zeta} d\zeta] \dot{\mathrm{E}}_{IJ} \\ &- \frac{1}{\rho_0 T} [\rho_0 \hat{s} - M^0 - \int_{-\infty}^{t} \beta_{IJ}(\mathbf{X}, t-\tau, 0) \frac{\partial \mathrm{E}_{IJ}(\mathbf{X},\tau)}{\partial \tau} d\tau \\ &- \frac{1}{T_0} \int_{-\infty}^{t} C_H(\mathbf{X}, t-\tau, 0) \frac{\partial \theta(\mathbf{X},\tau)}{\partial \tau} d\tau] \dot{T} \\ &+ \frac{1}{T} \hat{\Lambda} + \frac{1}{\rho_0} \mathbf{J}_q \cdot \nabla_R \frac{1}{T} + \frac{1}{\rho_0} \nabla_R \cdot (\mathbf{J}_s - \frac{\mathbf{J}_q}{T}) \geq 0,\end{aligned} \tag{3.35}$$

where

$$\rho_0\hat{\Lambda} = -\int_{-\infty}^{t} \frac{\partial}{\partial t} L_{IJ}(\mathbf{X}, t-\tau)\frac{\partial \mathrm{E}_{IJ}(\mathbf{X},\tau)}{\partial \tau} d\tau$$
$$+\int_{-\infty}^{t} \frac{\partial}{\partial t} M(\mathbf{X}, t-\tau)\frac{\partial \theta(\mathbf{X},\tau)}{\partial \tau} d\tau$$
$$-\frac{1}{2}\int_{-\infty}^{t}\int_{-\infty}^{t} \frac{\partial}{\partial t} G_{IJKL}(\mathbf{X}, t-\tau, t-\zeta)\frac{\partial \mathrm{E}_{IJ}(\mathbf{X},\tau)}{\partial \tau}\frac{\partial \mathrm{E}_{KL}(\mathbf{X},\zeta)}{\partial \zeta} d\tau d\zeta \tag{3.36}$$
$$+\int_{-\infty}^{t}\int_{-\infty}^{t} \frac{\partial}{\partial t} \beta_{IJ}(\mathbf{X}, t-\tau, t-\zeta)\frac{\partial \mathrm{E}_{IJ}(\mathbf{X},\tau)}{\partial \tau}\frac{\partial \theta(\mathbf{X},\zeta)}{\partial \zeta} d\tau d\zeta$$
$$+\frac{1}{2T_0}\int_{-\infty}^{t}\int_{-\infty}^{t} \frac{\partial}{\partial t} C_H(\mathbf{X}, t-\tau, t-\zeta)\frac{\partial \theta(\mathbf{X},\tau)}{\partial \tau}\frac{\partial \theta(\mathbf{X},\zeta)}{\partial \zeta} d\tau d\zeta.$$

Since the entropy production inequality (3.35) is always valid, state equations should fulfill the following conditions:

$$\Sigma_{IJ}(\mathbf{X},t) = L_{IJ}^0(\mathbf{X}) + \int_{-\infty}^{t} G_{IJKL}(\mathbf{X},0,t-\zeta)\frac{\partial \mathrm{E}_{KL}(\mathbf{X},\zeta)}{\partial \zeta} d\zeta$$
$$-\int_{-\infty}^{t} \beta_{IJ}(\mathbf{X},0,t-\zeta)\frac{\partial \theta(\mathbf{X},\zeta)}{\partial \zeta} d\zeta, \tag{3.37}$$

$$\rho_0\hat{s}(\mathbf{X},t) = M^0(\mathbf{X}) + \int_{-\infty}^{t} \beta_{IJ}(\mathbf{X},t-\tau,0)\frac{\partial \mathrm{E}_{IJ}(\mathbf{X},\tau)}{\partial \tau} d\tau$$
$$+\frac{1}{T_0}\int_{-\infty}^{t} C_H(\mathbf{X},t-\tau,0)\frac{\partial \theta(\mathbf{X},\tau)}{\partial \tau} d\tau, \tag{3.38}$$

$$\mathbf{J}_s = \frac{1}{T}\mathbf{J}_q, \tag{3.39}$$

$$\frac{d_i\hat{s}}{dt} = \frac{1}{\rho_0}\mathbf{J}_q \cdot \nabla_R \frac{1}{T} + \frac{\hat{\Lambda}}{T} \geq 0, \tag{3.40}$$

where $G_{IJKL}(\mathbf{X},0,t-\zeta)$, $C_H(\mathbf{X},t-\tau,0)$, $\beta_{IJ}(\mathbf{X},0,t-\zeta)$, and $\beta_{IJ}(\mathbf{X},t-\tau,0)$ are appropriate memory functions.

The first terms L_{IJ}^0 and M^0 on the right-hand sides of Eqs. (3.37) and (3.38) stand for the values of Σ_{IJ} and $\rho_0\hat{s}$ in the reference state, the second terms for mechanical contribution, and the third terms for thermal contribution. It is shown from (3.40) that the total dissipation is associated with heat conduction and time-dependent dissipation. Since the inequality (3.40) should always be satisfied, kinetic laws for specific irreversible processes may be determined accordingly.

The time-dependent dissipation rate per unit mass satisfies the following inequality:

$$\hat{\Lambda} \geq 0\,. \tag{3.41}$$

It is proposed that the thermodynamic flux for heat conduction depends linearly on the corresponding thermodynamic force, that is,

$$\mathbf{J}_q = \hat{\mathbf{L}}^{qq} \cdot \nabla_R \frac{1}{T}, \tag{3.42}$$

where $\hat{\mathbf{L}}^{qqT} = \hat{\mathbf{L}}^{qq}$ is positive definite.

Substituting (3.38) and (3.42) into (3.31) yields the following coupled heat transfer equation based on the Helmholtz free energy functional:

$$\begin{aligned} &\frac{d}{dt}[\int_{-\infty}^{t} \beta_{IJ}(\mathbf{X}, t-\tau, 0) \frac{\partial \mathrm{E}_{IJ}(\mathbf{X},\tau)}{\partial \tau} d\tau \\ &+ \frac{1}{T_0}\int_{-\infty}^{t} C_H(\mathbf{X}, t-\tau, 0) \frac{\partial \theta(\mathbf{X},\tau)}{\partial \tau} d\tau] \\ &= \frac{1}{T_0^3} \nabla_R \cdot \left(\hat{\mathbf{L}}^{qq} \cdot \nabla_R \theta\right) + \frac{1}{T_0} \rho_0 \hat{\Lambda}, \end{aligned} \tag{3.43}$$

where the integral involving the strain history gives rise to a coupling between thermal and mechanical effects.

3.3.3 *Formulation based on Gibbs free energy functional*

With Gibbs free energy $\hat{g} = \hat{e} - T\hat{s} - \boldsymbol{\Sigma} : \mathbf{E}/\rho_0$, the local energy balance equation (3.28) can be rewritten in the reference configuration V_R as

$$\frac{d\hat{s}}{dt} = -\frac{1}{\rho_0} \nabla_R \cdot \frac{\mathbf{J}_q}{T} + \frac{1}{\rho_0} \mathbf{J}_q \cdot \nabla_R \frac{1}{T} - \frac{1}{\rho_0 T} \dot{\boldsymbol{\Sigma}} : \mathbf{E} - \frac{1}{T} \hat{s}\dot{T} - \frac{1}{T}\frac{d\hat{g}}{dt}. \tag{3.44}$$

The Gibbs free energy is taken to be a functional of the histories of stress and temperature. Expansion of the Gibbs free energy functional for materials with fading memory up to the second order yields

$$
\begin{aligned}
-\rho_0 \hat{g} = &-\rho_0 \hat{g}_0 + \int_{-\infty}^{t} L_{IJ}(\mathbf{X}, t-\tau) \frac{\partial \Sigma_{IJ}(\mathbf{X},\tau)}{\partial \tau} d\tau \\
&+ \int_{-\infty}^{t} M(\mathbf{X}, t-\tau) \frac{\partial \theta(\mathbf{X},\tau)}{\partial \tau} d\tau \\
&+ \frac{1}{2} \int_{-\infty}^{t} \int_{-\infty}^{t} J_{IJKL}(\mathbf{X}, t-\tau, t-\zeta) \frac{\partial \Sigma_{IJ}(\mathbf{X},\tau)}{\partial \tau} \frac{\partial \Sigma_{KL}(\mathbf{X},\zeta)}{\partial \zeta} d\tau d\zeta \\
&+ \int_{-\infty}^{t} \int_{-\infty}^{t} \alpha_{IJ}(\mathbf{X}, t-\tau, t-\zeta) \frac{\partial \Sigma_{IJ}(\mathbf{X},\tau)}{\partial \tau} \frac{\partial \theta(\mathbf{X},\zeta)}{\partial \zeta} d\tau d\zeta \\
&+ \frac{1}{2T_0} \int_{-\infty}^{t} \int_{-\infty}^{t} C_G(\mathbf{X}, t-\tau, t-\zeta) \frac{\partial \theta(\mathbf{X},\tau)}{\partial \tau} \frac{\partial \theta(\mathbf{X},\zeta)}{\partial \zeta} d\tau d\zeta
\end{aligned}
\tag{3.45}
$$

where $\hat{g}_0$ is the value of the Gibbs free energy in the reference state (i.e., $\mathbf{\Sigma} = 0$, $T = T_0$), $J_{IJKL}(\mathbf{X}, t-\tau, t-\zeta) = J_{KLIJ}(\mathbf{X}, t-\zeta, t-\tau)$, and $C_G(\mathbf{X}, t-\tau, t-\zeta) = C_G(\mathbf{X}, t-\zeta, t-\tau)$.

Substituting Eqs. (3.44) and (3.45) into (3.30) and performing differentiation with respect to time using Leibnitz rule, we obtain

$$
\begin{aligned}
\frac{d_i \hat{s}}{dt} = &-\frac{1}{\rho_0 T}[\mathrm{E}_{IJ} - L_{IJ}^0 - \int_{-\infty}^{t} J_{IJKL}(\mathbf{X}, 0, t-\zeta) \frac{\partial \Sigma_{KL}(\mathbf{X},\zeta)}{\partial \zeta} d\zeta \\
&- \int_{-\infty}^{t} \alpha_{IJ}(\mathbf{X}, 0, t-\zeta) \frac{\partial \theta(\mathbf{X},\zeta)}{\partial \zeta} d\zeta] \dot{\Sigma}_{IJ} \\
&- \frac{1}{\rho_0 T}[\rho_0 \hat{s} - M^0 - \int_{-\infty}^{t} \alpha_{IJ}(\mathbf{X}, t-\tau, 0) \frac{\partial \Sigma_{IJ}(\mathbf{X},\tau)}{\partial \tau} d\tau \\
&- \frac{1}{T_0} \int_{-\infty}^{t} C_G(\mathbf{X}, t-\tau, 0) \frac{\partial \theta(\mathbf{X},\tau)}{\partial \tau} d\tau] \dot{T} \\
&+ \frac{1}{T} \hat{\Lambda} + \frac{1}{\rho_0} \mathbf{J}_q \cdot \nabla_R \frac{1}{T} + \frac{1}{\rho_0} \nabla_R \cdot (\mathbf{J}_s - \frac{\mathbf{J}_q}{T}) \geq 0,
\end{aligned}
\tag{3.46}
$$

where

$$\begin{aligned}
\rho_0\hat{\Lambda} &= \int_{-\infty}^{t} \frac{\partial L_{IJ}(\mathbf{X},t-\tau)}{\partial t}\frac{\partial \Sigma_{IJ}(\mathbf{X},\tau)}{\partial \tau}d\tau \\
&+\int_{-\infty}^{t} \frac{\partial M(\mathbf{X},t-\tau)}{\partial t}\frac{\partial \theta(\mathbf{X},\tau)}{\partial \tau}d\tau \\
&+\frac{1}{2}\int_{-\infty}^{t}\int_{-\infty}^{t} \frac{\partial J_{IJKL}(\mathbf{X},t-\tau,t-\zeta)}{\partial t}\frac{\partial \Sigma_{IJ}(\mathbf{X},\tau)}{\partial \tau}\frac{\partial \Sigma_{KL}(\mathbf{X},\zeta)}{\partial \zeta}d\tau d\zeta \\
&+\int_{-\infty}^{t}\int_{-\infty}^{t} \frac{\partial \alpha_{IJ}(\mathbf{X},t-\tau,t-\zeta)}{\partial t}\frac{\partial \Sigma_{IJ}(\mathbf{X},\tau)}{\partial \tau}\frac{\partial \theta(\mathbf{X},\zeta)}{\partial \zeta}d\tau d\zeta \\
&+\frac{1}{2T_0}\int_{-\infty}^{t}\int_{-\infty}^{t} \frac{\partial C_G(\mathbf{X},t-\tau,t-\zeta)}{\partial t}\frac{\partial \theta(\mathbf{X},\tau)}{\partial \tau}\frac{\partial \theta(\mathbf{X},\zeta)}{\partial \zeta}d\tau d\zeta.
\end{aligned} \tag{3.47}$$

Since the entropy production inequality (3.46) is always valid, the state equations should fulfill the following conditions:

$$\begin{aligned}
\mathrm{E}_{IJ}(\mathbf{X},t) &= L_{IJ}^0(\mathbf{X}) + \int_{-\infty}^{t} J_{IJKL}(\mathbf{X},0,t-\zeta)\frac{\partial \Sigma_{KL}(\mathbf{X},\zeta)}{\partial \zeta}d\zeta \\
&+\int_{-\infty}^{t} \alpha_{IJ}(\mathbf{X},0,t-\zeta)\frac{\partial \theta(\mathbf{X},\zeta)}{\partial \zeta}d\zeta,
\end{aligned} \tag{3.48}$$

$$\begin{aligned}
\rho_0\hat{s}(\mathbf{X},t) &= M^0(\mathbf{X}) + \int_{-\infty}^{t} \alpha_{IJ}(\mathbf{X},t-\tau,0)\frac{\partial \Sigma_{IJ}(\mathbf{X},\tau)}{\partial \tau}d\tau \\
&+\frac{1}{T_0}\int_{-\infty}^{t} C_G(\mathbf{X},t-\tau,0)\frac{\partial \theta(\mathbf{X},\tau)}{\partial \tau}d\tau,
\end{aligned} \tag{3.49}$$

$$\mathbf{J}_s = \frac{1}{T}\mathbf{J}_q, \tag{3.50}$$

$$\frac{d_i\hat{s}}{dt} = \frac{1}{\rho_0}\mathbf{J}_q \cdot \nabla_R \frac{1}{T} + \frac{\hat{\Lambda}}{T} \geq 0, \tag{3.51}$$

where $J_{IJKL}(\mathbf{X},0,t-\zeta)$, $C_G(\mathbf{X},t-\tau,0)$, $\alpha_{IJ}(\mathbf{X},0,t-\zeta)$, and $\alpha_{IJ}(\mathbf{X},t-\tau,0)$ are appropriate memory functions.

The first terms L_{IJ}^0 and M^0 on the right-hand sides of Eqs. (3.48) and (3.49) stand for the values of E_{IJ} and $\rho_0\hat{s}$ in the reference state, the second terms for mechanical contribution, and the third terms for thermal contribution. It is shown from (3.51) that the total dissipation is associated with heat conduction and time-dependent dissipation. Since the inequality (3.51) should always be satisfied, kinetic laws for specific irreversible processes may be determined accordingly.

The time-dependent dissipation rate associated with the hysteresis effect satisfies the inequality

$$\hat{\Lambda} \geq 0. \tag{3.52}$$

It is proposed that the thermodynamic flux for heat conduction depends linearly on the corresponding thermodynamic force, that is,

$$\mathbf{J}_q = \hat{\mathbf{L}}^{qq} \cdot \nabla_R \frac{1}{T}, \tag{3.53}$$

where $\mathbf{L}^{qqT} = \mathbf{L}^{qq}$ is positive definite.

Substituting (3.49) and (3.53) into (3.44) yields the following coupled heat transfer equation based on the Gibbs free energy functional:

$$\begin{aligned} &\frac{d}{dt}[\int_{-\infty}^{t} \alpha_{IJ}(\mathbf{X}, t-\tau, 0) \frac{\partial \Sigma_{IJ}(\mathbf{X}, \tau)}{\partial \tau} d\tau \\ &+ \frac{1}{T_0} \int_{-\infty}^{t} C_G(\mathbf{X}, t-\tau, 0) \frac{\partial \theta(\mathbf{X}, \tau)}{\partial \tau} d\tau] \\ &= \frac{1}{T_0^3} \nabla_R \cdot \left(\hat{\mathbf{L}}^{qq} \cdot \nabla_R \theta\right) + \frac{1}{T_0} \rho_0 \hat{\Lambda}, \end{aligned} \tag{3.54}$$

where the integral involving the stress history gives rise to a coupling between thermal and mechanical effects.

3.4 Thermoviscoelastic Boundary-Initial Value Problems

Coupled thermoviscoelastic boundary-initial value problems can be formulated with the balance equations of mass (3.25), linear momentum (3.26) and angular momentum (3.27), Green strain measure (3.32), constitutive relations (3.37) or (3.48), and heat transfer equation (3.43) or (3.54), as well as appropriate boundary and initial conditions.

In the small-strain formulation, the basic equations of coupled thermoviscoelasticity are summarized as follows:

Equation of continuity:

$$\dot{\rho} = -\rho v_{i,i}. \tag{3.55}$$

Equations of motion:

$$\rho \dot{v}_i = \sigma_{ij,j} + \rho \hat{f}_i . \tag{3.56}$$

Infinitesimal strain-displacement relation:

$$\varepsilon_{ij} = \frac{1}{2}(u_{i,j} + u_{j,i}) . \tag{3.57}$$

Constitutive relation based on Helmholtz free energy functional:

$$\begin{aligned}\sigma_{ij}(\mathbf{x},t) = L_{ij}^{0}(\mathbf{x}) + \int_{-\infty}^{t} G_{ijkl}(\mathbf{x},0,t-\zeta)\frac{\partial \varepsilon_{kl}(\mathbf{x},\zeta)}{\partial \zeta}d\zeta \\ -\int_{-\infty}^{t} \beta_{ij}(\mathbf{x},0,t-\zeta)\frac{\partial \theta(\mathbf{x},\zeta)}{\partial \zeta}d\zeta.\end{aligned} \tag{3.58a}$$

Constitutive relation based on Gibbs free energy functional:

$$\begin{aligned}\varepsilon_{ij}(\mathbf{x},t) = L_{ij}^{0}(\mathbf{x}) + \int_{-\infty}^{t} J_{ijkl}(\mathbf{x},0,t-\zeta)\frac{\partial \sigma_{KL}(\mathbf{x},\zeta)}{\partial \zeta}d\zeta \\ +\int_{-\infty}^{t} \alpha_{ij}(\mathbf{x},0,t-\zeta)\frac{\partial \theta(\mathbf{x},\zeta)}{\partial \zeta}d\zeta.\end{aligned} \tag{3.58b}$$

Heat transfer equation based on Helmholtz free energy functional:

$$\begin{aligned}&\frac{d}{dt}[\int_{-\infty}^{t} \beta_{ij}(\mathbf{x},t-\tau,0)\frac{\partial \varepsilon_{ij}(\mathbf{x},\tau)}{\partial \tau}d\tau \\ &+\frac{1}{T_0}\int_{-\infty}^{t} C_H(\mathbf{x},t-\tau,0)\frac{\partial \theta(\mathbf{x},\tau)}{\partial \tau}d\tau] \\ &=\frac{1}{T_0^3}\nabla\cdot\left(\hat{\mathbf{L}}^{qq}\cdot\nabla\theta\right)+\frac{1}{T_0}\rho_0\hat{\Lambda}.\end{aligned} \tag{3.59a}$$

Heat transfer equation based on Gibbs free energy functional:

$$\begin{aligned}&\frac{d}{dt}[\int_{-\infty}^{t} \alpha_{ij}(\mathbf{x},t-\tau,0)\frac{\partial \sigma_{ij}(\mathbf{x},\tau)}{\partial \tau}d\tau \\ &+\frac{1}{T_0}\int_{-\infty}^{t} C_G(\mathbf{x},t-\tau,0)\frac{\partial \theta(\mathbf{x},\tau)}{\partial \tau}d\tau] \\ &=\frac{1}{T_0^3}\nabla\cdot\left(\hat{\mathbf{L}}^{qq}\cdot\nabla\theta\right)+\frac{1}{T_0}\rho_0\hat{\Lambda}.\end{aligned} \tag{3.59b}$$

Initial conditions are taken as

$$\mathbf{u}(\mathbf{x},t)=\mathbf{u}_0 \ (t<0), \tag{3.60}$$

$$\dot{\mathbf{u}}(\mathbf{x},t)=\mathbf{v}_0 \ (t=0), \tag{3.61}$$

$$\theta(\mathbf{x},t)=0 \ (t<0). \tag{3.62}$$

Boundary conditions are given by

$$\mathbf{u}=\mathbf{u}_B(\mathbf{x},t) \text{ on } S_u \ (t\geq 0), \tag{3.63}$$

$$\mathbf{n}\cdot\boldsymbol{\sigma}=\mathbf{t}_B(\mathbf{x},t) \text{ on } S_\sigma \ (t\geq 0), \tag{3.64}$$

$$\theta=\theta_B(\mathbf{x},t) \text{ on } S_\theta \ (t\geq 0), \tag{3.65}$$

$$\mathbf{n}\cdot\mathbf{j}_q=q_B(\mathbf{x},t) \text{ on } S_q \ (t\geq 0), \tag{3.66}$$

where $S_{[\,]}$ refers to a certain part of the boundary: displacement is prescribed on S_u, traction on S_σ (the complement of S_u), temperature on S_θ, and heat flux on S_q (the complement of S_θ). Hence, we have $S_u \cup S_\sigma = S$ and $S_\theta \cup S_q = S$. Like thermoelastic problems, other mixed boundary conditions for thermoviscoelastic problems may also be used.

Integral transform methods provide a useful tool for solving such problems. After an integral transform, such as the Laplace transform, is applied to the basic equations for thermoviscoelastic boundary-initial value problems, the transformed boundary value problems can be solved in a manner similar to that for thermoelastic problems. The final thermoviscoelastic solution is then obtained upon inversion of the transformed solution. This analogy is often referred to as the correspondence principle. For establishment of the existence and uniqueness of solutions for linear viscoelastic and thermoviscoelstic boundary-initial value problems, the reader may refer to the paper by Onat and Breuer (1963) and the book by Christensen (1982).

Chapter 4

Overview on Fracture of Electromagnetic Materials

4.1 Introduction

Electromagnetic materials have broad civilian and defense applications such as infrastructure monitoring, electronic packaging, novel antenna designs, and biomedical devices, due to their remarkable multifunctional properties. Energy can be converted from one form to another due to interactions between magnetic, electric, thermal, and mechanical effects. However, a major concern of these materials is their susceptibility to cracking whilst in service (Fig. 4.1). Fracture of these smart material systems has become the subject of active research over the past few decades because of the rapid development of these new kinds of multifunctional materials for various engineering applications (for example, see review articles or book chapters by Qin, 2001; Trimarco and Maugin, 2001; Zhang *et al.*, 2002; Chen and Lu, 2003; Zhang and Gao, 2004; Chen and Hasebe, 2005; Schneider, 2007; Wang *et al.*, 2009; Kuna, 2010 including the references cited therein). A general consensus is that a major challenge is how to resolve the fundamental discrepancies between theoretical predictions and experimental results on crack propagation in piezoelectric materials. In this chapter, a summary on linear piezoelectric/piezomagnetic fracture mechanics is given, covering basic field equations, general solution procedures, crack-face boundary conditions, fracture criteria, and experimental observations. Some nonlinear problems for which the linear theory is not sufficient will also be discussed.

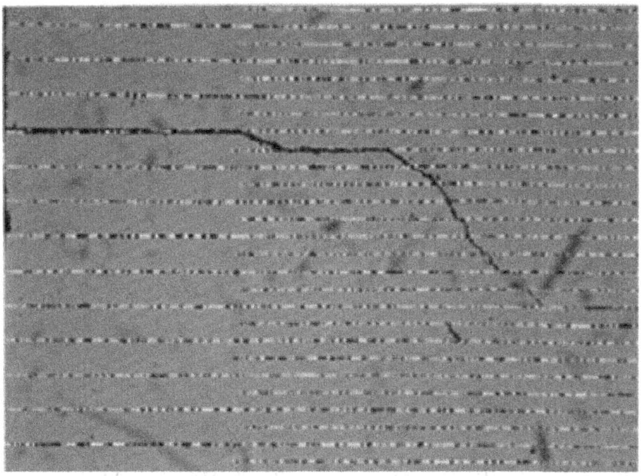

Fig. 4.1. A crack in a PZT multilayer actuator. (From Kuna, 2010, with permission from Elsevier).

4.2 Basic Field Equations

The fundamental concepts of fracture mechanics and the elements of electrodynamics are described in Chapters 1 and 2, respectively. The body of knowledge developed for describing the fracture behavior of piezoelectric materials within the framework of the classical linear theory of piezoelectricity (Voigt, 1910) is referred to as linear piezoelectric fracture mechanics (LPFM). The extension of LPFM to electromagnetic materials inherits the same limitations and drawbacks.

For a simple illustration, the basic field equations in the absence of temperature change ($\dot{T}=0$) and electricity conduction ($j_e = 0$) under the quasi-static approximation for the electromagnetic fields are summarized as follows:

$$\sigma_{ij} = c_{ijkl}\varepsilon_{kl} - e_{kij}E_k - h_{kij}H_k, \tag{4.1}$$

$$D_i = e_{ijk}\varepsilon_{jk} + \kappa_{ij}E_j + g_{ij}H_j, \tag{4.2}$$

$$B_i = h_{ijk}\varepsilon_{jk} + g_{ji}E_j + \mu_{ij}H_j, \tag{4.3}$$

$$\varepsilon_{ij} = \frac{1}{2}(u_{i,j} + u_{j,i}), \tag{4.4}$$

$$E_i = -\phi_{,i}, \tag{4.5}$$

$$H_i = -\psi_{,i}, \tag{4.6}$$

$$\sigma_{ij,j} + \rho\hat{f}_i = \rho\ddot{u}_i, \tag{4.7}$$

$$D_{i,i} = q_f, \tag{4.8}$$

$$B_{i,i} = 0, \tag{4.9}$$

where $c_{ijkl} = c_{klij} = c_{jikl} = c_{ijlk}$, $e_{kij} = e_{kji}$, $h_{kij} = h_{kji}$, $\kappa_{ij} = \kappa_{ji}$, and $\mu_{ij} = \mu_{ji}$.

The boundary conditions for a cracked body are given by

$$\mathbf{n}\cdot[[\mathbf{D}]] = \varpi_f, \quad \mathbf{n}\times[[\mathbf{E}]] = 0, \quad \mathbf{n}\cdot[[\mathbf{B}]] = 0, \quad \mathbf{n}\times[[\mathbf{H}]] = 0, \tag{4.10}$$

$$\mathbf{n}\cdot[[\boldsymbol{\sigma}]] = 0, \tag{4.11}$$

where $[[\cdots]]$ represents the jump of the field quantity inside the double square brackets across a surface of discontinuity (see Fig. 4.2), **n** is the unit normal vector, and ϖ_f is the surface charge density.

The initial conditions are taken as

$$\mathbf{u}\big|_{t=t_0} = \mathbf{u}_0, \tag{4.12}$$

$$\dot{\mathbf{u}}\big|_{t=t_0} = \mathbf{v}_0. \tag{4.13}$$

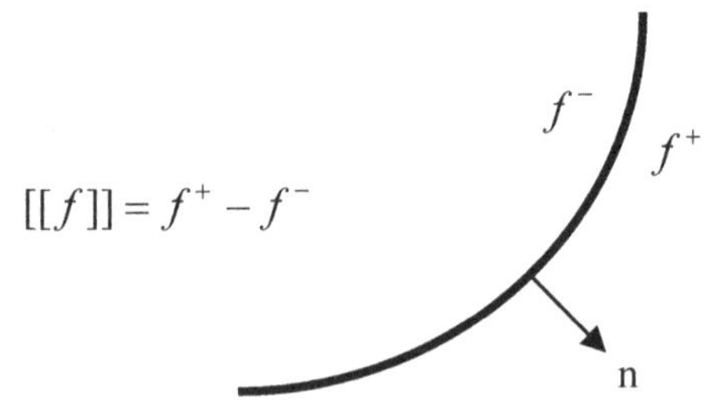

Fig. 4.2. Jump of field quantity f across a surface of discontinuity between two regions.

4.3 General Solution Procedures

Static and dynamic crack problems may be solved by a variety of techniques. Barnett and Lothe (1975) first extended the Stroh formalism in anisotropic elasticity to treatment of dislocations and line charges in anisotropic piezoelectric insulators. Shindo (1977, 1978) studied the distribution of mechanical and magnetic fields in an infinite body with a planar or penny-shaped crack by the integral transform following a linear theory for soft ferromagnetic elastic materials (Pao and Yeh, 1973). Deeg (1980) generalized the distributed dislocation approach to piezoelectric crack and inclusion problems. Sosa and Pak (1990) conducted an eigenfunction analysis of a crack in a piezoelectric material. Sosa (1991) started with stress functions and obtained solutions for plane problems in piezoelectric media with defects based on the extended Lekhnitskii formalism. Pak (1992) applied the dislocation approach to linear electro-elastic fracture. Suo *et al.* (1992) solved crack problems in piezoelectrics or on the interfaces between piezoelectrics and other materials based on the extended Stroh formalism. Dunn (1994) examined the fracture of piezoelectric solids based on the equivalent inclusion method of Eshelby (1957). Dascalu and Maugin (1995) analyzed dynamic fracture problems for piezoelectric materials with inertial effects using the extended Stroh formalism. Li and Mataga (1996a–b) described the Bleustein–Gulyaev surface wave phenomenon for the propagation of a semi-infinite crack in piezoelectric materials using the Laplace transform, Wiener–Hopf, and Cagniard–de Hoop techniques. Shindo *et al.* (1996) and Narita and Shindo (1998) reduced the problem of a finite crack subjected to longitudinal waves in a dielectric medium or horizontal shear waves in a piezoelectric medium to a Fredholm integral equation of the second kind by means of the Fourier transform. Qin and Mai (1998) explored the application of the thermoelectroelastic Green's function. Meguid and Wang (1998) and Wang and Meguid (2000) studied the dynamic behavior of piezoelectric materials containing interacting cracks using integral transform techniques and Chebyshev polynomial expansions. Finite element method (FEM) and boundary element method (BEM) have also been employed for solving complicated thermo-electro-mechanical boundary-initial value problems (Qin, 2001; Kuna, 2010).

These typical solution techniques can be readily extended to analysis of magneto-electro-thermo-elastic inclusion or crack problems (e.g., Alshits *et al.*, 1995; Chung and Ting, 1995; Kirchner and Alshits, 1996; Huang *et al.*, 1998; Li, 2000; Liu *et al.*, 2001; Sih *et al.*, 2003; Wang and Mai, 2003, 2007a; Gao *et al.*, 2004; Du *et al.*, 2004; Hu and Li, 2005; Zhong and Li, 2006; Hu *et al.*, 2007; Feng *et al.*, 2007; Zhong *et al.*, 2009). As one of the most commonly used techniques for a planar crack in anisotropic magneto-electro-elastic materials (Fig. 4.3), the solution procedure based on the extended Stroh formalism is illustrated below.

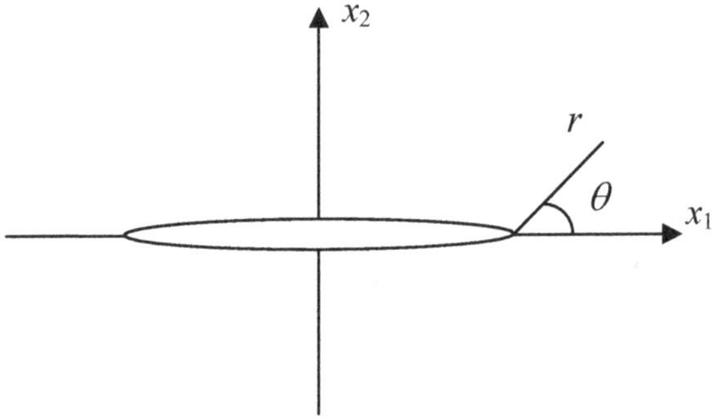

Fig. 4.3. A planar crack in an anisotropic magneto-electro-elastic material.

In the shorthand notation used by Barnett and Lothe (1975), a general solution is sought with consideration of displacement components, electric potential, and magnetic potential:

$$u_m = a_m f(z) \ (m = 1,2,\ldots,5), \tag{4.14}$$

where $u_4 = \phi$, $u_5 = \psi$, the function f is analytic in the complex variable $z = x_1 + px_2$, and p and a_m are complex numbers to be determined from the governing equations.

In the absence of mechanical body forces, inertial effects, and free electric charges, the basic field equations given in Section 4.2 lead to

$$[\mathbf{Q} + (\mathbf{R} + \mathbf{R}^T)p + \mathbf{T}p^2]\mathbf{a} = 0, \tag{4.15}$$

with $\mathbf{a} = (a_1, a_2, a_3, a_4, a_5)^T$ and the 5×5 matrices:

$$\mathbf{Q} = \begin{bmatrix} c_{1jk1} & e_{1j1} & h_{1j1} \\ e_{1k1}^T & -\kappa_{11} & -g_{11} \\ h_{1k1}^T & -g_{11} & -\mu_{11} \end{bmatrix}, \quad \mathbf{R} = \begin{bmatrix} c_{1jk2} & e_{2j1} & h_{2j1} \\ e_{1k2}^T & -\kappa_{12} & -g_{12} \\ h_{1k2}^T & -g_{21} & -\mu_{12} \end{bmatrix},$$

$$\mathbf{T} = \begin{bmatrix} c_{2jk2} & e_{2j2} & h_{2j2} \\ e_{2k2}^T & -\kappa_{22} & -g_{22} \\ h_{2k2}^T & -g_{22} & -\mu_{22} \end{bmatrix}. \tag{4.16}$$

Nontrivial solutions are obtained if p is a root of

$$\det[\mathbf{Q} + (\mathbf{R} + \mathbf{R}^T)p + \mathbf{T}p^2] = 0 . \tag{4.17}$$

There is a total of ten eigenvalues from Eq. (4.17) which consists of five pairs of complex conjugates. We suppose that p_α ($\alpha = 1,2,\cdots,5$) are five distinct roots with positive imaginary parts and construct a 5×5 matrix $\mathbf{A} = [A_{m\alpha}]$ with columns which are the associated eigenvectors. Then, the general solution is given by

$$u_m = \sum_{\alpha=1}^{5} A_{m\alpha} f_\alpha(z_\alpha) + \sum_{\alpha=1}^{5} \overline{A}_{m\alpha} \overline{f_\alpha(z_\alpha)}, \tag{4.18}$$

where $z_\alpha = x_1 + p_\alpha x_2$ and over-bars denote complex conjugates.

Hence, stress, electric displacement, and magnetic induction are expressed as

$${}_t\sigma_{i1} = -\sum_{\alpha=1}^{5} p_\alpha L_{i\alpha} f_\alpha'(z_\alpha) - \sum_{\alpha=1}^{5} \overline{p}_\alpha \overline{L}_{i\alpha} \overline{f_\alpha'(z_\alpha)}, \tag{4.19}$$

$${}_t\sigma_{i2} = \sum_{\alpha=1}^{5} L_{i\alpha} f_\alpha'(z_\alpha) + \sum_{\alpha=1}^{5} \overline{L}_{i\alpha} \overline{f_\alpha'(z_\alpha)}, \tag{4.20}$$

$$D_1 = -\sum_{\alpha=1}^{5} p_\alpha L_{4\alpha} f_\alpha'(z_\alpha) - \sum_{\alpha=1}^{5} \overline{p}_\alpha \overline{L}_{4\alpha} \overline{f_\alpha'(z_\alpha)}, \tag{4.21}$$

$$D_2 = \sum_{\alpha=1}^{5} L_{4\alpha} f_\alpha'(z_\alpha) + \sum_{\alpha=1}^{5} \overline{L}_{4\alpha} \overline{f_\alpha'(z_\alpha)}, \tag{4.22}$$

$$B_1 = -\sum_{\alpha=1}^{5} p_\alpha L_{5\alpha} f_\alpha'(z_\alpha) - \sum_{\alpha=1}^{5} \overline{p}_\alpha \overline{L}_{5\alpha} \overline{f_\alpha'(z_\alpha)}, \tag{4.23}$$

$$B_2 = \sum_{\alpha=1}^{5} L_{5\alpha} f_\alpha'(z_\alpha) + \sum_{\alpha=1}^{5} \overline{L}_{5\alpha} \overline{f_\alpha'(z_\alpha)}, \tag{4.24}$$

where $L_{n\alpha}=(R_{mn}+p_\alpha T_{nm})A_{m\alpha}=-(Q_{nm}+p_\alpha R_{nm})A_{m\alpha}/p_\alpha$ can be used to construct a 5×5 matrix $\mathbf{L}=[L_{n\alpha}]$, $f_\alpha'(z_\alpha)=df_\alpha(z_\alpha)/dz_\alpha$.

The analytical functions $f_\alpha(z_\alpha)$ can be determined for a given boundary value problem in magneto-electro-elasticity with a similar solution structure to that in piezoelectricity.

4.4 Debates on Crack-Face Boundary Conditions

Debates exist about the selection of crack-face boundary conditions in piezoelectric materials, as reviewed by Zhang *et al.* (2002). The electric boundary conditions are considered for "insulating" or "conducting" cracks. The former is compatible with the crack interior filled by vacuum, air, or oil, whereas the latter is compatible with the crack interior filled by a conducting medium. Four sets of electrical boundary conditions on insulating crack faces have been adopted – exact, electrically permeable, impermeable, and semi-permeable. Since the exact boundary-initial value problems need to be solved in both the cracked solid region and the interior vacuum region, approximations have often been made for analyzing slit crack problems.

Parton (1976) made the first attempt to define the electric boundary conditions over crack faces by considering both the electric displacement and the electric potential continuous across a traction-free slit:

$$D_n^+=D_n^-,\quad \phi^+=\phi^-, \tag{4.25}$$

where the subscript "n" denotes the component normal to the crack faces and the superscripts "+" and "–" denote the upper and lower crack faces.

However, there may be a potential drop across the piezoelectric crack, which can be assumed to be a low-capacitance medium. Deeg (1980), Pak (1990, 1992), and Suo *et al.* (1992), among others, imposed the electrically impermeable condition on the crack faces:

$$D_n^+=D_n^-=0. \tag{4.26}$$

Later, Hao and Shen (1994) provided the semi-permeable condition across the crack faces, that is,

$$D_n^+=D_n^-,\quad D_n^+(u_n^+-u_n^-)=-\kappa_c(\phi^+-\phi^-), \tag{4.27}$$

where κ_c is the dielectric permittivity of the crack.

The electric boundary condition on crack faces for conducting (electroded) crack problems is

$$\phi^+ = \phi^- = 0 . \tag{4.28}$$

To determine the effect of the dielectric medium inside a crack on the electric boundary condition, McMeeking (1989), Zhang and Tong (1996), and Zhang *et al.* (1998) introduced the parameter $(\kappa^m / \kappa^f)(b/a)$ to study an elliptical, flaw-like crack in electrostrictive and piezoelectric materials, where b/a is the flaw aspect ratio, a and b are the semi-axes of the ellipse (a>b), and κ^m / κ^f is the ratio of the dielectric permittivity of the surrounding material to that of the flaw interior. Likewise, Gao *et al.* (2004) used two parameters $\lambda_e = (\kappa^m / \kappa^f)(b/a)$ and $\lambda_m = (\mu^m / \mu^f)(b/a)$ for a mode-III, elliptical, flaw-like crack in a magneto-electro-elastic solid, where μ^m / μ^f is the ratio of the magnetic permeability of the surrounding material to that of the flaw interior. The crack is impermeable if $\lambda_e \to \infty$ and $\lambda_m \to \infty$, permeable if $\lambda_e \to 0$ and $\lambda_m \to 0$, and semi-permeable if λ_e and λ_m are nonzero finite constants.

Xu and Rajapaske (2001) discussed the influence of different electric boundary conditions on an arbitrarily oriented crack by reducing an elliptical void solution to a crack solution using the extended Lekhnitskii formalism. Wang and Jiang (2004) studied the nonlinear fracture behavior of an arbitrarily oriented crack in a piezoelectric medium considering the deformed crack geometry. Landis (2004) proposed energetically consistent boundary conditions for electromechanical fracture. Haug and McMeeking (2006) also investigated cracks with surface charge in poled ferroelectrics.

Furthermore, Wang and Mai (2007a) examined four ideal crack-face electromagnetic boundary condition assumptions: (i) electrically and magnetically impermeable, (ii) electrically impermeable and magnetically permeable, (iii) electrically permeable and magnetically impermeable, and (iv) electrically and magnetically permeable. In addition, Gao *et al.* (2008) studied the effects of applying only a magnetic field to a magnetically permeable crack in a soft ferromagnetic solid, including the Maxwell stresses in the boundary conditions, not only on the crack faces, but also at infinity. It is found that all the field variables are uniform, which means that there is no crack-tip field singularity when a mathematical slit crack is dealt with in this case.

In practice, the crack-face electromagnetic boundary condition may be expressed as

$$D_n^+ = D_n^- = D_n^0, \; B_n^+ = B_n^- = B_n^0, \tag{4.29}$$

where D_n^0 and B_n^0 are either prescribed for the impermeable condition or determined through the permeable or semi-permeable condition.

4.5 Fracture Criteria

A fracture criterion can be used to determine whether or not a crack advances by comparison of the crack driving force with the crack resistance. The extension of classical fracture criteria to combined magnetic, electric, and mechanical loadings is summarized below in terms of field intensity factors, path-independent integral, mechanical strain energy release rate, as well as global and local energy release rates.

4.5.1 *Field intensity factors*

Like linear elastic crack solutions, stress, electric displacement, and magnetic induction at the crack tip exhibit the traditional inverse square-root singularity (Fig. 4.4). The asymptotic near-tip fields can be expressed in terms of the crack-tip polar coordinates (r, θ) as

$$\begin{aligned}\sigma_{ij}(r,\theta) = {} & \frac{K_I}{\sqrt{2\pi r}}\Sigma_{ij}^{I}(\theta) + \frac{K_{II}}{\sqrt{2\pi r}}\Sigma_{ij}^{II}(\theta) + \frac{K_{III}}{\sqrt{2\pi r}}\Sigma_{ij}^{III}(\theta) \\ & + \frac{K_D}{\sqrt{2\pi r}}\Sigma_{ij}^{IV}(\theta) + \frac{K_B}{\sqrt{2\pi r}}\Sigma_{ij}^{V}(\theta),\end{aligned} \tag{4.30}$$

$$\begin{aligned}D_i(r,\theta) = {} & \frac{K_I}{\sqrt{2\pi r}}\Sigma_{i4}^{I}(\theta) + \frac{K_{II}}{\sqrt{2\pi r}}\Sigma_{i4}^{II}(\theta) + \frac{K_{III}}{\sqrt{2\pi r}}\Sigma_{i4}^{III}(\theta) \\ & + \frac{K_D}{\sqrt{2\pi r}}\Sigma_{i4}^{IV}(\theta) + \frac{K_B}{\sqrt{2\pi r}}\Sigma_{i4}^{V}(\theta),\end{aligned} \tag{4.31}$$

$$B_i(r,\theta) = \frac{K_I}{\sqrt{2\pi r}}\Sigma_{i5}^{I}(\theta) + \frac{K_{II}}{\sqrt{2\pi r}}\Sigma_{i5}^{II}(\theta) + \frac{K_{III}}{\sqrt{2\pi r}}\Sigma_{i5}^{III}(\theta) + \frac{K_D}{\sqrt{2\pi r}}\Sigma_{i5}^{IV}(\theta) + \frac{K_B}{\sqrt{2\pi r}}\Sigma_{i5}^{V}(\theta), \tag{4.32}$$

where K_I, K_{II}, K_{III} are the mode-I, mode-II, and mode-III stress intensity factors, K_D is the electric displacement intensity factor, K_B is the magnetic induction intensity factor, and the functions $\Sigma_{ij}^{I}(\theta)$, $\Sigma_{ij}^{II}(\theta)$ $\Sigma_{ij}^{III}(\theta)$, $\Sigma_{ij}^{IV}(\theta)$, and $\Sigma_{ij}^{V}(\theta)$ prescribe angular variations (Wang and Mai, 2003; Kuna, 2010).

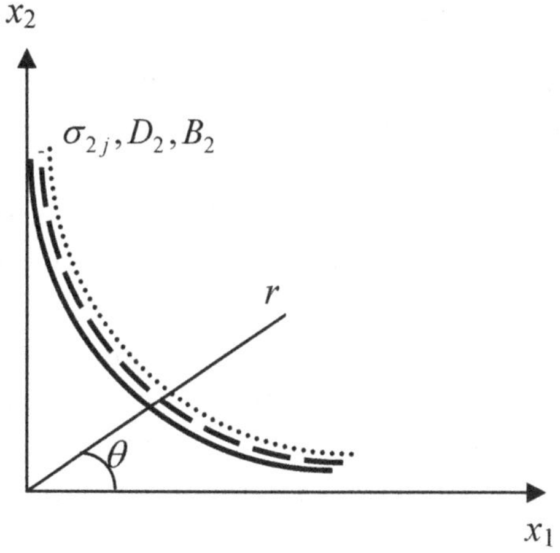

Fig. 4.4. Singular crack-tip fields.

At a distance r ahead of the crack tip along the crack plane ($\theta = 0$), the near-tip fields can be expressed as

$$(\sigma_{21}, \sigma_{22}, \sigma_{23}, D_2, B_2)^T = \frac{1}{\sqrt{2\pi r}}\mathbf{k}, \tag{4.33}$$

where $\mathbf{k} = (K_{II}, K_I, K_{III}, K_D, K_B)^T$ is the field intensity factor vector.

For a conventional Griffith-type crack of length $2a$, the crack-tip field intensity factor vector is related to the remote and crack-face loadings by

$$\mathbf{k} = (\sigma_{21}^{\infty} - \sigma_{21}^{0}, \sigma_{22}^{\infty} - \sigma_{22}^{0}, \sigma_{23}^{\infty} - \sigma_{23}^{0}, D_2^{\infty} - D_2^{0}, B_2^{\infty} - B_2^{0})^T \sqrt{\pi a}\,. \tag{4.34}$$

It can be seen that, at the crack tip, the stress fields are decoupled from the electromagnetic fields within the framework of the linear theory of piezoelectricity and its extension. It should also be noted that the internal stress field induced by domain switching under cyclic electric loading may cause fatigue damage in ferroelectric materials, even in the absence of an applied mechanical load. Thus, the local stress intensity factor based on domain switching models (e.g., Lynch *et al.*, 1995; Zhu and Yang, 1997, 1999; Yang and Zhu, 1998) has been used in a fracture criterion to predict crack growth in ferroelectrics under small-scale switching conditions, similar to small-scale yielding conditions for metals.

4.5.2 *Path-independent integral*

Many efforts (e.g., Cherepanov, 1979; Pak and Hermann, 1986a–b; Pak, 1990; McMeeking, 1990; Maugin and Epstein, 1991; Suo *et al.*, 1992; Maugin *et al.*, 1992; Dascalu and Maugin, 1994; Maugin, 1994; Trimarco and Maugin, 2001; Wang and Mai, 2003) have been devoted to extend the basic concepts of the energy-momentum tensor and the path-independent integral of Eshelby (1951, 1956, 1970, 1975), Cherepanov (1967, 1968) and Rice (1968) to linear and nonlinear electromechanical and magneto-electro-mechanical problems. It should be noted that all the theoretical treatments of these fracture mechanics models take the path-independent integral constructed with use of the electric enthalpy or the electromagnetic enthalpy as the crack extension force.

A straightforward derivation for linear magneto-electro-elastic media (Wang and Mai, 2003) starts with the differentiation of the electromagnetic enthalpy density $\tilde{f} = (\sigma_{ij}\varepsilon_{ij} - D_i E_i - B_i H_i)/2$ with respect to the spatial coordinate x_k:

$$\begin{aligned} &\frac{\partial}{\partial x_k}\tilde{f}(u_{i,j}, E_i, H_i; x_i) \\ &= \frac{\partial \tilde{f}}{\partial u_{i,j}} u_{i,jk} + \frac{\partial \tilde{f}}{\partial E_i} E_{i,k} + \frac{\partial \tilde{f}}{\partial H_i} H_{i,k} + \left.\frac{\partial \tilde{f}}{\partial x_k}\right|_{\text{expl}}, \end{aligned} \tag{4.35}$$

where the subscript "expl" denotes the explicit dependence of $\tilde{f}$ on x_k.

With the use of the basic field equations (4.1)–(4.9) in the absence of inertial force, mechanical body force, and free electric charge, Eq. (4.35) can be rewritten as

$$\left.\frac{\partial \tilde{f}}{\partial x_k}\right|_{\text{expl}} = b_{jk,j}, \tag{4.36}$$

where the energy-momentum tensor b_{jk} is defined as

$$b_{jk} = \tilde{f}\delta_{jk} - \sigma_{ji}u_{i,k} + D_j E_k + B_j H_k . \tag{4.37}$$

If there is no discontinuity in a material, the above divergence becomes zero. The energy-momentum tensor provides a general method for establishing invariant integrals. For example, the J_k-integral vector for a three-dimensional body bounded by the closed surface S with the outer unit normal vector **n** can be easily deduced using the divergence theorem as

$$J_k = \int_S (\tilde{f}\delta_{jk} - \sigma_{ji}u_{i,k} + D_j E_k + B_j H_k) n_j dS . \tag{4.38}$$

The J-integral is the first component of the J_k-integral vector. For a generalized plane magneto-electro-elastic crack problem with the crack line along the x_1-axis, it follows that

$$J = \int_\Gamma (\tilde{f}n_1 - n_j \sigma_{ji} u_{i,1} + n_j D_j E_1 + n_j B_j H_1) ds , \tag{4.39}$$

where Γ is a contour surrounding the crack tip.

As a generalization of the crack closure analysis by Irwin (1957), the path-independent J-integral constructed with the electromagnetic enthalpy can be evaluated from the field intensity factors using the crack closure integral, that is,

$$\begin{aligned} J = \lim_{\delta a \to 0} \frac{1}{2\delta a} \int_a^{a+\delta a} & [\sigma_{2i}(x_1,0)\Delta u_i(x_1 - \delta a) \\ & + D_2(x_1,0)\Delta\phi(x_1 - \delta a) + B_2(x_1,0)\Delta\psi(x_1 - \delta a)] dx_1, \end{aligned} \tag{4.40}$$

where δa is the virtual crack extension and Δ denotes the jump between the upper and lower crack faces.

Hence, the Irwin-type relation is obtained as

$$J = \frac{1}{4}(K_{II}, K_I, K_{III}, K_D, K_B) \cdot \mathbf{H}^i \cdot (K_{II}, K_I, K_{III}, K_D, K_B)^T , \tag{4.41}$$

where $\mathbf{H}^i = 2\,\mathrm{Re}(i\mathbf{A}\mathbf{L}^{-1})$ is indefinite and sometimes referred to as the Irwin matrix (Lothe and Barnett, 1976; Suo *et al.*, 1992; Kuna, 2010), $i = \sqrt{-1}$, and the matrices **A** and **L** are given in Section 4.3.

As the magnetic field becomes zero, the above formulae are reduced to those for piezoelectric materials. Consider a piezoelectric material with hexagonal crystal structure (class $C_{6v} = 6\mathrm{mm}$, see notations in Section 2.5) with the x_3-axis in the poling direction. Pak (1990) evaluated the *J*-integral for an antiplane crack problem for which a finite crack of length $2a$ is embedded in an infinite piezoelectric medium subjected to far-field mechanical and electric loads:

$$\text{Case 1: } J = \frac{\pi a}{2}\left[\frac{\kappa_{11}\tau_\infty^2 + 2e_{15}\tau_\infty D_\infty - c_{44}D_\infty^2}{\kappa_{11}c_{44} + e_{15}^2}\right]; \tag{4.42}$$

$$\text{Case 2: } J = \frac{\pi a}{2}\left[c_{44}\gamma_\infty^2 - 2e_{15}E_\infty\gamma_\infty - \kappa_{11}E_\infty^2\right]; \tag{4.43}$$

$$\text{Case 3: } J = \frac{\pi a}{2}\left[\frac{\tau_\infty^2 - (\kappa_{11}c_{44} + e_{15}^2)E_\infty^2}{c_{44}}\right]; \tag{4.44}$$

$$\text{Case 4: } J = \frac{\pi a}{2}\left[\frac{(\kappa_{11}c_{44} + e_{15}^2)\gamma_\infty^2 - D_\infty^2}{\kappa_{11}}\right], \tag{4.45}$$

where c_{44}, e_{15}, and κ_{11} are, respectively, shear modulus, piezoelectric constant, and dielectric constant, τ_∞ is the far-field shear stress, γ_∞ is the far-field shear strain, D_∞ is the far-field electric displacement, and E_∞ is the far-field electric field.

Pak (1990) noticed that the *J*-integral constructed with the electric enthalpy is always negative in the absence of mechanical loading and the electric field would impede crack propagation, regardless of its direction, which is essentially different from the traditional role of the *J*-integral as the crack extension force. The more complicated in-plane crack solution predicted the same trend (Pak, 1992; Suo *et al.*, 1992). However, this theoretical prediction is contradictory to experimental evidence (e.g., Pak and Tobin, 1993; Tobin and Pak, 1993; Cao and Evans, 1994; Lynch *et al.* 1995; Park and Sun, 1995a–b; Zhang *et al.*, 2002; Chen and Lu, 2003).

A fracture criterion based on the J-integral formulated from the electromagnetic enthalpy indicates that the presence of an electric/magnetic field, either positive or negative, should elevate the critical load. Hence, an even dependence exists between the critical load and the applied electric/magnetic field. As remarked by Chen and Lu (2003), whether or not the J-integral thus formulated can be a candidate for a piezoelectric fracture criterion has been a long-standing controversial issue and some have even used chaos to describe the contentious situation. The fundamental discrepancy between experimental observations and theoretical predictions has hindered the development of the piezoelectric fracture theory, instigating numerous attempts to resolve this controversy (e.g., Park and Sun, 1995a–b; Gao *et al.*, 1997; Sih and Zuo, 2000; Fulton and Gao, 2001; McMeeking, 2001, 2004; Li, 2003a–b; Landis, 2004; Zhang *et al.*, 2005).

4.5.3 *Mechanical strain energy release rate*

Park and Sun (1995a–b) first pointed out that the path-independent integral formulated from the electric enthalpy could not be directly used as a fracture criterion for piezoelectric materials. Instead, they proposed that the mechanical strain energy release rate (MSERR) is a dominant parameter governing piezoelectric fracture.

By definition, the mechanical strain energy release rate is expressed as the mechanical part of the crack closure integral:

$$G^M = \lim_{\delta a \to 0} \frac{1}{2\delta a} \int_a^{a+\delta a} \sigma_{2i}(x_1, 0) \Delta u_i (x_1 - \delta a) dx_1 \, . \tag{4.46}$$

Thus, the mechanical strain energy release rate is related to the field intensity factors by

$$G^M = \frac{1}{4}(K_{II}, K_I, K_{III}, 0, 0) \cdot \mathbf{H}^i \cdot (K_{II}, K_I, K_{III}, K_D, K_B)^T \, . \tag{4.47}$$

Consider a Griffith-type crack of length $2a$ in PZT-4 piezoelectric ceramics poled along the x_2-axis under remote loads σ_{22}^{∞} and D_2^{∞} (Fig. 4.5). The crack plane is perpendicular to the poling axis. The mechanical strain energy release rate and the potential energy release rate are obtained as (Park and Sun, 1995a–b)

$$G^M = \frac{\pi a}{2}(1.48\times 10^{-11}\sigma_{22}^{\infty\,2} + 2.67\times 10^{-2}\sigma_{22}^{\infty}D_2^{\infty}), \quad (4.48)$$

$$G = J = \frac{\pi a}{2}(1.48\times 10^{-11}\sigma_{22}^{\infty\,2} + 2\times 2.67\times 10^{-2}\sigma_{22}^{\infty}D_2^{\infty} - 8.56\times 10^{7}D_2^{\infty\,2}). \quad (4.49)$$

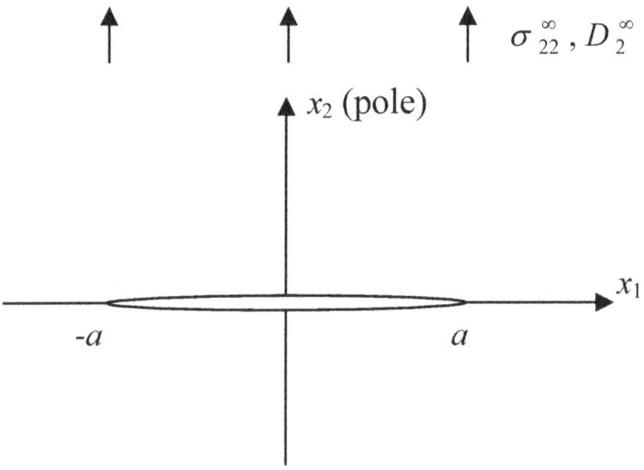

Fig. 4.5. A Griffith-type crack perpendicular to the poling axis under remote loads.

It can be seen that the mechanical strain energy release rate G^M is essentially different from the potential energy release rate G. Since G^M is an odd function of the electric displacement intensity factor, an applied electric field may either promote or retard crack propagation, depending on its direction. In their landmark study (Park and Sun, 1995a–b), the MSERR criterion agreed roughly with experimental measurement of the critical load for a crack perpendicular to the poling axis in simple tension and three-point bend PZT-4 specimens. They argued that it may be more suitable to take only the mechanical strain energy released during crack extension as the fracture criterion, since fracture is a mechanical process.

4.5.4 *Global and local energy release rates*

For a better understanding of the fracture mechanisms in piezoelectric ceramics under combined mechanical and electrical loadings, Gao *et al.* (1997) proposed the concept of global and local energy release rates based on a strip saturation model, via a simplified electroelasticity formulation. Figure 4.6 illustrates a view of multiscale singularity fields in piezoelectric fracture. Inspired by the classic Dugdale model for

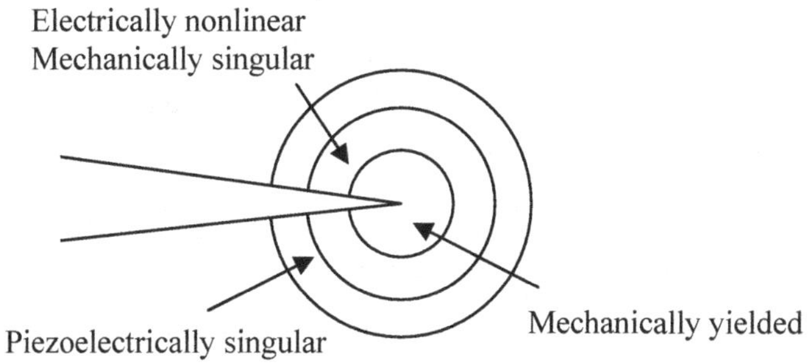

Fig. 4.6. Multiscale singularity fields in piezoelectric fracture. (After Gao *et al.*, 1997, with permission from Elsevier).

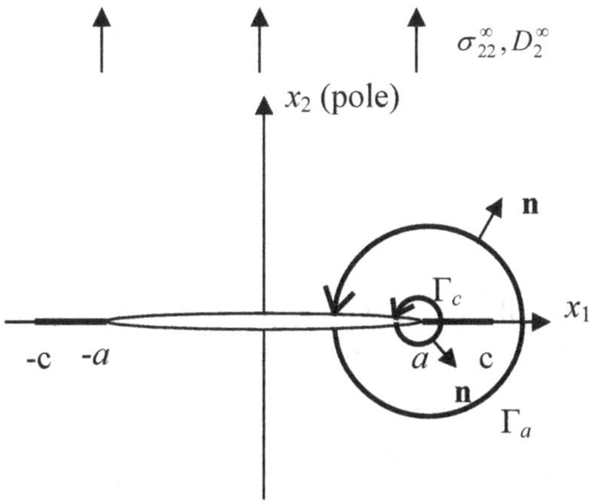

Fig. 4.7. A Dugdale-type electrically nonlinear crack perpendicular to the poling axis under remote loads. (After Gao *et al.*, 1997, with permission from Elsevier).

plastic yielding of metal sheets containing slits (Dugdale, 1960), Gao *et al.* (1997) assumed that the electric polarization is saturated only in a line segment of length $r_s = c - a$ in front of each tip of a crack of length $2a$ under remote loads σ_{22}^{∞} and D_2^{∞} (Fig. 4.7).

The boundary conditions along the electrically impermeable crack faces and the electrically nonlinear line segments are

$$\sigma_{22} = 0,\ D_2 = 0,\ \text{at}\ |x_1| < a, \tag{4.50}$$

$$u_2^+ = u_2^-,\ D_2 = D_s,\ \text{at}\ a < |x_1| < c, \tag{4.51}$$

where D_s is the saturation electric displacement.

A local energy release rate is obtained by evaluating the *J*-integral along an infinitesimal local contour Γ_c, that is,

$$J_c = \frac{\pi a}{2M}\left(1 + \frac{e^2}{M\kappa}\right)(eE_\infty + \sigma_\infty)^2, \tag{4.52}$$

where the material constants *M*, *e*, and κ represent, qualitatively, the elastic, piezoelectric, and dielectric properties.

An "apparent" or global energy release rate is obtained by evaluating the *J*-integral along a global contour Γ_a, that is,

$$\begin{aligned} J_a &= J_c + D_s(\phi_+ - \phi_-)\big|_{x=a} \\ &= J_c - \frac{4D_s^2 a}{\pi\kappa}\ln\left[\sec\left(\frac{\pi}{2}\frac{(M\kappa + e^2)E_\infty + e\sigma_\infty}{MD_s}\right)\right]. \end{aligned} \tag{4.53}$$

The strip saturation size is

$$r_s = c - a = a\left[\sec\left(\frac{\pi D_\infty}{2D_s}\right) - 1\right] << a. \tag{4.54}$$

Like the small-scale yielding condition for metals, the global energy release rate is reduced to that of a linear piezoelectric crack under the small-scale saturation condition $r_s << a$, that is,

$$(J_a)_{SSY} = \frac{\pi a}{2M}\left[\sigma_\infty^2 - (M\kappa + e^2)E_\infty^2\right]. \tag{4.55}$$

Thus, a fracture criterion based on the local energy release rate indicates that the electric field can positively influence fracture of piezoelectric materials, which is notably different from a fracture criterion based on the global energy release rate. Nonetheless, the major difficulty is that the theoretical treatments are all incomplete – for example, *ad hoc* neglect the electric contribution to the energy release rate (Park and Sun, 1995a–b) or the energy dissipation by the saturation of the electrical polarization (Gao *et al.*, 1997).

4.6 Experimental Observations

Despite the lack of experimental study on dynamic fracture of piezoelectric materials, increasing work has been done over the past few decades on experimental investigations of quasistatic crack propagation under combined electric and mechanical loadings. As reviewed by Zhang *et al.* (2002), there are discrepancies in experimental results, especially when the applied electric and mechanical fields are comparable in amplitude. Moreover, experimental data presented by different researchers sometimes contradict each other. It is a very important task to compare theoretical predictions with experimental results for various material systems, crack geometries, and loading combinations. A brief description is given in this section with a focus on the application of fracture mechanics concepts for explaining experimental observations.

4.6.1 *Indentation test*

The indentation technique has often been used for fracture toughness characterization of brittle materials due to its simplicity and economy (Anstis *et al.*, 1981; Chantikul *et al.*, 1981). A schematic illustration of the Vickers indentation technique is shown in Fig. 4.8. Inelastic deformation under the indenter would give rise to residual tensile stress at the crack front and, thus, propagates the radial crack to its final dimension as the indenter is unloaded.

For isotropic and homogeneous materials, the toughness K_c may be expressed in terms of the indentation load P and the induced crack length c as (Anstis *et al.*, 1981)

$$K_c = \xi \left(\frac{Y}{H} \right)^{1/2} \frac{P}{c^{3/2}}, \tag{4.56}$$

$$H = \frac{P}{2a^2} \sin(\alpha/2), \tag{4.57}$$

where $\xi = 0.016 \pm 0.004$ is an empirically determined calibration constant, Y is Young's modulus, H is the hardness, a is the impression half-diagonal, and α is the apex angle of the indenter.

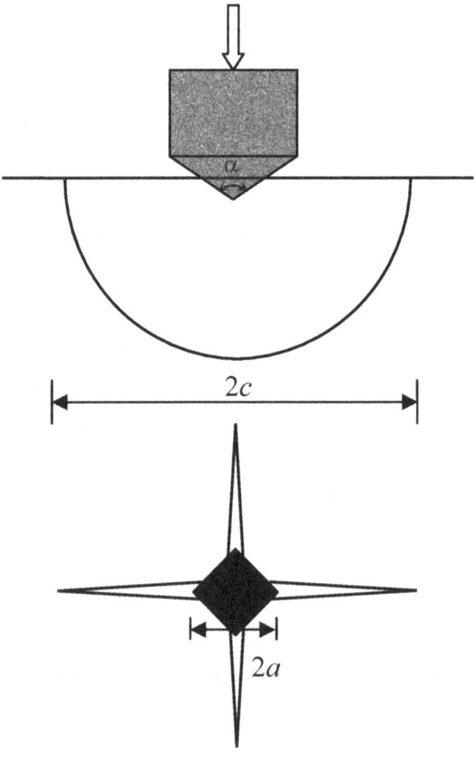

Fig. 4.8. Schematic illustration of the Vickers indentation technique.

Many researchers have observed that the electric field has significant effects on the crack behavior of piezoelectric ceramics from indentation tests. For instance, Pak and Tobin (1993) and Tobin and Pak (1993) found from the Vickers indentation tests on PZT-8 samples that the cracks perpendicular to the poling direction were longer with an electric field aligned with the poling direction and shorter with an electric field opposite to the poling direction compared to the case without an applied electric field (Fig. 4.9). This trend is consistent with the fracture test results on PIC 151 (similar to PZT-5H) Vickers indentation under electric loading by Zhang *et al.* (2004) and PZT-4 compact tension and three-point bending under combined mechanical and electric loadings by Park and Sun (1995b).

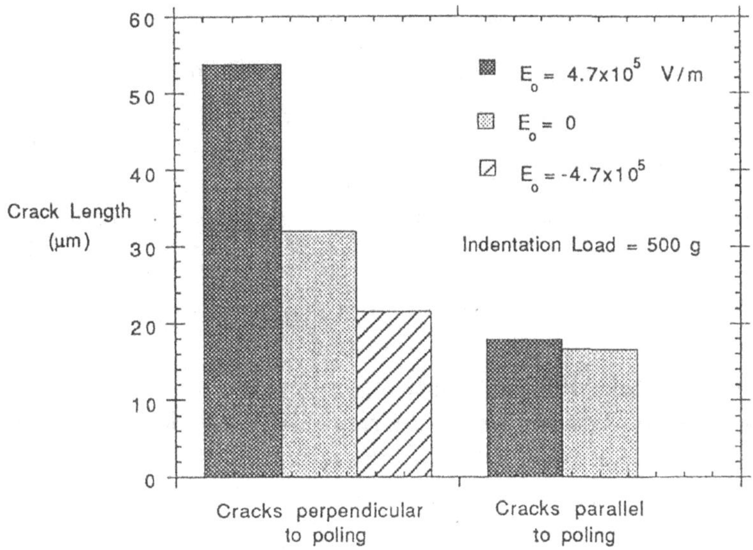

Fig. 4.9. Influence of electric field on the indentation fracture behavior of poled PZT-8. (From Pak and Tobin, 1993, with permission from ASME).

With precracks produced by indentation, Cao and Evens (1994), Lynch *et al.* (1995), Jiang and Sun (1999), and Zhang *et al.* (2004) further studied the crack growth behavior of piezoelectric ceramics under cyclic loads. Stable crack growth perpendicular to the poling direction has been observed for both PZT and PLZT samples under alternating electric loading only, for which the linear piezoelectric fracture theory predicts a negative potential energy release rate. Jiang and Sun (2001) also derived an approximate analytical solution for a half penny-shaped crack (semicircular surface crack) in a piezoceramic half space and then used it in conjunction with the mechanical strain energy release rate to interpret the effect of the electric field on indentation crack length.

Using the wedge model, the stress intensity factor and the electric displacement intensity factor for the semicircular surface crack in piezoelectric materials are (Jiang and Sun, 2001)

$$K_I = \frac{2f(\theta)}{(\pi c)^{3/2}}\left[P_0^o\left(\frac{c}{c_0}\right) - \alpha_{piezo} k_E c^2 E_z^\infty\right], \tag{4.58}$$

$$K_D = 2E_z^\infty \chi_E \sqrt{c/\pi}\,, \tag{4.59}$$

where c_0 is the indentation-induced crack length in the absence of the electric field, α_{piezo} is a piezoelectric constant reduction factor, E_z^∞ is the applied electric field, $f(\theta) = 1 + 0.2(1 - 2\theta/\pi)^2$, $P_0^o = (0.089/f_0)(Y/H)^{1/2} P$, and $f_0 = f(0) = 1.2$.

For a mode-I crack in PZT-4, the mechanical strain energy release rate can be expressed as

$$G_I^M = \frac{1}{2}(1.75\times10^{-11} K_I^2 + 2.21\times10^{-2} K_I K_D) \text{ (Nm}^{-1}\text{)}. \tag{4.60}$$

Sun and Park (1995) obtained the critical mechanical strain energy release rates of 3.68 Nm^{-1} and 4.63 Nm^{-1} for indentation loads of 9.8 N and 49 N, respectively. In terms of the critical mechanical strain energy release rate and the given indentation load, the relation between the crack length and the electric field can be determined iteratively using Eq. (4.60). The piezoelectric constant reduction factor α_{piezo} depends on the degree of completion of domain switching. It was found that the solution in conjunction with the mechanical strain energy release rate was able to explain the electric field effects on indentation crack growth.

Zhang *et al.* (2004) used a sphere cavity model in dielectrics to explain the growth of indentation cracks due to cyclic electric loading. It is found that low electric field intensity does not promote fatigue crack growth in PIC 151 but, at high applied electric field, the indentation cracks initially grow quickly and are then arrested. Electrostrictive strain drives cyclic fatigue crack growth and domain switching is the main fatigue mechanism. These results have significant consequences for the long-term durability of piezoceramics.

4.6.2 *Compact tension test*

Compact tension specimens have been used to study the fracture behavior of piezoelectric materials, together with finite element analysis (e.g., Park and Sun, 1995b; Kuna, 2010). An experimental setup for compact tension specimens under combined mechanical and electric loadings is shown in Fig. 4.10. Electrodes were coated in silver on the top and bottom surfaces of the specimens. The procedure of testing using compact tension specimens is to increase the tensile load applied by the crosshead displacement control of the MTS machine under a certain electric field generated by a D.C. power supplier until fracture occurs. Since electric discharging between electrodes through the air was observed when the electric field exceeded 5 kV/cm during initial exploratory tests, the specimen was immersed in a tub filled with silicone oil to enforce an insulated crack surface boundary condition.

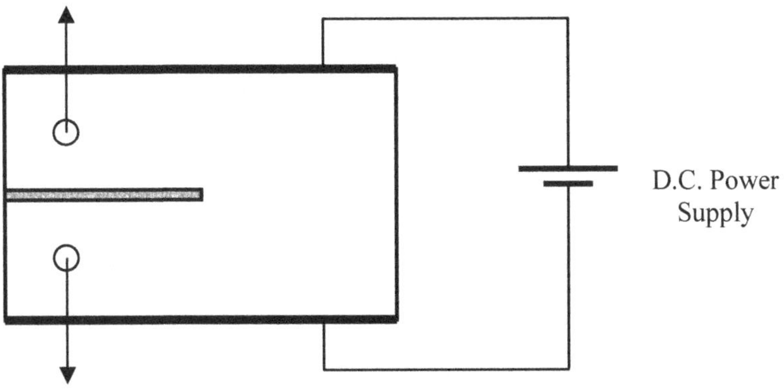

Fig. 4.10. Experimental setup for compact tension specimen under combined mechanical and electrical loadings.

Park and Sun (1995b) performed compact tension tests on PZT-4 specimens poled along the axis perpendicular to the crack plane. The material properties for poled PZT-4 in the principal material coordinate system (X_1, X_2, X_3) are listed in Table 4.1. The poling direction is along the X_3-axis. Figure 4.11 shows the fracture initiation loads under different electric fields from fracture testing on PZT-4 compact tension specimens. It appears that a positive electric field reduces the fracture load while a negative electric field increases the fracture load. It turns out that the mechanical strain energy release rate criterion is superior to the total potential energy release rate criterion at predicting the effect of the electric field on the fracture load.

Table 4.1 Material constants for poled PZT-4 in the principal material coordinate system. (After Park and Sun, 1995b, with permission from John Wiley and Sons).

c_{11} (N/m^2)	c_{12} (N/m^2)	c_{13} (N/m^2)	c_{33} (N/m^2)	c_{44} (N/m^2)
13.9×10^{10}	7.78×10^{10}	7.43×10^{10}	11.3×10^{10}	2.56×10^{10}
e_{31} (C/m^2)	e_{33} (C/m^2)	e_{15} (C/m^2)	κ_{11} (C/Vm)	κ_{33} (C/Vm)
−6.98	13.84	13.44	6.00×10^{-9}	5.47×10^{-9}

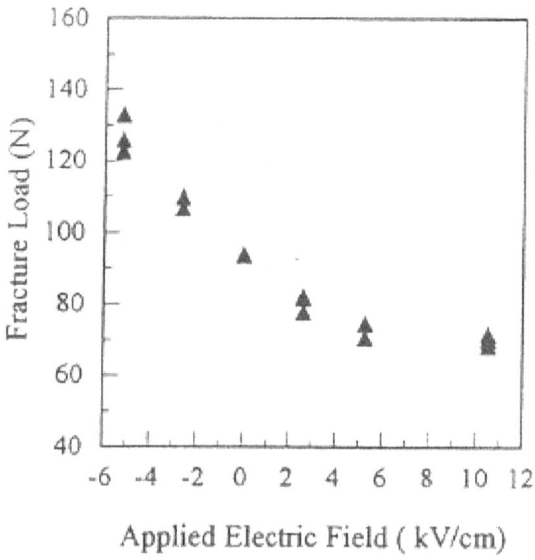

Fig. 4.11. Fracture loads under applied electric fields for PZT-4 compact tension specimens. (From Park and Sun, 1995b, with permission from John Wiley and Sons).

4.6.3 *Bending test*

To further verify the validity of the mechanical strain energy release rate as a fracture criterion, Park and Sun (1995b) also conducted fracture tests on PZT-4 piezoelectric ceramics using three-point bending specimens with a symmetric crack for mode-I fracture and an asymmetric crack for mixed-mode fracture (Fig. 4.12). The entire setup, including the indenter, was made of Plexiglas to avoid electric discharge. The prepared specimen was placed in the silicone oil tub that was mounted on the MTS machine. The poling direction is parallel to the span of the bending setup. Fracture loads versus applied electric fields were obtained for various crack locations (Fig. 4.13).

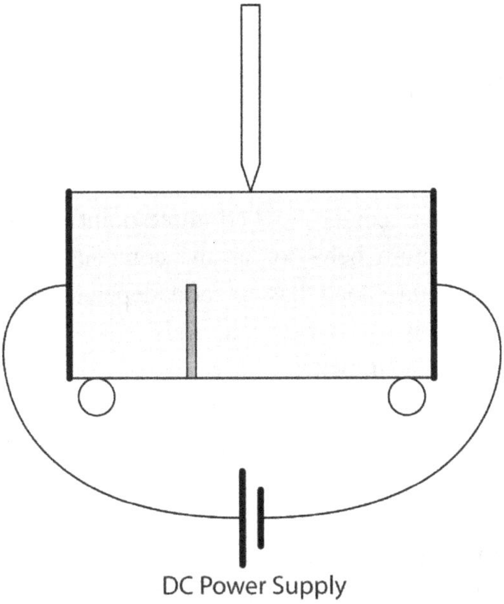

Fig. 4.12. Experimental setup for three-point bending specimens under combined mechanical and electrical loadings.

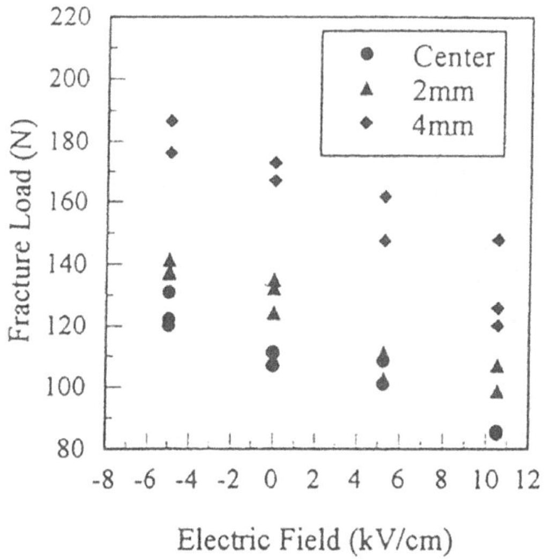

Fig. 4.13. Fracture loads under applied electric fields for various crack locations in PZT three-point bending specimens. (From Park and Sun, 1995b, with permission from John Wiley and Sons).

It appears that the center-cracked three-point bending specimens exhibit the same fracture behavior as the compact tension specimens. Specifically, the fracture load has an odd dependence on the applied electric field – positive electric field aids crack propagation, while negative electric field impedes crack propagation. The three-point bending specimens with an off-center crack also exhibit the same trend.

Later, Soh *et al.* (2003) used central crack specimens to study the effects of an applied electric field on the fracture toughness of poled piezoelectric ceramics and demonstrated that changing the applied electric field from negative to positive reduced the fracture toughness of poled PZT-5 ceramic, which is consistent with the observations by Park and Sun (1995b).

4.7 Nonlinear Studies

Linear piezoelectric/piezomagnetic fracture mechanics analysis is an important first step towards understanding the fracture behavior of electromagnetic materials. Nevertheless, experimental findings and microstructural diagnostics give evidence that there may exist factors beyond the scope of the linear theory of piezoelectricity and its extension that would affect fracture such as electrostriction/magnetostriction, polarization/magnetization saturation, domain switching, and domain wall motion, although these activities may be confined to a small region near the crack tip. The studies on these nonlinear effects are briefly summarized in this section.

4.7.1 *Electrostriction/magnetostriction*

Electrostriction is a property of dielectric materials whose shape is changed under an applied electric field, with the resulting strain proportional to the square of the polarization. Since the deformation remains unchanged with reversal of the electric field, the electrostrictive effect is obviously different from the piezoelectric effect that is characterized by a reversal in the direction of deformation when the electric field is reversed. Piezomagnetism and magnetostriction can be taken as magnetic analogues of piezoelectricity and electrostriction.

Smith and Warren (1966, 1968) studied an elliptical inclusion in an infinite isotropic dielectric material with electrostriction in consideration of the Maxwell stress. McMeeking (1989) investigated electrostrictive stresses near crack-like flaws in terms of a comparison between the fields of an elliptical flaw and a slit crack. Yang and Suo (1994) estimated the magnitude of the stress intensity factors for the flaws around the electrode edges in ceramic actuators caused by electrostriction under small-scale saturation conditions.

Based on the rotationally invariant (finite-strain) quasi-magnetostatic theory of elastic paramagnets and soft ferromagnets without magnetic spin-ordering effects, Sabir and Maugin (1996) constructed two path-independent integrals with the use of the magnetic enthalpy, including or excluding the contribution of the energy of the free magnetic field, and yielded essentially the same results as the canonical field-theoretic approach using the notions of Eshelby stress and material forces (Maugin *et al.*, 1992). The expressions obtained were applied to an antiplane crack problem of an isotropic magnetostrictive body in an axial bias magnetic field. In this case, the near-tip solution has the inverse square-root singularity like those obtained by Shindo (1977) based on a linear theory for soft ferromagnetic elastic materials (Pao and Yeh, 1973). Sabir and Maugin (1996) concluded that the magnetic field has a negative contribution to the energy release rate so that its presence is beneficial from the viewpoint of fracture toughness. However, there is no experimental support for this conclusion.

Recently, Gao *et al.* (2010a–b, 2011) obtained the solutions of a single crack and collinear cracks in an electrostrictive solid under pure electric loads based on the complex variable method. It is found that the total stresses exhibit the classical inverse square-root singularity at the crack tip and the applied electric field may either enhance or retard crack growth depending on the electric boundary conditions adopted on the crack faces and the Maxwell stresses on the crack faces and at infinity.

4.7.2 *Polarization/magnetization saturation*

Polarization saturation or magnetization saturation is the state which is reached when an increase in applied electric or magnetic field cannot increase the polarization or magnetization of the material further. As discussed in Section 4.5.4, Gao *et al.* (1997), inspired by the classical Dugdale model for plastic yielding near the crack tip in metals (Dugdale, 1960), developed a strip saturation model for an electrically insulating crack perpendicular or parallel to the poling axis of an infinite poled piezoelectric medium via a simplified electroelasticity formulation. Since the linear piezoelectric model gives singular electrical displacement distribution near the crack tip, it is assumed by this idealized strip saturation model that electric polarization reaches a saturation limit along a line segment in front of the crack tip.

Subsequently, Ru (1999) examined the effect of polarization saturation on stress intensity factors for a general piezoelectric medium. Ru and Mao (1999) also studied conducting cracks in a poled ferroelectric of limited electrical polarization based on a strip saturation model of the Dugdale type. Beom (2001) further analyzed an electrically impermeable crack in an electrostrictive ceramic with a strip saturation model. The strip saturation model may be applicable to magnetization saturation due to its similarity to polarization saturation.

As remarked by McMeeking (2001), the polarization saturation model may not correspond to the classical Dugdale model from the energy point of view because the electric displacement behaves like strain and the electric field behaves like stress. Zhang *et al.* (2005) proposed a strip dielectric breakdown model for an electrically impermeable crack in a piezoelectric medium with the assumption that the electric field in a strip ahead of the crack tip is equal to the dielectric breakdown strength, which is analogous to the classical Dugdale model from the energy point of view. The dielectric breakdown strength is defined as the critical electric field at which dielectric discharge occurs, leading to dielectric breakdown. Motivated by the similarities in electricity and magnetism, Zhao and Fan (2008) extended the strip dielectric breakdown model to magneto-electro-elastic media.

4.7.3 *Domain switching*

One class of widely used piezoelectric materials exhibit the ferroelectric effect. They possess spontaneous electric polarization that can be reversed by the application of an external electric field, yielding a hysteresis loop. This term is used by analogy to ferromagnetism because of the similarity between this hysteresis process and the corresponding process involving ferromagnetic materials. Typically, materials demonstrate ferroelectricity only below a certain characteristic temperature, T_c, called the Curie temperature. That is, spontaneous polarization disappears above this temperature.

A ferroelectric domain, in which all dipole moments of neighboring unit cells are oriented in the same direction, can switch its orientation to align itself in the direction of an applied external electric field as close as possible. This phenomenon is called "domain switching". Consequently, not only the local state of polarization is rotated but also the local state of strain is changed, which is described by the polarization switch vector

ΔP_i and the switching strain tensor $\Delta\varepsilon_{ij}$. A sufficiently strong electric field may rotate the polarization direction of an individual domain by $\pm 90^{\circ}$ or 180°, that is, the new polarization can be -90°, 90° or 180° rotated from the original direction.

As reviewed by Kuna (2010), models based on domain switching for ferroelectric materials have been developed to describe the polarization hysteresis loop and the strain butterfly loop (see Fig. 4.14) as well as the internal stress field induced by domain switching. For example, Hwang *et al.* (1995) proposed an energetic switching criterion for combined loadings:

$$E_i \Delta P_i + \sigma_{ij} \Delta\varepsilon_{ij} \geq 2P_s E_c. \tag{4.61}$$

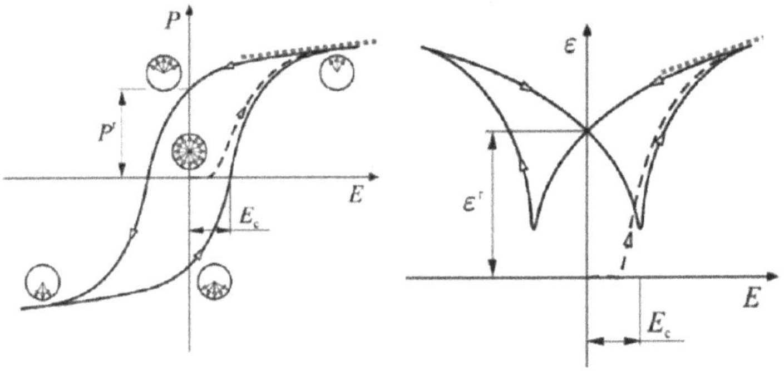

Fig. 4.14. Ferroelectric hysteresis of polarization and deformation (P^r – remanent polarization, ε^r – remanent strain, E_c – coercive field strength). (From Kuna, 2010, with permission from Elsevier).

It has been realized that domain switching plays an important role in the apparent fracture toughness variation for ferroelectric materials (e.g., Lynch *et al.*, 1995; Zhu and Yang, 1997, 1999; Yang and Zhu, 1998; Fulton and Gao, 2001; Zhang *et al.*, 2002; Beom and Atluri, 2003; Chen and Hasebe, 2005; Schneider, 2007; Kuna, 2010). The re-orientation of the polarization direction could significantly affect the solution for the corresponding boundary value problem and, consequently, the stress and electric displacement intensity factors. With an assumption like small-scale yielding, investigations on the nonlinear influence of domain switching on the fracture of ferroelectric materials have been confined to the near-tip process zone.

By analogy with the phase-transformation toughening mechanism (McMeeking and Evans, 1982), Zhu and Yang (1997) and Yang and Zhu (1998) studied switching toughening of ferroelectrics by adopting the switching criterion of Hwang *et al.* (1995) and derived the change in stress intensity factor ΔK as a result of stress redistribution induced by 90° polarization switching. In the case of small-scale switching, where the switching zone size is considerably smaller than the specimen size, the stress field near the switching boundary may be approximated by the remote K-field. The near-tip stress intensity factor K_{tip} is related to the applied stress intensity factor K_a by

$$K_{tip} = K_a + \Delta K \,. \tag{4.62}$$

A fracture criterion may be defined in terms of the near-tip stress intensity factor as

$$K_{tip} = K_{IC} \,. \tag{4.63}$$

Depending on the sign of ΔK for shielding or anti-shielding effects caused by domain switching, the apparent fracture load measured in experiments can be either enhanced or reduced. Based on the small-scale domain switching model, Zhu and Yang (1999) provided a mechanistic explanation of fatigue crack growth in ferroelectrics driven by cyclic electric loading. Recently, Kalyanam and Sun (2009) modeled the fracture behavior of piezoelectric materials using a gradual polarization switching model with the internal energy density as the parameter to estimate the amount of domain switching and the resulting gradual change in the polarization direction.

4.7.4 *Domain wall motion*

Domain switching may be regarded as a result of domain wall motion caused by the growth of domains with low-energy orientations and the shrinkage of domains with high-energy orientations (Zhang *et al.*, 2002). The domain wall may be treated either as a sharp or diffuse interface. The configuration (material) force method has been used to study the motion of a ferroelectric or ferromagnetic domain wall as a sharp interface. For example, Fomethe and Maugin (1997) studied the propagation of phase-transition fronts and domain walls in thermoelastic ferromagnets by exploiting the notion of the material forces. Fu and Zhang (2000) proposed a domain wall kinetic model to explain the effects of temperature and electric field on the bending strength of PZT-941 ceramics. Shilo *et al.* (2007) developed a model for large electrostrictive actuation in ferroelectric crystals by assuming a reasonable arrangement of domain walls and formulating equations of motion for these walls.

By contrast, phase-field simulation provides a powerful method for studying the evolution of ferroelectric or ferromagnetic domain structure as a diffuse interface. The major advantage of this approach lies in that the well-accepted Ginzburg–Landau equation is used to govern the time dependence of a spatial inhomogeneous order parameter without any preset transformation criterion. Phase transformation is a direct consequence of the minimization process of the total free energy of an entire simulated system. Wang and Zhang (2007) simulated polarization switching-induced toughening in a ferroelectric material with the original polarization direction perpendicular to an electrically permeable crack by a phase field model, accounting for the domain wall energy and the long-range mechanical and electrical interactions. Based on a local J-integral as a fracture criterion, the result indicates that an applied uniform electric field parallel to the original polarization direction reduces the apparent fracture toughness while an applied uniform electric field anti-parallel to the original polarization direction enhances it.

4.8 Status and Prospects

The fracture mechanics approaches within the framework of the linear theory of piezoelectricity and its extension outlined in this chapter cover the great majority of current applications. Simplicity is generally associated with the linear piezoelectric/piezomagnetic fracture mechanics methodology and so it is useful for a first approach at achieving a solution for a given crack problem. While nonlinear investigations beyond those of linear piezoelectric/piezomagnetic fracture mechanics are increasingly attempted, they are still confined to the small-scale region in the vicinity of a crack tip such as small-scale saturation conditions or small-scale switching conditions.

The major challenges in the current understanding of the complex fracture behavior of electromagnetic materials are:

- Discrepancy between theoretical predictions and experimental observations
- Various nonlinear effects
- Magneto-electro-thermo-mechanical coupling
- Large-scale dissipation
- Fully coupled dynamic framework
- Functionally graded materials (FGMs)
- Damage and failure at multiscales (nano, micro, meso, and macroscales)

At this transition between the elementary aspects and the more advanced treatments of the subject to come, it is worth stating that a highly important question in the development of a fracture mechanics theory for electromagnetic materials is whether there is any particular thermodynamic quantity of a cracked body that can be interpreted as the "driving force" for crack propagation under combined magnetic, electric, thermal, and mechanical loadings. The answer to this question has been pursued for decades, but no satisfactory agreement has yet been reached. Thus, the establishment of a physically sound fracture criterion becomes the hallmark of an advanced fracture mechanics treatment for electromagnetic materials. The objective of this book is to further the progress with development of a nonlinear field theory of fracture mechanics for electromagnetic materials by inclusion of magneto-electro-thermo-mechanical coupling and dissipative effects.

Chapter 5

Crack Driving Force in Electro-Thermo-Elastodynamic Fracture

5.1 Introduction

As reviewed in Chapter 4, it is theoretically predicted that an even dependence should exist between critical load and applied electric field based on the path-independent integral constructed with electric enthalpy (Pak and Hermann, 1986; Pak, 1990; Maugin and Epstein, 1991; Suo *et al.*, 1992; Dascalu and Maugin, 1994, 1995). On the contrary, however, it is found experimentally that the critical load for piezoelectric fracture is an odd function of the applied electric field (e.g., Pak and Tobin, 1993; Tobin and Pak, 1993; Cao and Evans, 1994; Lynch *et al.*, 1995). Hence, a major challenge in the fracture mechanics of piezoelectric materials is how to resolve the fundamental discrepancy between theoretical predictions and experimental observations.

A great advance in this area is owed to Park and Sun (1995a–b) who first pointed out that the path-independent integral thus formulated cannot be used directly as a fracture criterion for piezoelectric materials. Instead, they proposed that the mechanical part of the crack closure integral, i.e., the mechanical strain energy release rate (MSERR), is the dominant parameter governing piezoelectric fracture. In their landmark study, the Park–Sun semi-empirical fracture criterion could be reconciled with experimental measurement of the critical load for a crack perpendicular to the poling axis in simple tension and three-point bending PZT-4 specimens. Nevertheless, the difficulty is that the theoretical treatments are incomplete, with omissions such as an *ad hoc* neglect of the electric contribution to the energy release rate by Park and Sun (1995a–b).

In the sections to follow, the crack driving force in electro-thermo-elastodynamic fracture is evaluated based on the fundamental principles of thermodynamics within the framework of the nonlinear theory of coupled electric, thermal, and mechanical fields (Chen, 2009a). The presentation is restricted to the quasi-electrostatic approximation for a simple formulation, which implies the near absence of a time-varying magnetic field.

5.2 Fundamental Principles of Thermodynamics

As shown in Chapter 2, the physical laws in electrodynamics consist of the conservation of mass, conservation of linear momentum, conservation of angular momentum, conservation of energy, and entropy production inequality, in addition to the Maxwell equations.

Using the notations in Chapter 2, the first principle of thermodynamics leads to the local energy balance equation in electro-thermo-elastodynamics:

$$\rho \frac{d\hat{e}}{dt} = -\nabla \cdot \mathbf{j}_q + \boldsymbol{\sigma} : \mathbf{v}\nabla + \rho \mathbf{E} \cdot \dot{\boldsymbol{\pi}} + \mathbf{E} \cdot (\mathbf{j}_e - q_f \mathbf{v}) . \tag{5.1}$$

Substituting the internal energy $\hat{e}$ by the Helmholtz free energy $\hat{h} = \hat{e} - T\hat{s}$ and using a series of transformations, Eq. (5.1) becomes

$$\begin{aligned} \frac{d\hat{s}}{dt} &= -\frac{1}{\rho_0} \nabla_R \cdot \frac{\mathbf{J}_q}{T} + \frac{1}{\rho_0} \mathbf{J}_q \cdot \nabla_R \frac{1}{T} + \frac{1}{\rho_0 T} \hat{\mathbf{E}} \cdot \boldsymbol{J}_e \\ &+ \frac{1}{2\rho_0 T} {}_t\boldsymbol{\Sigma} : \dot{\mathbf{C}} + \frac{1}{\rho_0 T} \hat{\mathbf{E}} \cdot \dot{\hat{\mathbf{D}}} - \frac{1}{T} \hat{s} \dot{T} - \frac{1}{T} \frac{d}{dt} \left(\hat{h} + \frac{{}_e u^f}{\rho} \right), \end{aligned} \tag{5.2}$$

where ${}_t\boldsymbol{\Sigma} = j\mathbf{F}^{-1} {}_t\boldsymbol{\sigma} \mathbf{F}^{-T}$ is the second Piola–Kirchhoff total stress tensor, ${}_t\boldsymbol{\sigma} = \boldsymbol{\sigma} + {}_e\boldsymbol{\sigma}$ is the total stress tensor (which is the sum of the Cauchy stress tensor $\boldsymbol{\sigma}$ and the electric stress tensor ${}_e\boldsymbol{\sigma} = \mathbf{D} \otimes \mathbf{E} - {}_e u^f \mathbf{I}$),

${}_{e}u^{f} = \varepsilon_0 \mathbf{E}\cdot\mathbf{E}/2$ is the energy density of the free electric field, $\hat{\mathbf{E}} = \mathbf{E}\cdot\mathbf{F}$, $\hat{\mathbf{D}} = j\mathbf{F}^{-1}\cdot\mathbf{D}$, $\mathbf{J}_q = j\mathbf{F}^{-1}\cdot\mathbf{j}_q$, $\boldsymbol{J}_e = j\mathbf{F}^{-1}\cdot\boldsymbol{j}_e$, and $\boldsymbol{j}_e = \mathbf{j}_e - q_f\mathbf{v}$.

The augmented Helmholtz free energy, including the contribution of the energy of the free electric field, is introduced by

$$\tilde{h} \equiv \hat{h} + \frac{{}_{e}u^{f}}{\rho}. \tag{5.3}$$

The second principle of thermodynamics leads to the entropy production inequality

$$\frac{d_i\hat{s}}{dt} \equiv \frac{d\hat{s}}{dt} + \frac{1}{\rho}\nabla\cdot\mathbf{j}_s \geq 0. \tag{5.4}$$

In the reference configuration, the entropy production inequality can be rewritten as

$$\frac{d_i\hat{s}}{dt} = \frac{d\hat{s}}{dt} + \frac{1}{\rho_0}\nabla_R\cdot\mathbf{J}_s \geq 0. \tag{5.5}$$

The augmented Helmholtz free energy, including the contribution of the energy of the free electric field, is assumed to be a function of deformation, temperature, temperature gradient, and electric displacement in the reference configuration V_R, with respect to which the deformation gradient $\mathbf{F}$ is measured, that is,

$$\tilde{h} = \tilde{h}(\mathbf{C}, T, \nabla_R T, \hat{\mathbf{D}}; \mathbf{X}). \tag{5.6}$$

Since the entropy production inequality (5.5) should be always valid, it is necessary and sufficient that the state equations fulfill the following conditions:

$$\frac{\partial\tilde{h}}{\partial T_{,K}} = 0, \tag{5.7}$$

$${}_{t}\Sigma_{KL} = 2\rho_0\frac{\partial\tilde{h}}{\partial C_{KL}}, \tag{5.8}$$

$$\hat{s} = -\frac{\partial \tilde{h}}{\partial T}, \tag{5.9}$$

$$\hat{E}_K = \rho_0 \frac{\partial \tilde{h}}{\partial \hat{D}_K}, \tag{5.10}$$

$$\mathbf{J}_s = \frac{1}{T}\mathbf{J}_q, \tag{5.11}$$

$$\frac{d_i \hat{s}}{dt} = \frac{1}{\rho_0}\mathbf{J}_q \cdot \nabla_R \frac{1}{T} + \frac{1}{\rho_0 T}\hat{\mathbf{E}} \cdot \boldsymbol{J}_e \geq 0. \tag{5.12}$$

From Eq. (5.7), it is shown that the augmented Helmholtz free energy does not depend on the temperature gradient.

5.3 Energy Flux and Dynamic Contour Integral

Consider a two-dimensional body $\tilde{B}$ that contains an extending crack (Fig. 5.1). The boundary of the cracked body $\tilde{B}$ is denoted by $\partial\tilde{B}$. A contour $\tilde{\Gamma}$ enclosing the crack tip translates with the crack tip moving at instantaneous speed V_C. When the energy balance is written in global form, the energy flux through $\tilde{\Gamma}$ can be expressed as

$$\begin{aligned} F(\tilde{\Gamma}) &\equiv \int_{\tilde{\Gamma}} [\mathbf{n} \cdot (\boldsymbol{\sigma} + {}_e\boldsymbol{\sigma}) \cdot \mathbf{v} - \mathbf{n} \cdot \boldsymbol{S} + (\tilde{\rho}\tilde{h} + \tilde{\rho}\hat{k})\mathbf{n} \cdot \boldsymbol{V}_C] d\tilde{\Gamma} \\ &= \int_{\partial\tilde{B}} [\mathbf{n} \cdot (\boldsymbol{\sigma} + {}_e\boldsymbol{\sigma}) \cdot \mathbf{v} - \mathbf{n} \cdot \boldsymbol{S}] d\tilde{s} - \int_{\tilde{B}-\tilde{A}_{\tilde{\Gamma}}} \frac{\tilde{\partial}}{\tilde{\partial} t}(\tilde{\rho}\tilde{h} + \tilde{\rho}\hat{k}) d\tilde{A} \\ &\quad + \int_{\tilde{B}-\tilde{A}_{\tilde{\Gamma}}} \tilde{\rho}\hat{\mathbf{f}} \cdot \mathbf{v} d\tilde{A} - \int_{\tilde{B}-\tilde{A}_{\tilde{\Gamma}}} \tilde{\rho}\hat{s}\dot{T} d\tilde{A} - \int_{\tilde{B}-\tilde{A}_{\tilde{\Gamma}}} \frac{\tilde{\rho}}{\rho_0} \hat{\mathbf{E}} \cdot \boldsymbol{J}_e d\tilde{A}, \end{aligned} \tag{5.13}$$

where the Poynting vector in the co-moving frame is given by

$$\boldsymbol{S} = -\mathbf{E} \times (\mathbf{v} \times \mathbf{D}). \tag{5.14}$$

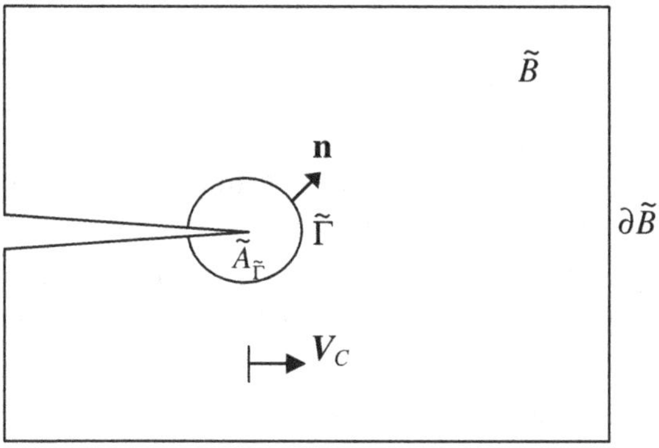

Fig. 5.1. A contour translating with the crack tip moving at instantaneous speed V_C.

The dynamic contour integral is related to the energy flux integral by

$$\tilde{J}_{\tilde{\Gamma}} \equiv \frac{F(\tilde{\Gamma})}{V_C} = \frac{1}{V_C}\int_{\tilde{\Gamma}}[\mathbf{n}\cdot(\boldsymbol{\sigma}+{}_e\boldsymbol{\sigma})\cdot\mathbf{v} - \mathbf{n}\cdot\boldsymbol{S} + (\tilde{\rho}\tilde{h} + \tilde{\rho}\hat{k})\mathbf{n}\cdot\boldsymbol{V}_C]d\tilde{\Gamma}, \quad (5.15)$$

where $V_C = |\boldsymbol{V}_C|$ is the magnitude of the crack speed.

In general, the dynamic contour integral is not path independent. The difference in the energy flux through two contours $\tilde{\Gamma}_1$ and $\tilde{\Gamma}_2$ is

$$\begin{aligned} F(\tilde{\Gamma}_2) - F(\tilde{\Gamma}_1) = \int_{\tilde{A}_{12}} \frac{\tilde{\partial}}{\tilde{\partial} t}(\tilde{\rho}\tilde{h} + \tilde{\rho}\hat{k})d\tilde{A} - \int_{\tilde{A}_{12}} \tilde{\rho}\hat{\mathbf{f}}\cdot\mathbf{v}d\tilde{A} \\ + \int_{\tilde{A}_{12}} \tilde{\rho}\hat{s}\dot{T}d\tilde{A} + \int_{\tilde{A}_{12}} \frac{\tilde{\rho}}{\rho_0}\hat{\mathbf{E}}\cdot\boldsymbol{J}_e d\tilde{A}, \end{aligned} \quad (5.16)$$

where $\tilde{A}_{12}$ is the difference in the areas enclosed by the contours $\tilde{\Gamma}_1$ and $\tilde{\Gamma}_2$, including the crack faces.

The energy flux integral can be extended to the three-dimensional case. If $\tilde{\Gamma}$ is interpreted as a surface in the reference configuration that moves at speed V_C with respect to the material particles instantaneously on it, the energy flux through the surface $\tilde{\Gamma}$ is

$$F(\tilde{\Gamma}) = \int_{\tilde{\Gamma}}[\mathbf{n}\cdot(\boldsymbol{\sigma}+{}_e\boldsymbol{\sigma})\cdot\mathbf{v} - \mathbf{n}\cdot\boldsymbol{S} + (\tilde{\rho}\tilde{h} + \tilde{\rho}\hat{k})\mathbf{n}\cdot\boldsymbol{V}_C]d\tilde{\Gamma}. \quad (5.17)$$

The first term of the energy flux integral expression is the rate of work done by the total traction acting on $\tilde{\Gamma}$, the second term represents the contribution due to the traveling of electromagnetic waves through $\tilde{\Gamma}$, and the third term represents the contribution due to the transport of material through $\tilde{\Gamma}$. It is noted that the associated energy density includes the kinetic energy density and the augmented Helmholtz free energy density, rather than the electric enthalpy density. As a general expression for the energy flux through a surface translating through a deformable solid, expression (5.17) does not depend on the existence or absence of a moving crack. Like its counterpart in elastodynamics (Freund, 1990), the energy flux integral expression (5.17) is valid for large deformation applications.

5.4 Dynamic Energy Release Rate Serving as Crack Driving Force

The dynamic energy release rate is defined as the rate of energy flow out of the body and into the crack tip per unit crack advance, that is,

$$\tilde{J}_0 \equiv \lim_{\tilde{\Gamma}\to 0}\left\{\frac{1}{V_C}\int_{\tilde{\Gamma}}[\mathbf{n}\cdot(\boldsymbol{\sigma}+{}_e\boldsymbol{\sigma})\cdot\mathbf{v}-\mathbf{n}\cdot \boldsymbol{S}+(\tilde{\rho}\tilde{h}+\tilde{\rho}\hat{k})\mathbf{n}\cdot \boldsymbol{V}_C]d\tilde{\Gamma}\right\}, \quad (5.18)$$

where the limit implies that $\tilde{\Gamma}$ is shrunk onto the crack tip.

In view of its definition, the dynamic energy release rate provides a unique characterization of the near-tip fields and thus plays a central role in the fracture criteria. The quantity $\tilde{J}_0$ introduced here cannot be related to mechanical energy variation alone. Instead, $\tilde{J}_0$ refers to total energy variation due to work done by total traction, traveling of electromagnetic waves, and transport of material with its associated energy density.

5.5 Configuration Force and Energy-Momentum Tensor

For steady-state crack propagation in the absence of mechanical body force ($\hat{\mathbf{f}} = 0$), temperature change ($\dot{T} = 0$), and electricity conduction ($\boldsymbol{J}_e = 0$), it can be seen from Eq. (5.16) that the dynamic contour integral becomes path independent as the contour including the crack faces is chosen, that is,

$$\tilde{J}_g = \tilde{J}_l = \tilde{J}_0, \tag{5.19}$$

where $\tilde{\Gamma}_g$ and $\tilde{\Gamma}_l$ are the global and local contours.

If a field quantity is invariant in the reference frame affixed to the crack tip moving at a uniform speed $V_C = V_C\overline{\mathbf{E}}_1$, the field quantity depends on t through the combination $\tilde{\mathbf{X}} = \mathbf{X} - V_C t$ only, where $\overline{\mathbf{E}}_1$ is the unit vector along the crack advance direction. Expression (5.15) for the dynamic contour integral takes the special form

$$\tilde{J} = \int_{\tilde{\Gamma}} \mathbf{n} \cdot [-(\boldsymbol{\sigma} + {}_e\boldsymbol{\sigma}) \cdot \mathbf{u}\tilde{\nabla} + \mathbf{D} \otimes \mathbf{E} \cdot \mathbf{u}\tilde{\nabla} - (\mathbf{D} \cdot \mathbf{E})\mathbf{u}\tilde{\nabla} + (\tilde{\rho}\tilde{h} + \tilde{\rho}\hat{k})\mathbf{I}]d\tilde{\Gamma} \cdot \overline{\mathbf{E}}_1 . \tag{5.20}$$

Using $(\mathbf{n} \times \mathbf{E}) \times \mathbf{D} = (\mathbf{n} \cdot \mathbf{D})\mathbf{E} - (\mathbf{D} \cdot \mathbf{E})\mathbf{n}$, (5.20) becomes

$$\begin{aligned} \tilde{J} &= -\int_{\tilde{\Gamma}} \mathbf{n} \cdot (\boldsymbol{\sigma} + {}_e\boldsymbol{\sigma}) \cdot \mathbf{u}\tilde{\nabla} d\tilde{\Gamma} \cdot \overline{\mathbf{E}}_1 + \int_{\tilde{\Gamma}} [(\mathbf{n} \times \mathbf{E}) \times \mathbf{D}] \cdot \mathbf{u}\tilde{\nabla} d\tilde{\Gamma} \cdot \overline{\mathbf{E}}_1 \\ &\quad + \int_{\tilde{\Gamma}} (\tilde{\rho}\tilde{h} + \tilde{\rho}\hat{k})\mathbf{n} d\tilde{\Gamma} \cdot \overline{\mathbf{E}}_1 . \end{aligned} \tag{5.21}$$

Introducing the energy-momentum tensor

$$\tilde{\mathbf{b}} \equiv -(\boldsymbol{\sigma} + {}_e\boldsymbol{\sigma}) \cdot \mathbf{u}\tilde{\nabla} + \mathbf{D} \otimes \mathbf{E} \cdot \mathbf{u}\tilde{\nabla} - (\mathbf{D} \cdot \mathbf{E})\mathbf{u}\tilde{\nabla} + (\tilde{\rho}\tilde{h} + \tilde{\rho}\hat{k})\mathbf{I} , \tag{5.22}$$

expression (5.20) can be rewritten as

$$\tilde{J} = \int_{\tilde{\Gamma}} \mathbf{n} \cdot \tilde{\mathbf{b}} d\tilde{\Gamma} \cdot \overline{\mathbf{E}}_1 = \tilde{\mathbf{J}} \cdot \overline{\mathbf{E}}_1 , \tag{5.23}$$

where $\tilde{\mathbf{J}} = \int_{\tilde{\Gamma}} \mathbf{n} \cdot \tilde{\mathbf{b}} d\tilde{\Gamma}$ is the configuration force (material force) on the singularity as an extension to the notation by Eshelby (1951, 1970).

Hence, the dynamic contour integral can be taken as the projection of the configuration force on the crack advance direction, which is consistent with the physical interpretation of being the dynamic energy release rate. The dynamic contour integral thus formulated is related to the energy-momentum tensor in the same way as given by Pak and Hermann (1986), Maugin and Epstein (1991) and Dascalu and Maugin

(1994). Nevertheless, distinct from previous work, the dynamic contour integral constructed with the use of the augmented Helmholtz free energy within the framework of the nonlinear theory of coupled electric, thermal, and mechanical fields fully satisfies the thermodynamic requirements and hence can be used in a physically sound fracture criterion.

5.6 Coupled Electromechanical Jump/Boundary Conditions

There are debates in the literature about the selection of the electric boundary conditions on crack faces, as discussed in Section 4.4. Dascalu and Maugin (1995) studied the dynamic fracture problem for piezoelectric materials with the impermeable crack-face condition. Li and Mataga (1996a–b) imposed electrode- and vacuum-type of electric boundary conditions on the crack surfaces, respectively, in their analysis of semi-infinite antiplane crack propagation in a hexagonal piezoelectric medium. Chen and Yu (1997), Chen and Karihaloo (1999) and Wang and Yu (2000) investigated dynamic crack problems in piezoelectric materials subjected to mechanical and electrical impacts for two kinds of crack-face conditions: impermeable and electrical contact. We discuss below the application of coupled electromechanical jump/boundary conditions for a cracked body within the framework of the nonlinear theory of coupled electric and mechanical fields.

For an inclusion problem, the coupled electromechanical jump conditions across the interface are given by

$$\mathbf{n}\cdot[[\mathbf{D}]]=\varpi_f\,, \tag{5.24}$$

$$\mathbf{n}\times[[\mathbf{E}]]=0\,, \tag{5.25}$$

$$\mathbf{n}\cdot[[\boldsymbol{\sigma}+{}_e\boldsymbol{\sigma}]]=0\,, \tag{5.26}$$

$$[[\mathbf{u}]]=0\,. \tag{5.27}$$

For a slit crack problem, the crack-face boundary conditions may be expressed as

$$\mathbf{n}^+ \cdot (\boldsymbol{\sigma} +_e \boldsymbol{\sigma})^+ = -\mathbf{n}^- \cdot (\boldsymbol{\sigma} +_e \boldsymbol{\sigma})^- \\ = -\{\sigma_{21}^0 +_e \sigma_{21}^0, \sigma_{22}^0 +_e \sigma_{22}^0, \sigma_{23}^0 +_e \sigma_{23}^0\}^T, \tag{5.28}$$

$$\mathbf{n}^+ \cdot \mathbf{D}^+ = -\mathbf{n}^- \cdot \mathbf{D}^- = -D_2^0. \tag{5.29}$$

Conditions (5.24)–(5.27) are exact, but the corresponding boundary-initial value problem needs to be solved in both the cracked solid region and the interior fluid (vacuum, air, etc.) region. It is noted that the total traction should be considered in the coupled electromechanical boundary conditions along the crack faces and the remainder surfaces of the solid. D_2^0 in (5.29) can be either prescribed for the impermeable crack-face condition or determined through solving the boundary-initial value problem with the permeable or semi-permeable crack-face condition.

5.7 Asymptotic Near-Tip Field Solution

The dynamic energy release rate serves as the crack driving force for any electro-thermo-elastic boundary-initial value problem and can be evaluated as long as the solution for the propagation of a crack, either electrically insulating or conducting, is given. Different from a stationary or quasi-static crack problem, a dynamic crack problem is concerned with fracture phenomena for which inertia effects arising from either rapidly applied loads or rapid crack propagation become significant. The influence of material inertia on the distribution of near-tip fields is of great importance because these fields represent the environment in which the mechanisms of crack advance are operative. Since it is rare to obtain closed-form exact solutions for this class of complicated problems, asymptotic solutions are often sought.

It emerges from the analysis by Yang (2004) and Li and Yang (2005) that the difference between the solutions for the fully dynamic antiplane unelectroded crack problem of polarized ceramics and the dynamic antiplane unelectroded crack problem based on the quasi-electrostatic approximation is small, since the crack speed is much lower than the speed of light. Thus, the quasi-electrostatic approximation can still be adopted for studying dynamic crack propagation so that the electric field may be expressed by the gradient of a scalar function called the electric potential.

Consider a generalized plane crack problem with the crack tip of primary interest advancing at instantaneous speed V_C along the X_1-axis. The displacement component u_m (m=1,2,3) and the electric potential ϕ are independent of X_3. To derive the asymptotic expansion, the scaled variables $\hat{X}_1 = \tilde{X}_1/\varepsilon$ and $\hat{X}_2 = \tilde{X}_2/\varepsilon$ are introduced, where ε is a small parameter, $\tilde{X}_1 = X_1 - V_C t$, and $\tilde{X}_2 = X_2$. If ε is taken to be indefinitely small, all points in the plane ($\tilde{X}_1, \tilde{X}_2$) except those near the crack tip are pushed out of the field of observation in the plane ($\hat{X}_1, \hat{X}_2$). Furthermore, as viewed from the scaled reference coordinate system affixed to the moving crack tip, the crack appears to be semi-infinite along the negative $\hat{X}_1$-axis.

As an extension of a standard solution procedure for asymptotic fields near a moving crack tip in elastodynamic fracture mechanics (Freund, 1990), the displacement components and the electric potential are expanded in powers of ε of the form

$$u_m(X_1, X_2, t) = \varepsilon^{q_0} \hat{u}_m^{(0)}(\hat{X}_1, \hat{X}_2, t) + \varepsilon^{q_1} \hat{u}_m^{(1)}(\hat{X}_1, \hat{X}_2, t) + \cdots, \quad (5.30)$$

$$\phi(X_1, X_2, t) = \varepsilon^{q_0} \hat{\phi}^{(0)}(\hat{X}_1, \hat{X}_2, t) + \varepsilon^{q_1} \hat{\phi}^{(1)}(\hat{X}_1, \hat{X}_2, t) + \cdots, \quad (5.31)$$

where $\hat{u}_m^{(0)}$ and $\hat{\phi}^{(0)}$ represent the dominant contribution, $\hat{u}_m^{(1)}$ and $\hat{\phi}^{(1)}$ represent the first-order correction, and so on. This implies that the exponents are ordered such that $q_0 < q_1 < q_2 < \cdots$.

The above expansion is essentially an assumption that the near-tip fields can be represented as a series of homogeneous functions of increasing degree. The assumed form of expansion is substituted into the governing equations and the coefficient of each power of ε is set equal to zero. The coefficient of the lowest power of ε vanishes if the dominant asymptotic solution satisfies

$$c_{ijkl} \hat{u}_{k,jl}^{(0)} + e_{lij} \hat{\phi}_{,jl}^{(0)} = \rho V_C^2 \hat{u}_{i,11}^{(0)}, \quad (5.32)$$

$$e_{jkl} \hat{u}_{k,jl}^{(0)} - \kappa_{jl} \hat{\phi}_{,jl}^{(0)} = 0. \quad (5.33)$$

Based on the Stroh-type formalism discussed extensively in Section 4.3, the solution is sought in the form

$$\hat{u}_m^{(0)} = a_m f(\hat{z}) \ (m{=}1,2,3), \ \hat{\phi}^{(0)} = a_4 f(\hat{z}), \ \hat{z} = \hat{X}_1 + p\hat{X}_2, \quad (5.34)$$

where the function f is analytic in the complex variable $\hat{z} = \hat{X}_1 + p\hat{X}_2$ and the complex numbers p and a_m must be determined from Eqs. (5.32) and (5.33).

Substituting Eq. (5.34) into Eqs. (5.32) and (5.33) yields

$$[\mathbf{Q} - \rho V_C^2 \mathbf{U} + (\mathbf{R} + \mathbf{R}^T)p + \mathbf{T}p^2]\mathbf{a} = 0, \quad (5.35)$$

with $\mathbf{a} = (a_1, a_2, a_3, a_4)^T$, $\mathbf{U} = \text{diag}(1,1,1,0)$ and the 4×4 matrices:

$$\mathbf{Q} = \begin{bmatrix} c_{1jk1} & e_{1j1} \\ e_{1k1}^T & -\kappa_{11} \end{bmatrix}, \ \mathbf{R} = \begin{bmatrix} c_{1jk2} & e_{2j1} \\ e_{1k2}^T & -\kappa_{12} \end{bmatrix}, \mathbf{T} = \begin{bmatrix} c_{2jk2} & e_{2j2} \\ e_{2k2}^T & -\kappa_{22} \end{bmatrix}. \quad (5.36)$$

Nontrivial solutions are obtained if p is a root of

$$\det[\mathbf{Q} - \rho V_C^2 \mathbf{U} + (\mathbf{R} + \mathbf{R}^T)p + \mathbf{T}p^2] = 0. \quad (5.37)$$

The eight roots of Eq. (5.37) depend on the crack velocity V_C, that is, $p_\alpha = p_\alpha(V_C)$. A real root p of Eq. (5.37) corresponds to a value of V_C equal to the velocity of bulk waves propagating in the direction (1, p) in the ($\hat{X}_1, \hat{X}_2$) plane. Following the treatment by Lothe and Barnett (1976) in the study of surface waves in piezoelectric crystals, c_L is introduced to denote the inferior limit of such bulk wave velocities. Then, Eq. (5.37) has no real roots for $V_C < c_L$. Since the coefficients of Eq. (5.37) are real, the eigenvalues and the eigenvectors form two sets of complex quantities with one set being conjugate to the other. We suppose p_α ($\alpha = 1,2,3,4$) are four distinct roots with positive imaginary parts and construct the matrix **A** with columns that are the associated eigenvectors. Then, the solution of Eqs. (5.32)–(5.33) is expressed as

$$\hat{u}_m^{(0)} = \sum_{\alpha=1}^{4} A_{m\alpha} f_\alpha(\hat{z}_\alpha) + \sum_{\alpha=1}^{4} \overline{A}_{m\alpha} \overline{f_\alpha(\hat{z}_\alpha)}, \quad (5.38)$$

$$\hat{\phi}^{(0)} = \sum_{\alpha=1}^{4} A_{4\alpha} f_\alpha(\hat{z}_\alpha) + \sum_{\alpha=1}^{4} \overline{A}_{4\alpha} \overline{f_\alpha(\hat{z}_\alpha)}, \tag{5.39}$$

where $\hat{z}_\alpha = \hat{X}_1 + p_\alpha \hat{X}_2$ ($\alpha = 1,2,3,4$) and the over-bars denote complex conjugates.

Hence, the total stress and electric displacement in the vicinity of the crack tip moving at instantaneous speed V_C are given by

$$\begin{aligned} {}_t\sigma_{i1} &= \varepsilon^{(q_0-1)} \sum_{\alpha=1}^{4} (\rho V_C^2 A_{i\alpha} - p_\alpha L_{i\alpha}) f_\alpha'(\hat{z}_\alpha) \\ &+ \varepsilon^{(q_0-1)} \sum_{\alpha=1}^{4} (\rho V_C^2 \overline{A}_{i\alpha} - \overline{p}_\alpha \overline{L}_{i\alpha}) \overline{f_\alpha'(\hat{z}_\alpha)}, \end{aligned} \tag{5.40}$$

$${}_t\sigma_{i2} = \varepsilon^{(q_0-1)} \sum_{\alpha=1}^{4} L_{i\alpha} f_\alpha'(\hat{z}_\alpha) + \varepsilon^{(q_0-1)} \sum_{\alpha=1}^{4} \overline{L}_{i\alpha} \overline{f_\alpha'(\hat{z}_\alpha)}, \tag{5.41}$$

$$D_1 = -\varepsilon^{(q_0-1)} \sum_{\alpha=1}^{4} p_\alpha L_{4\alpha} f_\alpha'(\hat{z}_\alpha) - \varepsilon^{(q_0-1)} \sum_{\alpha=1}^{4} \overline{p}_\alpha \overline{L}_{4\alpha} \overline{f_\alpha'(\hat{z}_\alpha)}, \tag{5.42}$$

$$D_2 = \varepsilon^{(q_0-1)} \sum_{\alpha=1}^{4} L_{4\alpha} f_\alpha'(\hat{z}_\alpha) + \varepsilon^{(q_0-1)} \sum_{\alpha=1}^{4} \overline{L}_{4\alpha} \overline{f_\alpha'(\hat{z}_\alpha)}, \tag{5.43}$$

where $L_{n\alpha} = (R_{mn} + p_\alpha T_{nm}) A_{m\alpha} = -(Q_{nm} + p_\alpha R_{nm} - \rho V_C^2 U_{nm}) A_{m\alpha} / p_\alpha$.

Let us introduce

$$\mathbf{f} = (f_1(\hat{z}_1), f_2(\hat{z}_2), f_3(\hat{z}_3), f_4(\hat{z}_4))^T, \tag{5.44}$$

$$\mathbf{h} = \mathbf{L}\mathbf{f}'. \tag{5.45}$$

The singular solution that gives bounded displacements and electric potential is

$$\mathbf{h}(\hat{z}) = \frac{1}{\sqrt{8\pi\hat{z}}} \tilde{\mathbf{k}}, \tag{5.46}$$

where $\tilde{\mathbf{k}}$ is the dynamic field intensity factor vector.

It is evident that the parameter q_0 in expressions (5.30) and (5.31) has the value of one-half. The total stress and electric displacement have the classical inverse square-root singularities at the crack tip. Accordingly, at a distance r ahead of the crack tip,

$$({}_t\sigma_{21}, {}_t\sigma_{22}, {}_t\sigma_{23}, D_2)^T = \frac{1}{\sqrt{2\pi r}}\tilde{\mathbf{k}}, \tag{5.47}$$

where $\tilde{\mathbf{k}} = (\tilde{K}_{II}, \tilde{K}_I, \tilde{K}_{III}, \tilde{K}_D)^T$, $\tilde{K}_I$, $\tilde{K}_{II}$, $\tilde{K}_{III}$ are mode-I, mode-II, and mode-III dynamic total stress intensity factors and $\tilde{K}_D$ is the dynamic electric displacement intensity factor.

To evaluate the dynamic energy release rate, we choose the contour $\tilde{\Gamma}_0$ in the reference frame ($\tilde{X}_1 = X_1 - V_C t$, $\tilde{X}_2 = X_2$) as shown in Fig. 5.2. This is a convenient choice because $n_1 = 0$ along the segments parallel to the $\tilde{X}_1$-axis. The contour is shrunk onto the crack tip by first letting $\delta_2 \to 0$ and then $\delta_1 \to 0$. As in the purely elastodynamic case discussed in Section 1.5, there is no contribution to $\tilde{J}_0$ from the segments parallel to the $\tilde{X}_2$-axis. Furthermore, the second and third terms on the right-hand side of Eq. (5.18) along the segments parallel to the $\tilde{X}_1$-axis vanish. Consequently, $\tilde{J}_0$ can be computed by evaluating only the first term on the right-hand side of Eq. (5.18) along the segments parallel to the $\tilde{X}_1$-axis, that is,

$$\tilde{J}_0 = \frac{2}{V_C} \lim_{\delta_1 \to 0} \{ \lim_{\delta_2 \to 0} \int_{-\delta_1}^{\delta_1} [\sigma_{2j}(\tilde{X}_1, \delta_2, t) + {}_e\sigma_{2j}(\tilde{X}_1, \delta_2, t)] \dot{u}_j(\tilde{X}_1, \delta_2, t) d\tilde{X}_1 \}. \tag{5.48}$$

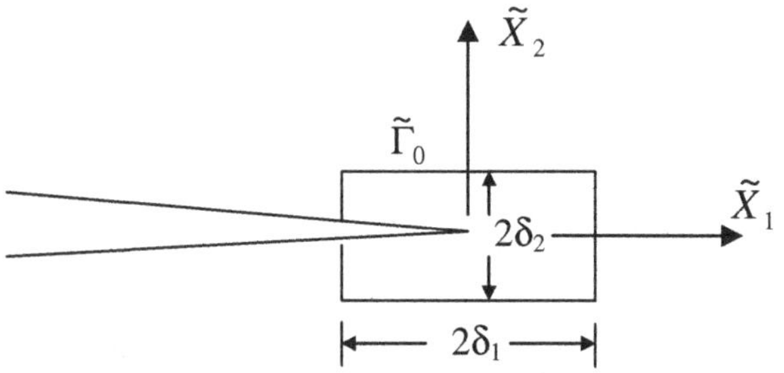

Fig. 5.2. A particular choice of the contour for evaluating the dynamic energy release rate.

Hence, the dynamic energy release rate is equal to the following crack closure integral:

$$\tilde{J}_0 = \lim_{\delta a \to 0} \frac{1}{2\delta a} \int_0^{\delta a} [\sigma_{2j}(\tilde{X}_1, 0, t) + {}_e\sigma_{2j}(\tilde{X}_1, 0, t)] \Delta u_j(\tilde{X}_1 - \delta a, 0^{\pm}, t) d\tilde{X}_1 \,, \tag{5.49}$$

where $\Delta u_j(\tilde{X}_1 - \delta a, 0^{\pm}, t) = u_j(\tilde{X}_1 - \delta a, 0^{+}, t) - u_j(\tilde{X}_1 - \delta a, 0^{-}, t)$ is the crack opening displacement at a distance $\delta a - \tilde{X}_1$ behind the crack tip.

Consequently, the dynamic energy release rate is calculated as

$$\tilde{J}_0 = \frac{1}{4}(\tilde{K}_{II}, \tilde{K}_I, \tilde{K}_{III}, 0) \cdot \tilde{\mathbf{H}}^i \cdot (\tilde{K}_{II}, \tilde{K}_I, \tilde{K}_{III}, \tilde{K}_D)^T \,, \tag{5.50}$$

where $\tilde{\mathbf{H}}^i = 2\,\mathrm{Re}(i\mathbf{A}\mathbf{L}^{-1})$ is the dynamic counterpart of the Irwin matrix described in Section 4.5.2, which depends on material properties and crack speed.

Remarkably, Eq. (5.50) shows that the dynamic energy release rate is an odd function of the electric displacement intensity factor, which is in agreement with experimental observations (Pak and Tobin, 1993; Tobin and Pak, 1993; Cao and Evans, 1994; Lynch *et al.*, 1995; Park and Sun, 1995a–b; Jiang and Sun, 1999, 2001; Qin, 2001; Zhang *et al.*, 2002; Soh *et al.*, 2003; Chen and Lu, 2003). As explained by Dascalu and Maugin (1995), the behavior of the dynamic Irwin matrix as a function of the crack velocity is intimately related to the existence of surface waves on the crack faces.

For the mode-I dynamic crack problem, the dynamic energy release rate and the dynamic crack opening displacement intensity factor are given by

$$\tilde{J}_0 = \frac{1}{4}(0, \tilde{K}_I, 0, 0) \cdot \tilde{\mathbf{H}}^i \cdot (0, \tilde{K}_I, 0, \tilde{K}_D)^T \,, \tag{5.51}$$

$$\tilde{K}_I^{COD} = 4\tilde{J}_0 / \tilde{K}_I = (0,1,0,0) \cdot \tilde{\mathbf{H}}^i \cdot (0, \tilde{K}_I, 0, \tilde{K}_D)^T \,. \tag{5.52}$$

For the mode-II dynamic crack problem, the dynamic energy release rate and the dynamic crack opening displacement intensity factor are given by

$$\tilde{J}_0 = \frac{1}{4}(\tilde{K}_{II}, 0, 0, 0) \cdot \tilde{\mathbf{H}}^i \cdot (\tilde{K}_{II}, 0, 0, \tilde{K}_D)^T, \tag{5.53}$$

$$\tilde{K}_{II}^{COD} = 4\tilde{J}_0 / \tilde{K}_{II} = (1, 0, 0, 0) \cdot \tilde{\mathbf{H}}^i \cdot (\tilde{K}_{II}, 0, 0, \tilde{K}_D)^T. \tag{5.54}$$

For the mode-III dynamic crack problem, the dynamic energy release rate and the dynamic crack opening displacement intensity factor are given by

$$\tilde{J}_0 = \frac{1}{4}(0, 0, \tilde{K}_{III}, 0) \cdot \tilde{\mathbf{H}}^i \cdot (0, 0, \tilde{K}_{III}, \tilde{K}_D)^T, \tag{5.55}$$

$$\tilde{K}_{III}^{COD} = 4\tilde{J}_0 / \tilde{K}_{III} = (0, 0, 1, 0) \cdot \tilde{\mathbf{H}}^i \cdot (0, 0, \tilde{K}_{III}, \tilde{K}_D)^T. \tag{5.56}$$

For complete evaluation of the crack driving force, total stress and electric displacement intensity factors should be obtained from the solution of a particular boundary-initial value problem. Quasi-static propagation of a crack perpendicular to the poling axis under the impermeable crack-face condition studied by Park and Sun (1995a–b) can be taken as a special case of the mode-I dynamic crack problem as the crack velocity tends to zero.

The antiplane dynamic crack problem studied by Dascalu and Maugin (1995) corresponds to the mode-III dynamic crack problem with the crack front parallel to the poling axis. With the replacement of the Cauchy stress by the total stress in their solution, the dynamic energy release rate and the dynamic crack opening displacement intensity factor are

$$\tilde{J}_0 = \frac{1}{2\alpha\rho c_T^2}\left(\tilde{K}_{III}^2 + \frac{e_{15}}{\kappa_{11}}\tilde{K}_{III}\tilde{K}_D\right), \tag{5.57}$$

$$\tilde{K}_{III}^{COD} = \frac{2}{\alpha\rho c_T^2}\left(\tilde{K}_{III} + \frac{e_{15}}{\kappa_{11}}\tilde{K}_D\right), \tag{5.58}$$

where $\alpha^2 = 1 - (V_C^2 / c_T^2)$ and c_T is the piezoelectrically stiffened bulk shear wave speed given by

$$c_T^2 = \frac{c_{44}}{\rho}\left(1 + \frac{e_{15}^2}{c_{44}\kappa_{11}}\right). \tag{5.59}$$

5.8 Remarks

This formulation successfully captures the singularity of coupled fields, offers the right expression for the crack driving force, and reconciles the fundamental discrepancy between theoretical predictions and experimental observations. The important features are summarized below:

- The dynamic total stress intensity factors describe the inverse square-root singularity of the near-tip total stress as the sum of the Cauchy stress and the electric stress, and the electric displacement intensity factor describes the inverse square-root singularity of the near-tip electric displacement. The crack-face conditions affect dynamic crack propagation by changing the field intensity factors.
- The definition of the dynamic contour integral originated from the energy flux integral, which is generally path dependent. For steady-state crack propagation in the absence of mechanical body force, temperature change, and electricity conduction, the dynamic contour integral becomes path independent as the contour including the crack faces is chosen.
- The dynamic energy release rate can be evaluated by the "mechanical" part of the crack closure integral with the replacement of the Cauchy stress by the total stress, which is consistent with the semi-empirical fracture criterion proposed by Park and Sun (1995a–b). Nevertheless, the difference lies in the replacement of the Cauchy stress tensor by the total stress tensor and the equivalence of the crack-tip dynamic contour integral to the dynamic energy release rate instead of the mechanical strain energy release rate.
- The dynamic energy release rate serving as the crack driving force is an odd function of the electric displacement intensity factor, which is

in agreement with experimental evidence (Pak and Tobin, 1993; Tobin and Pak, 1993; Park and Sun, 1995a–b; Jiang and Sun, 2001; Qin, 2001; Zhang *et al.*, 2002; Soh *et al.*, 2003; Chen and Lu, 2003).

- The application of a purely electric load can drive crack growth in the absence of a mechanical load due to its contribution to the dynamic energy release rate through the dynamic total stress and electric displacement intensity factors, which is in agreement with the experimental observations on fatigue crack growth under cyclic electric loading (e.g., Cao and Evans, 1994; Lynch *et al.*, 1995; Jiang and Sun, 1999; Zhang *et al.*, 2004).
- In addition to the dynamic energy release rate, the dynamic crack opening displacement intensity factor may be taken as an important parameter to monitor electro-elastodynamic fracture.

Chapter 6

Dynamic Fracture Mechanics of Magneto-Electro-Thermo-Elastic Solids

6.1 Introduction

In Chapter 5, electro-thermo-elastodynamic fracture was investigated under the quasi-electrostatic approximation, that is, the near-absence of a time-varying magnetic field. Since the early work by Van Suchtelen (1972), piezoelectric/piezomagnetic composites have been developed for various engineering applications as a result of the emergence of a new product property, i.e., magnetoelectric coupling, which is absent in single-phase piezoelectric or piezomagnetic materials. The co-existence of piezoelectric, piezomagnetic and magnetoelectric coupling effects (i.e., magneto-electro-elastic coupling effects) in composite materials consisting of piezoelectric and piezomagnetic phases introduces many complexities to multiphysics analysis.

Research on the deformation and fracture behavior of magneto-electro-thermo-elastic solids has drawn considerable attention (e.g., Harshe *et al.*, 1993; Nan, 1994; Maugin, 1994; Alshits *et al.*, 1995; Kirchner and Alshits, 1996; Huang *et al.*, 1998; Li, 2000; Trimarco and Maugin, 2001; Liu *et al.*, 2001; Sih *et al.*, 2003; Song and Sih, 2003; Wang and Mai, 2003, 2007a–b; Gao *et al.*, 2003, 2004; Du *et al.*, 2004; Hu and Li, 2005; Zhong and Li, 2006; Niraula and Wang, 2006; Hu *et al.*, 2007; Feng, *et al.*, 2007; Zhong *et al.*, 2009; Wang *et al.*, 2009) because of the safety and reliability requirements for their service in actuators, sensors, waveguides, electronic packaging, and biomedical devices. Recently, Chen (2009b) studied the energy release rate and the path-independent integral in dynamic fracture of magneto-electro-thermo-elastic solids, which is an extension of the new formation of the crack driving force and the energy-momentum tensor in electro-elastodynamic fracture (Chen, 2009a).

This chapter begins with the thermodynamic formulation of a fully coupled dynamic fracture mechanics framework for crack propagation in nonlinear magneto-electro-thermo-elastic solids, followed by evaluation of the dynamic energy release rate through seeking the complex variable solution based on the Stroh-type formalism. After that, magneto-electro-elastostatic crack problems are discussed as special cases. Finally, a summary is given.

6.2 Thermodynamic Formulation of Fully Coupled Dynamic Framework

The elements of the non-relativistic electrodynamics of continua have been discussed in Chapter 2. We now focus on developing a fully coupled dynamic framework for crack propagation in nonlinear magneto-electro-thermo-elastic solids based on the fundamental principles of thermodynamics. The thermodynamic formulation enables us to deal with complex material and fracture behaviors in a unified way and requires only that constitutive equations should be derived from an explicitly defined free energy function and transport laws conform to the requirement of non-negative dissipation.

6.2.1 *Field equations and jump conditions*

For a cracked body $\tilde{B}$ under combined magnetic, electric, thermal, and mechanical loadings, the basic field equations and associated jump conditions are summarized below, following the localization of the fundamental physical laws.

Gauss's law (in $\tilde{B}$):

$$\nabla \cdot \mathbf{D} = q_f . \tag{6.1}$$

Gauss's law for magnetism (in $\tilde{B}$):

$$\nabla \cdot \mathbf{B} = 0 . \tag{6.2}$$

Faraday's law (in $\tilde{B}$):

$$\nabla \times \mathbf{E} + \frac{\partial \mathbf{B}}{\partial t} = 0 . \tag{6.3}$$

Ampere's law (in $\tilde{B}$):

$$\nabla \times \mathbf{H} - \frac{\partial \mathbf{D}}{\partial t} = \mathbf{j}_e . \tag{6.4}$$

Conservation law of electric charges (in $\tilde{B}$):

$$\frac{\partial q_f}{\partial t} + \nabla \cdot \mathbf{j}_e = 0 . \tag{6.5}$$

Conservation law of mass (in $\tilde{B}$):

$$\frac{d\rho}{dt} + \rho \nabla \cdot \mathbf{v} = 0 . \tag{6.6}$$

Conservation law of momentum (in $\tilde{B}$):

$$\rho \frac{d\mathbf{v}}{dt} = \nabla \cdot (\boldsymbol{\sigma} + {}_{em}\boldsymbol{\sigma}) + \rho \hat{\mathbf{f}} - \frac{\partial \mathbf{G}}{\partial t} . \tag{6.7}$$

Conservation law of angular momentum (in $\tilde{B}$):

$$\varepsilon_{kij}(\sigma_{ij} + {}_{em}\sigma_{ij}) = 0 . \tag{6.8}$$

Conservation law of energy (in $\tilde{B}$):

$$\rho \frac{d\hat{e}}{dt} = -\nabla \cdot \mathbf{j}_q + \boldsymbol{\sigma} : \mathbf{v}\nabla + \rho \boldsymbol{E} \cdot \dot{\boldsymbol{\pi}} - \boldsymbol{M} \cdot \dot{\mathbf{B}} + \boldsymbol{E} \cdot \boldsymbol{j}_e . \tag{6.9}$$

Entropy production inequality (in $\tilde{B}$):

$$\frac{d\hat{s}}{dt}+\frac{1}{\rho}\nabla\cdot\mathbf{j}_s\geq 0\,. \tag{6.10}$$

Constitutive equations (in $\widetilde{B}$):

$$_{E}\Sigma_{KL}=2\rho_0\frac{\partial\hat{h}}{\partial C_{KL}}, \tag{6.11}$$

$$\hat{s}=-\frac{\partial\hat{h}}{\partial T}, \tag{6.12}$$

$$\hat{E}_K=\rho_0\frac{\partial\hat{h}}{\partial\hat{\Pi}_K}, \tag{6.13}$$

$$\hat{M}_K=-\rho_0\frac{\partial\hat{h}}{\partial\hat{B}_K}. \tag{6.14}$$

Transport laws (in $\widetilde{B}$):

$$\mathbf{J}_q=\hat{\mathbf{L}}^{qq}\cdot\nabla_R\frac{1}{T}+\frac{1}{T}\hat{\mathbf{L}}^{qe}\cdot\hat{\boldsymbol{E}}\,, \tag{6.15}$$

$$\boldsymbol{J}_e=\hat{\mathbf{L}}^{eq}\cdot\nabla_R\frac{1}{T}+\frac{1}{T}\hat{\mathbf{L}}^{ee}\cdot\hat{\boldsymbol{E}}\,. \tag{6.16}$$

Jump conditions (across $\partial\widetilde{B}$):

$$\mathbf{n}\cdot[[\mathbf{D}]]=\varpi_f\,, \tag{6.17}$$

$$\mathbf{n}\cdot[[\mathbf{B}]]=0\,, \tag{6.18}$$

$$\mathbf{n}\times[[\mathbf{E}+\mathbf{v}\times\mathbf{B}]]=0\,, \tag{6.19}$$

$$\mathbf{n}\times[[\mathbf{H}-\mathbf{v}\times\mathbf{D}]]=0\,, \tag{6.20}$$

$$\mathbf{n}\cdot[[\mathbf{j}_e-q_f\mathbf{v}]]+\frac{\bar{\delta}\varpi_f}{\bar{\delta}t}=0\,, \tag{6.21}$$

$$\mathbf{n}\cdot[[\boldsymbol{\sigma}+{}_{em}\boldsymbol{\sigma}+\mathbf{v}\otimes\mathbf{G}]]=0\,, \tag{6.22}$$

$$\mathbf{n}\cdot[[\mathbf{j}_q-(\boldsymbol{\sigma}+{}_{em}\boldsymbol{\sigma}+\mathbf{v}\otimes\mathbf{G})\cdot\mathbf{v}+S]]=0\,. \tag{6.23}$$

6.2.2 *Dynamic energy release rate*

Consider a three-dimensional deformable electromagnetic body $\tilde{B}$ containing a propagating crack of arbitrary shape (Fig. 6.1). The $\tilde{X}_3$-axis is tangent to the crack front at the observation point P attached to the reference frame translating with the crack front moving at instantaneous speed V_C along the $\tilde{X}_1$-axis. A surface $\tilde{\Gamma}$ surrounding the crack front is fixed relative to the reference frame.

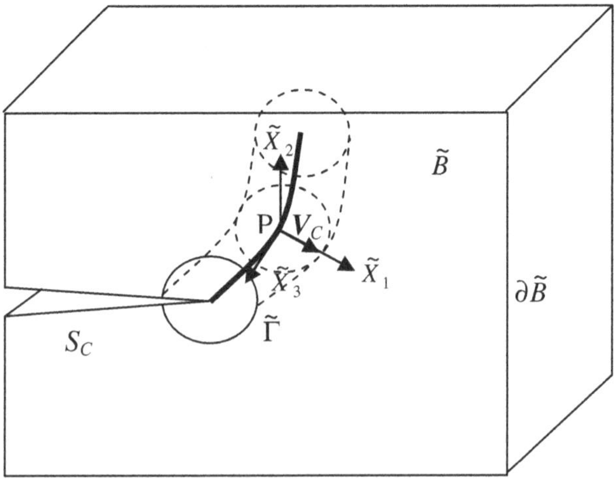

Fig. 6.1. A three-dimensional deformable electromagnetic body containing a propagating crack of arbitrary shape.

A global form of the energy balance leads to the following expression for the energy flux integral:

$$\begin{aligned}
F(\tilde{\Gamma}) &\equiv \int_{\tilde{\Gamma}} [\mathbf{n}\cdot(\boldsymbol{\sigma}+{}_{em}\boldsymbol{\sigma}+\mathbf{v}\otimes\mathbf{G})\cdot\mathbf{v}-\mathbf{n}\cdot\boldsymbol{S}+(\tilde{\rho}\hat{k}+\tilde{\rho}\hat{h}+{}_{em}u^{f})\mathbf{n}\cdot\boldsymbol{V}_{C}]d\tilde{\Gamma} \\
&= \int_{\partial\tilde{B}} [\mathbf{n}\cdot(\boldsymbol{\sigma}+{}_{em}\boldsymbol{\sigma}+\mathbf{v}\otimes\mathbf{G})\cdot\mathbf{v}-\mathbf{n}\cdot\boldsymbol{S}]d\tilde{S} \\
&\quad - \int_{\tilde{B}-\tilde{V}_{\tilde{\Gamma}}} \frac{\tilde{\partial}}{\tilde{\partial}t}(\tilde{\rho}\hat{k}+\tilde{\rho}\hat{h}+{}_{em}u^{f})d\tilde{V} + \int_{\tilde{B}-\tilde{V}_{\tilde{\Gamma}}} \tilde{\rho}\hat{\mathbf{f}}\cdot\mathbf{v}d\tilde{V} \\
&\quad - \int_{\tilde{B}-\tilde{V}_{\tilde{\Gamma}}} \frac{\tilde{\rho}}{\rho_0}\hat{\boldsymbol{E}}\cdot\boldsymbol{J}_e d\tilde{V} - \int_{\tilde{B}-\tilde{V}_{\tilde{\Gamma}}} \tilde{\rho}\hat{s}\dot{T}d\tilde{V},
\end{aligned} \tag{6.24}$$

where $\tilde{V}_{\tilde{\Gamma}}$ is the volume bounded by $\tilde{\Gamma}$, $\partial\tilde{B}$ is the boundary of the cracked body $\tilde{B}$, and $\boldsymbol{S}=\boldsymbol{E}\times\boldsymbol{H}$ is the Poynting vector in the co-moving frame R_C.

Hence, the total energy flux is caused by work done by the total traction, traveling of electromagnetic waves, and transport of material with its associated energy density. It is noted that the associated energy density includes the kinetic energy density, the Helmholtz free energy density, and the energy density of the free electromagnetic fields.

The dynamic energy release rate is defined as the rate of energy flow out of the body and into the crack front per unit crack advance, that is,

$$\tilde{J}_0 = \lim_{\tilde{\Gamma}\to 0}\frac{1}{\dot{A}}\int_{\tilde{\Gamma}}[\mathbf{n}\cdot(\boldsymbol{\sigma}+{}_{em}\boldsymbol{\sigma}+\mathbf{v}\otimes\mathbf{G})\cdot\mathbf{v}-\mathbf{n}\cdot\boldsymbol{S}+(\tilde{\rho}\hat{k}+\tilde{\rho}\hat{h}+{}_{em}u^{f})\mathbf{n}\cdot\boldsymbol{V}_{C}]d\tilde{\Gamma}, \tag{6.25}$$

where the limit implies that $\tilde{\Gamma}$ is shrunk onto the crack front and $\dot{A}$ is the crack area growth rate.

The above definition is reduced to Eq. (1.27) as the electromagnetic fields are shut off and to Eq. (5.18) under the quasi-electrostatic approximation.

6.2.3 *Invariant integral*

In view of its definition, the dynamic energy release rate serves as the crack driving force for any boundary-initial value problem and can be evaluated when the solution becomes available. Since it is rare to obtain closed-form full-field solutions under combined loading conditions, numerical techniques are often used to determine coupled magnetic, electric, thermal, and mechanical fields. However, it is difficult to proceed numerically to the limit required in (6.25) for definition of the dynamic energy release rate due to inaccuracy of numerical solutions for points very close to the crack front where gradients are severe. Therefore, an equivalent representation that is less sensitive to numerical inaccuracy in the crack-front region is needed for efficient evaluation of the dynamic energy release rate.

The relationship between the energy fluxes through two surfaces $\tilde{\Gamma}_1$ and $\tilde{\Gamma}_2$ is obtained as

$$\begin{aligned}
&\int_{\tilde{\Gamma}_2}[\mathbf{n}\cdot(\boldsymbol{\sigma}+{}_{em}\boldsymbol{\sigma}+\mathbf{v}\otimes\mathbf{G})\cdot\mathbf{v}-\mathbf{n}\cdot\boldsymbol{S}+(\tilde{\rho}\hat{k}+\tilde{\rho}\hat{h}+{}_{em}u^f)\mathbf{n}\cdot\boldsymbol{V}_C]d\tilde{\Gamma}\\
&-\lim_{\tilde{\Gamma}_0\to 0}\int_{\tilde{V}_{\tilde{\Gamma}_2}-\tilde{V}_{\tilde{\Gamma}_0}}\frac{\tilde{\partial}}{\tilde{\partial}t}(\tilde{\rho}\hat{k}+\tilde{\rho}\hat{h}+{}_{em}u^f)d\tilde{V}+\lim_{\tilde{\Gamma}_0\to 0}\int_{\tilde{V}_{\tilde{\Gamma}_2}-\tilde{V}_{\tilde{\Gamma}_0}}\tilde{\rho}\hat{\mathbf{f}}\cdot\mathbf{v}d\tilde{V}\\
&-\lim_{\tilde{\Gamma}_0\to 0}\int_{\tilde{V}_{\tilde{\Gamma}_2}-\tilde{V}_{\tilde{\Gamma}_0}}\frac{\tilde{\rho}}{\rho_0}\hat{\boldsymbol{E}}\cdot\boldsymbol{J}_e d\tilde{V}-\lim_{\tilde{\Gamma}_0\to 0}\int_{\tilde{V}_{\tilde{\Gamma}_2}-\tilde{V}_{\tilde{\Gamma}_0}}\tilde{\rho}\hat{s}\dot{T}d\tilde{V}\\
&=\int_{\tilde{\Gamma}_1}[\mathbf{n}\cdot(\boldsymbol{\sigma}+{}_{em}\boldsymbol{\sigma}+\mathbf{v}\otimes\mathbf{G})\cdot\mathbf{v}-\mathbf{n}\cdot\boldsymbol{S}+(\tilde{\rho}\hat{k}+\tilde{\rho}\hat{h}+{}_{em}u^f)\mathbf{n}\cdot\boldsymbol{V}_C]d\tilde{\Gamma}\\
&-\lim_{\tilde{\Gamma}_0\to 0}\int_{\tilde{V}_{\tilde{\Gamma}_1}-\tilde{V}_{\tilde{\Gamma}_0}}\frac{\tilde{\partial}}{\tilde{\partial}t}(\tilde{\rho}\hat{k}+\tilde{\rho}\hat{h}+{}_{em}u^f)d\tilde{V}+\lim_{\tilde{\Gamma}_0\to 0}\int_{\tilde{V}_{\tilde{\Gamma}_1}-\tilde{V}_{\tilde{\Gamma}_0}}\tilde{\rho}\hat{\mathbf{f}}\cdot\mathbf{v}d\tilde{V}\\
&-\lim_{\tilde{\Gamma}_0\to 0}\int_{\tilde{V}_{\tilde{\Gamma}_1}-\tilde{V}_{\tilde{\Gamma}_0}}\frac{\tilde{\rho}}{\rho_0}\hat{\boldsymbol{E}}\cdot\boldsymbol{J}_e d\tilde{V}-\lim_{\tilde{\Gamma}_0\to 0}\int_{\tilde{V}_{\tilde{\Gamma}_1}-\tilde{V}_{\tilde{\Gamma}_0}}\tilde{\rho}\hat{s}\dot{T}d\tilde{V}.
\end{aligned}\tag{6.26}$$

It can be seen that the above integral becomes invariant because of the added domain integral terms. Consequently, the dynamic energy release rate can be represented alternatively by

$$\begin{aligned}
\tilde{J}_0&=\hat{J}\\
&=\frac{1}{\dot{A}}\int_{\tilde{\Gamma}}[\mathbf{n}\cdot(\boldsymbol{\sigma}+{}_{em}\boldsymbol{\sigma}+\mathbf{v}\otimes\mathbf{G})\cdot\mathbf{v}-\mathbf{n}\cdot\boldsymbol{S}+(\tilde{\rho}\hat{k}+\tilde{\rho}\hat{h}+{}_{em}u^f)\mathbf{n}\cdot\boldsymbol{V}_C]d\tilde{\Gamma}\\
&\quad-\lim_{\tilde{\Gamma}_0\to 0}\frac{1}{\dot{A}}\int_{\tilde{V}_{\tilde{\Gamma}}-\tilde{V}_{\tilde{\Gamma}_0}}\frac{\tilde{\partial}}{\tilde{\partial}t}(\tilde{\rho}\hat{k}+\tilde{\rho}\hat{h}+{}_{em}u^f)d\tilde{V}+\lim_{\tilde{\Gamma}_0\to 0}\frac{1}{\dot{A}}\int_{\tilde{V}_{\tilde{\Gamma}}-\tilde{V}_{\tilde{\Gamma}_0}}\tilde{\rho}\hat{\mathbf{f}}\cdot\mathbf{v}d\tilde{V}\\
&\quad-\lim_{\tilde{\Gamma}_0\to 0}\frac{1}{\dot{A}}\int_{\tilde{V}_{\tilde{\Gamma}}-\tilde{V}_{\tilde{\Gamma}_0}}\frac{\tilde{\rho}}{\rho_0}\hat{\boldsymbol{E}}\cdot\boldsymbol{J}_e d\tilde{V}-\lim_{\tilde{\Gamma}_0\to 0}\frac{1}{\dot{A}}\int_{\tilde{V}_{\tilde{\Gamma}}-\tilde{V}_{\tilde{\Gamma}_0}}\tilde{\rho}\hat{s}\dot{T}d\tilde{V}.
\end{aligned}\tag{6.27}$$

If a field quantity is invariant in a reference frame traveling with the crack front at a uniform speed $V_C = V_C\overline{\mathbf{E}}_1$, the field quantity depends on t only through the combination $\tilde{\mathbf{X}} = \mathbf{X} - V_C t$. Hence, for steady-state crack propagation in the absence of mechanical body force, temperature change, and electricity conduction, the path-domain independent $\hat{J}$-integral becomes path independent, that is,

$$\begin{aligned}\hat{J} &= \tilde{J}\\ &= -\frac{1}{B}\int_{\tilde{\Gamma}} \mathbf{n}\cdot(\boldsymbol{\sigma}+{}_{em}\boldsymbol{\sigma})\cdot\mathbf{u}\tilde{\nabla}d\tilde{\Gamma}\cdot\overline{\mathbf{E}}_1 + \frac{1}{B}\int_{\tilde{\Gamma}} \mathbf{n}\cdot(\tilde{\rho}\hat{k} + \tilde{\rho}\hat{h} + {}_{em}u^f)\mathbf{I}d\tilde{\Gamma}\cdot\overline{\mathbf{E}}_1\\ &\quad + \frac{1}{B}\int_{\tilde{\Gamma}}[(\mathbf{n}\times\mathbf{E})\times\mathbf{D}]\cdot\mathbf{u}\tilde{\nabla}d\tilde{\Gamma}\cdot\overline{\mathbf{E}}_1 + \frac{1}{B}\int_{\tilde{\Gamma}}[(\mathbf{n}\times\mathbf{H})\times\mathbf{B}]\cdot\mathbf{u}\tilde{\nabla}d\tilde{\Gamma}\cdot\overline{\mathbf{E}}_1\\ &\quad + \frac{1}{B}\int_{\tilde{\Gamma}} \mathbf{n}\cdot\mathbf{v}\otimes(\mathbf{P}\times\mathbf{B})\cdot\mathbf{u}\tilde{\nabla}d\tilde{\Gamma}\cdot\overline{\mathbf{E}}_1,\end{aligned} \tag{6.28}$$

where B is the thickness along the crack front.

With the introduction of the energy-momentum tensor

$$\begin{aligned}\tilde{\mathbf{b}} = -[&\boldsymbol{\sigma}+{}_{em}\boldsymbol{\sigma} + (\mathbf{D}\cdot\mathbf{E})\mathbf{I} - \mathbf{D}\otimes\mathbf{E} + (\mathbf{B}\cdot\mathbf{H})\mathbf{I} - \mathbf{B}\otimes\mathbf{H}\\ &- \mathbf{v}\otimes(\mathbf{P}\times\mathbf{B})]\cdot\mathbf{u}\tilde{\nabla} + (\tilde{\rho}\hat{k} + \tilde{\rho}\hat{h} + {}_{em}u^f)\mathbf{I},\end{aligned} \tag{6.29}$$

the $\tilde{J}$-integral can be expressed as the first component of the $\tilde{J}_K$-integral vector as an extension of the configuration force (material force) notation (Eshelby, 1951, 1956, 1970, 1975; Maugin and Trimarco, 1992; Gurtin, 2000), that is,

$$\tilde{J} = \frac{1}{B}\int_{\tilde{\Gamma}} \mathbf{n}\cdot\tilde{\mathbf{b}}d\tilde{\Gamma}\cdot\overline{\mathbf{E}}_1 = \tilde{\mathbf{J}}\cdot\overline{\mathbf{E}}_1. \tag{6.30}$$

Nevertheless, the $\tilde{J}_K$-integral vector and the energy-momentum tensor $\tilde{\mathbf{b}}$ derived in this formulation are different from those obtained with use of the electromagnetic enthalpy (Maugin *et al.*, 1992; Maugin, 1994). The physical meaning of the crack-front $\tilde{J}$-integral is the dynamic energy release rate, which represents the rate of energy flow out of the body and into the crack front per unit crack advance. Unlike other path-independent integrals, the $\tilde{J}$-integral thus formulated fully satisfies thermodynamic requirements and, hence, can be used as a physically sound fracture criterion. When the added domain integral terms in Eq. (6.27) are nonzero, the $\tilde{J}$-integral becomes path dependent and, thus, the invariant $\hat{J}$-integral is used as an alternative representation. The invariant $\hat{J}$-integral method is not only generally applicable to various material systems and loading conditions but also relatively easy for finite element implementation due to its path-domain independency.

6.3 Stroh-Type Formalism for Steady-State Crack Propagation under Coupled Magneto-Electro-Mechanical Jump/Boundary Conditions

6.3.1 *Generalized plane crack problem*

To illustrate the application of the developed theory, consider a conventional planar crack extending in a magneto-electro-elastic solid (Fig. 6.2). A reference frame is affixed to the crack tip advancing at instantaneous speed V_C. The $\tilde{X}_3$-axis is along the crack front and the crack faces are on the half-plane containing the negative $\tilde{X}_1$-axis. For a generalized plane crack problem, the field quantities do not depend on $\tilde{X}_3$ but may have components in the $\tilde{X}_3$-direction.

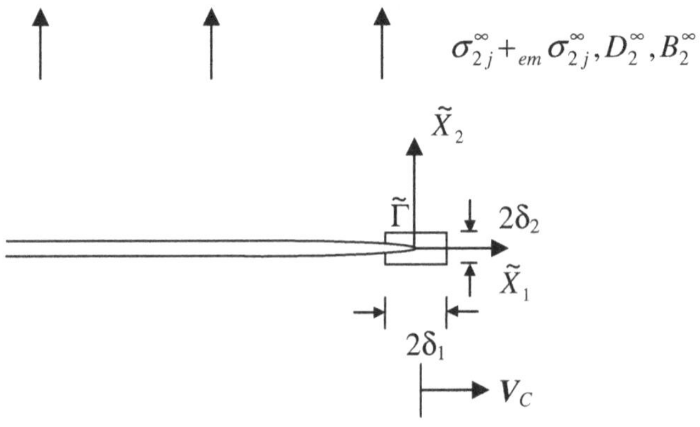

Fig. 6.2. A conventional planar crack extending in a magneto-electro-elastic solid.

For an elliptical cavity-like crack problem, the jump conditions across a material surface of discontinuity are given by

$$\mathbf{n}\cdot[[\boldsymbol{\sigma}+{}_{em}\boldsymbol{\sigma}]]=0\,,\ [[\mathbf{u}]]=0\,, \tag{6.31}$$

$$\mathbf{n}\cdot[[\mathbf{D}]]=\varpi_f\,,\ \mathbf{n}\times[[\mathbf{E}]]=0,\ \mathbf{n}\cdot[[\mathbf{B}]]=0\,,\ \mathbf{n}\times[[\mathbf{H}]]=0\,. \tag{6.32}$$

For a slit crack problem, the crack-face conditions may be expressed as

$$\begin{aligned}\mathbf{n}^+\cdot(\boldsymbol{\sigma}+{}_{em}\boldsymbol{\sigma})^+ &= -\mathbf{n}^-\cdot(\boldsymbol{\sigma}+{}_{em}\boldsymbol{\sigma})^- \\ &= -\{\sigma_{21}^0+{}_{em}\sigma_{21}^0,\sigma_{22}^0+{}_{em}\sigma_{22}^0,\sigma_{23}^0+{}_{em}\sigma_{23}^0\},\end{aligned} \tag{6.33}$$

$$\mathbf{n}^+\cdot\mathbf{D}^+=-\mathbf{n}^-\cdot\mathbf{D}^-=-D_2^0\,,\ \mathbf{n}^+\cdot\mathbf{B}^+=-\mathbf{n}^-\cdot\mathbf{B}^-=-B_2^0\,. \tag{6.34}$$

The total traction should be considered in the coupled magneto-electro-mechanical boundary conditions along the crack faces and the remainder surface of the solid. The jump conditions (6.31) and (6.32) are exact, but the corresponding boundary value problem needs to be solved in both the solid region and the cavity region. For a slit crack problem, D_2^0 and B_2^0 in (6.34) are either prescribed under the impermeable crack-face condition or determined through solving the boundary value problem with the permeable or semi-permeable crack-face condition. The crack-face boundary conditions may also involve crack opening, surface charge or discharge.

Since the basic equations in anisotropic magneto-electro-elasticity have the same structure as those in anisotropic electroelasticity, general solution techniques for linearized problems like the Stroh-type formalism remain valid. As an illustration, a steady crack growth problem is dealt with below.

6.3.2 *Steady-state solution*

A steady-state solution for dynamic crack propagation at constant speed may be achieved in some limiting sense. The steady-state condition permits further reduction of the number of independent variables from three to two so that the analysis is considerably simplified. If a field quantity is an element of a steady-state solution, the field quantity depends on t only through the combination $\tilde{\mathbf{X}} = \mathbf{X} - V_C t$, that is,

$$f(X_1, X_2, t) = f(\tilde{X}_1, \tilde{X}_2), \tag{6.35}$$

$$\frac{\partial}{\partial t} f(X_1, X_2, t) = -V_C \frac{\partial}{\partial \tilde{X}_1} f(\tilde{X}_1, \tilde{X}_2), \tag{6.36}$$

$$\frac{\partial}{\partial X_1} f(X_1, X_2, t) = \frac{\partial}{\partial \tilde{X}_1} f(\tilde{X}_1, \tilde{X}_2), \tag{6.37}$$

$$\frac{\partial}{\partial X_2} f(X_1, X_2, t) = \frac{\partial}{\partial \tilde{X}_2} f(\tilde{X}_1, \tilde{X}_2), \tag{6.38}$$

where the forms of the functions on the left-hand and right-hand sides of the equations are different, but the values of the functions represent the same physical quantity, and so they may be represented by the same symbol with little risk of confusion.

Under quasi-electrostatic and quasi-magnetostatic approximations, the governing equations for an anisotropic magneto-electro-elastic solid in the absence of mechanical body force ($\hat{\mathbf{f}} = 0$), electricity conduction ($\boldsymbol{j}_e = 0$), and free electric charge ($q_f = 0$) can be rewritten in the rectangular reference coordinate system affixed to the moving crack tip as

$$c_{ijkl} u_{k,jl} + e_{lij} \phi_{,jl} + h_{lij} \psi_{,jl} = \rho V_C^2 u_{i,11}, \tag{6.39}$$

$$e_{jkl} u_{k,jl} - \kappa_{jl} \phi_{,jl} - g_{jl} \psi_{,jl} = 0, \tag{6.40}$$

$$h_{jkl} u_{k,jl} - g_{lj} \varphi_{,jl} - \mu_{jl} \psi_{,jl} = 0, \tag{6.41}$$

$$E_i = -\phi_{,i}, \tag{6.42}$$

$$H_i = -\psi_{,i}, \tag{6.43}$$

where ϕ is electric potential, ψ is magnetic potential, e_{ijk}, h_{ijk}, and g_{ij} are piezoelectric, piezomagnetic, and magnetoelectric coupling coefficients, respectively, c_{ijkl}, κ_{ij}, and μ_{ij} are elastic stiffness, dielectric permittivity, and magnetic permeability coefficients, respectively.

Based on the Stroh-type formalism, a general solution is sought of the form

$$u_m = a_m f(z) \ \ (m = 1,2,3), \tag{6.44}$$

$$\phi = a_4 f(z), \tag{6.45}$$

$$\psi = a_5 f(z), \tag{6.46}$$

where the function f is analytic in the complex variable $z = \tilde{X}_1 + p\tilde{X}_2$ and the complex numbers p and a_m must be determined from Eqs. (6.39)–(6.41).

Substitution of (6.44)–(6.46) into (6.39)–(6.41) yields

$$[\mathbf{Q} - \rho V_C^2 \mathbf{U} + (\mathbf{R} + \mathbf{R}^T)p + \mathbf{T}p^2]\mathbf{a} = 0, \tag{6.47}$$

with $\mathbf{a} = (a_1, a_2, a_3, a_4, a_5)^T$, $\mathbf{U} = \mathrm{diag}(1,1,1,0,0)$, and the 5×5 matrices:

$$\mathbf{Q} = \begin{bmatrix} c_{1jk1} & e_{1j1} & h_{1j1} \\ e_{1k1}^T & -\kappa_{11} & -g_{11} \\ h_{1k1}^T & -g_{11} & -\mu_{11} \end{bmatrix}, \tag{6.48}$$

$$\mathbf{R} = \begin{bmatrix} c_{1jk2} & e_{2j1} & h_{2j1} \\ e_{1k2}^T & -\kappa_{12} & -g_{12} \\ h_{1k2}^T & -g_{21} & -\mu_{12} \end{bmatrix}, \tag{6.49}$$

$$\mathbf{T} = \begin{bmatrix} c_{2jk2} & e_{2j2} & h_{2j2} \\ e_{2k2}^T & -\kappa_{22} & -g_{22} \\ h_{2k2}^T & -g_{22} & -\mu_{22} \end{bmatrix}. \tag{6.50}$$

Nontrivial solutions are obtained if p is a root of

$$\det[\mathbf{Q}-\rho V_C^2\mathbf{U}+(\mathbf{R}+\mathbf{R}^T)p+\mathbf{T}p^2]=0. \tag{6.51}$$

The ten roots of Eq. (6.51) depend on the crack velocity V_C, that is, $p_\alpha = p_\alpha(V_C)$. A real root p of Eq. (6.51) corresponds to a value of V_C equal to the velocity of bulk wave propagating in the direction (1, p) in the ($\tilde{X}_1, \tilde{X}_2$) plane. Let c_L denote the inferior limit of such bulk-wave velocities. Then, Eq. (6.51) has no real roots for $V_C < c_L$. Since the coefficients of Eq. (6.51) are real, the eigenvalues p_α and the eigenvectors a_α form two sets of complex quantities, with one set being conjugate to the other. We suppose p_α ($\alpha = 1,2,3,4,5$) are five distinct roots with positive imaginary parts and construct the matrix **A** with columns that are the associated eigenvectors. Thus, the general solution of Eqs. (6.39)–(6.41) is given by

$$u_m = \sum_{\alpha=1}^{5} A_{m\alpha} f_\alpha(z_\alpha) + \sum_{\alpha=1}^{5} \bar{A}_{m\alpha} \overline{f_\alpha(z_\alpha)}, \tag{6.52}$$

$$\phi = \sum_{\alpha=1}^{5} A_{4\alpha} f_\alpha(z_\alpha) + \sum_{\alpha=1}^{5} \bar{A}_{4\alpha} \overline{f_\alpha(z_\alpha)}, \tag{6.53}$$

$$\psi = \sum_{\alpha=1}^{5} A_{5\alpha} f_\alpha(z_\alpha) + \sum_{\alpha=1}^{5} \bar{A}_{5\alpha} \overline{f_\alpha(z_\alpha)}, \tag{6.54}$$

where $z_\alpha = \tilde{X}_1 + p_\alpha \tilde{X}_2$ ($\alpha = 1,2,3,4,5$) and over-bars denote complex conjugates.

Hence, total stress, electric displacement, and magnetic induction are readily expressed as

$${}_t\sigma_{i1} = \sum_{\alpha=1}^{5} (\rho V_C^2 A_{i\alpha} - p_\alpha L_{i\alpha}) f'_\alpha(z_\alpha) + \sum_{\alpha=1}^{5} (\rho V_C^2 \bar{A}_{i\alpha} - \bar{p}_\alpha \bar{L}_{i\alpha}) \overline{f'_\alpha(z_\alpha)}, \tag{6.55}$$

$${}_t\sigma_{i2} = \sum_{\alpha=1}^{5} L_{i\alpha} f'_\alpha(z_\alpha) + \sum_{\alpha=1}^{5} \bar{L}_{i\alpha} \overline{f'_\alpha(z_\alpha)}, \tag{6.56}$$

$$D_1 = -\sum_{\alpha=1}^{5} p_\alpha L_{4\alpha} f'_\alpha(z_\alpha) - \sum_{\alpha=1}^{5} \bar{p}_\alpha \bar{L}_{4\alpha} \overline{f'_\alpha(z_\alpha)}, \tag{6.57}$$

$$D_2 = \sum_{\alpha=1}^{5} L_{4\alpha} f_\alpha'(z_\alpha) + \sum_{\alpha=1}^{5} \overline{L}_{4\alpha} \overline{f_\alpha'(z_\alpha)}, \tag{6.58}$$

$$B_1 = -\sum_{\alpha=1}^{5} p_\alpha L_{5\alpha} f_\alpha'(z_\alpha) - \sum_{\alpha=1}^{5} \overline{p}_\alpha \overline{L}_{5\alpha} \overline{f_\alpha'(z_\alpha)}, \tag{6.59}$$

$$B_2 = \sum_{\alpha=1}^{5} L_{5\alpha} f_\alpha'(z_\alpha) + \sum_{\alpha=1}^{5} \overline{L}_{5\alpha} \overline{f_\alpha'(z_\alpha)}, \tag{6.60}$$

where $L_{n\alpha} = (R_{mn} + p_\alpha T_{nm})A_{m\alpha} = -(Q_{nm} + p_\alpha R_{nm} - \rho V_C^2 U_{nm})A_{m\alpha} / p_\alpha$ is used to construct the matrix **L**. The unknown functions f_α can be determined by the boundary conditions for a given crack growth problem.

Let us introduce

$$\mathbf{f} = (f_1(z_1), f_2(z_2), f_3(z_3), f_4(z_4), f_5(z_5))^T. \tag{6.61}$$

The asymptotic solution is

$$\mathbf{h}(z) = \mathbf{L}\mathbf{f}'(z) = \frac{1}{\sqrt{8\pi z}} \tilde{\mathbf{k}}. \tag{6.62}$$

The dynamic field intensity factor vector is defined in terms of the total stress, electric displacement and magnetic induction at a distance r ahead of the crack tip as

$$\tilde{\mathbf{k}} = \lim_{r \to 0} \sqrt{2\pi r} ({}_t\sigma_{21}, {}_t\sigma_{22}, {}_t\sigma_{23}, D_2, B_2)^T, \tag{6.63}$$

where $\tilde{\mathbf{k}} = (\tilde{K}_{II}, \tilde{K}_I, \tilde{K}_{III}, \tilde{K}_D, \tilde{K}_B)^T$, $\tilde{K}_I$, $\tilde{K}_{II}$, $\tilde{K}_{III}$ are mode-I, mode-II, and mode-III dynamic total stress intensity factors, $\tilde{K}_D$ is the dynamic electric displacement intensity factor, and $\tilde{K}_B$ is the dynamic magnetic induction intensity factor.

The jumps of the displacements, electric potential, and magnetic potential across the crack faces at a distance r behind the crack tip are

$$(\Delta u_1, \Delta u_2, \Delta u_3, \Delta\phi, \Delta\psi)^T = \sqrt{\frac{2r}{\pi}} \tilde{\mathbf{H}}^i \cdot \tilde{\mathbf{k}}, \tag{6.64}$$

where $\tilde{\mathbf{H}}^i = 2\,\mathrm{Re}(i\mathbf{A}\mathbf{L}^{-1})$ is the dynamic counterpart of the Irwin matrix described in Section 4.5.2, which depends on material properties and crack speed.

Hence, the dynamic field intensity factor vector in terms of the crack opening displacement (COD), electric potential difference, and magnetic potential difference across the crack faces at a distance r behind the crack tip is defined as

$$\tilde{\mathbf{k}}^* = \lim_{r\to 0}\sqrt{\frac{\pi}{2r}}(\Delta u_1, \Delta u_2, \Delta u_3, \Delta\phi, \Delta\psi)^T, \tag{6.65}$$

where $\tilde{\mathbf{k}}^* = (\tilde{K}_{II}^{COD}, \tilde{K}_{I}^{COD}, \tilde{K}_{III}^{COD}, \tilde{K}_{\phi}, \tilde{K}_{\psi})^T$, $\tilde{K}_{I}^{COD}$, $\tilde{K}_{II}^{COD}$, $\tilde{K}_{III}^{COD}$ are mode-I, mode-II, and mode-III dynamic crack opening displacement intensity factors, $\tilde{K}_{\phi}$ is the electric potential intensity factor, and $\tilde{K}_{\psi}$ is the magnetic potential intensity factor.

As a result, the two dynamic field intensity factor vectors $\tilde{\mathbf{k}}^*$ and $\tilde{\mathbf{k}}$ are related by

$$\tilde{\mathbf{k}}^* = \tilde{\mathbf{H}}^i \cdot \tilde{\mathbf{k}}. \tag{6.66}$$

6.3.3 *Path-independent integral for steady crack growth*

For steady-state propagation of a planar crack without mechanical body force ($\hat{\mathbf{f}} = 0$), temperature change ($\dot{T} = 0$) and electricity conduction ($\boldsymbol{J}_e = 0$), the dynamic energy release rate can be represented by the path-independent dynamic contour integral as the closed contour including the crack faces is chosen, that is,

$$\tilde{J}_0 = \tilde{J}_g = \tilde{J}_l\,. \tag{6.67}$$

Choose the contour as shown in Fig. 6.2. This is a convenient choice because $n_1 = 0$ along the segments parallel to the $\tilde{X}_1$-axis. The contour is shrunk onto the crack tip by first letting $\delta_2 \to 0$ and then $\delta_1 \to 0$. By analogy to the purely elastodynamic case (Freund, 1990), there is no contribution to $\tilde{J}_0$ from the segments parallel to the $\tilde{X}_2$-axis and the segments along the crack faces. Moreover, the second, third, fourth, and fifth terms on the right-hand side of Eq. (6.28) along the segments parallel to the $\tilde{X}_1$-axis vanish. Consequently, $\tilde{J}_0$ can be computed by evaluating only the first term on the right-hand side of Eq. (6.28) along the segments parallel to the $\tilde{X}_1$-axis, that is,

$$\tilde{J}_0 = -2 \lim_{\delta_1 \to 0} \{ \lim_{\delta_2 \to 0} \int_{-\delta_1}^{\delta_1} [\sigma_{2j}(\tilde{X}_1, \delta_2) + {}_{em}\sigma_{2j}(\tilde{X}_1, \delta_2)] \frac{\partial u_j(\tilde{X}_1, \delta_2)}{\partial \tilde{X}_1} d\tilde{X}_1 \}. \tag{6.68}$$

Consequently, the dynamic energy release rate is equal to the following crack closure integral:

$$\tilde{J}_0 = \lim_{\delta a \to 0} \frac{1}{2\delta a} \int_0^{\delta a} [\sigma_{2j}(\tilde{X}_1, 0) + {}_{em}\sigma_{2j}(\tilde{X}_1, 0)] \Delta u_j(\tilde{X}_1 - \delta a, 0^{\pm}) d\tilde{X}_1 , \tag{6.69}$$

where $\Delta u_j(\tilde{X}_1 - \delta a, 0^{\pm}) = u_j(\tilde{X}_1 - \delta a, 0^{+}) - u_j(\tilde{X}_1 - \delta a, 0^{-})$ is the crack opening displacement at a distance $\delta a - \tilde{X}_1$ behind the crack tip.

The dynamic energy release rate is thus calculated as

$$\tilde{J}_0 = \frac{1}{4} (\tilde{K}_{II}, \tilde{K}_I, \tilde{K}_{III}, 0, 0) \cdot \tilde{\mathbf{H}}^i \cdot (\tilde{K}_{II}, \tilde{K}_I, \tilde{K}_{III}, \tilde{K}_D, \tilde{K}_B)^T . \tag{6.70}$$

Equation (6.70) shows that the dynamic energy release rate is an odd function of the electric displacement intensity factor and the magnetic induction intensity factor, which is consistent with experimental evidence (Pak and Tobin, 1993; Tobin and Pak, 1993; Cao and Evans, 1994; Lynch *et al.*, 1995; Park and Sun, 1995a–b; Jiang and Sun, 1999, 2001; Qin, 2001; Zhang *et al.*, 2002; Soh *et al.*, 2003; Chen and Lu, 2003). The axisymmetric dynamic crack problem under the electromagnetically impermeable or permeable conditions studied by Feng *et al.* (2007) is analogous to the mode-I dynamic crack problem with the crack plane perpendicular to the poling direction.

6.4 Magneto-Electro-Elastostatic Crack Problem as a Special Case

As the crack velocity tends to zero, the near-tip field formulae in the previous section are reduced to the quasi-static case discussed in Chapter 4 with the replacement of the Cauchy stress by the total stress. For a conventional Griffith-type crack of length $2a$, the crack-tip field intensity factor vector is obtained as

$$\begin{aligned}\tilde{\mathbf{k}} &= (\tilde{K}_{II}, \tilde{K}_{I}, \tilde{K}_{III}, \tilde{K}_{D}, \tilde{K}_{B})^{T} \\ &=({}_t\sigma_{21}^{\infty} - {}_t\sigma_{21}^{0}, {}_t\sigma_{22}^{\infty} - {}_t\sigma_{22}^{0}, {}_t\sigma_{23}^{\infty} - {}_t\sigma_{23}^{0}, D_2^{\infty} - D_2^{0}, B_2^{\infty} - B_2^{0})^{T}\sqrt{\pi a},\end{aligned} \tag{6.71}$$

where ${}_t\sigma_{2j}^{\infty}$, D_2^{∞} and B_2^{∞} are, respectively, total traction, electric displacement, and magnetic induction components in the far field, ${}_t\sigma_{2j}^{0}$, D_2^{0} and B_2^{0} are, respectively, total traction, electric displacement, and magnetic induction components at the crack surface.

Since the total stress tensor is the sum of the Cauchy stress tensor and the electromagnetic stress tensor, the total stress fields are coupled with the electromagnetic fields, which is fundamentally different from the decoupled prediction based on the linear theory of piezoelectricity and its extension.

From Eq. (6.28), the energy release rate for a quasi-static or stationary planar crack in the absence of mechanical body force ($\hat{\mathbf{f}} = 0$), temperature change ($\dot{T} = 0$), and electricity conduction ($\boldsymbol{J}_e = 0$) can be expressed by a path-independent integral constructed with the augmented Helmholtz free energy, including the contribution of the energy of the free electromagnetic fields, that is,

$$\begin{aligned}\tilde{J} = \frac{1}{B}\int_{\tilde{\Gamma}} \mathbf{n}\cdot\{&-[(\boldsymbol{\sigma} + {}_{em}\boldsymbol{\sigma}) + (\mathbf{D}\cdot\mathbf{E})\mathbf{I} - \mathbf{D}\otimes\mathbf{E} + (\mathbf{B}\cdot\mathbf{H})\mathbf{I} \\ &- \mathbf{B}\otimes\mathbf{H}]\cdot\mathbf{u}\tilde{\nabla} + (\tilde{\rho}\hat{h} + {}_{em}u^{f})\mathbf{I}\}d\tilde{\Gamma}\cdot\overline{\mathbf{E}}_1.\end{aligned} \tag{6.72}$$

As the electric field or the magnetic field is shut off, the above expression becomes

$$\tilde{J} = \frac{1}{B}\int_{\tilde{\Gamma}} \mathbf{n} \cdot \{-[(\boldsymbol{\sigma} + {}_e\boldsymbol{\sigma}) + (\mathbf{D} \cdot \mathbf{E})\mathbf{I} - \mathbf{D} \otimes \mathbf{E}] \cdot \mathbf{u}\tilde{\nabla} + (\tilde{\rho}\hat{h} + {}_e u^f)\mathbf{I}\} d\tilde{\Gamma} \cdot \overline{\mathbf{E}}_1, \tag{6.73}$$

$$\tilde{J} = \frac{1}{B}\int_{\tilde{\Gamma}} \mathbf{n} \cdot \{-[(\boldsymbol{\sigma} + {}_m\boldsymbol{\sigma}) + (\mathbf{B} \cdot \mathbf{H})\mathbf{I} - \mathbf{B} \otimes \mathbf{H}] \cdot \mathbf{u}\tilde{\nabla} + (\tilde{\rho}\hat{h} + {}_m u^f)\mathbf{I}\} d\tilde{\Gamma} \cdot \overline{\mathbf{E}}_1. \tag{6.74}$$

6.5 Summary

The thermodynamic approach provides a uniform treatment of nonlinear constitutive and fracture behaviors of deformable electromagnetic materials involving multifield coupling effects. The elements of dynamic fracture mechanics for nonlinear magneto-electro-thermo-elastic solids are summarized in Table 6.1. The dynamic energy release rate representing the rate of energy flow out of the body and into the crack front per unit crack advance under combined magnetic, electric, thermal, and mechanical loadings can be expressed as the crack-front $\tilde{J}$-integral or, alternatively, the invariant $\hat{J}$-integral. The $\hat{J}$-integral including both path and domain integral terms is invariant (i.e., path-domain independent) as a whole, but in general neither path independent nor domain independent separately. For steady-state crack propagation in the absence of mechanical body force, temperature change, and electricity conduction, the path-domain independent $\hat{J}$-integral is equivalent to the $\tilde{J}$-integral, which becomes path independent as the closed surface including the crack faces is chosen. Unlike other invariant integrals in the literature, the invariant $\hat{J}$-integral thus formulated can be used as a physically sound fracture criterion for magneto-electro-thermo-elastic solids so as to provide guidelines for design and analysis of smart material and structure systems. Remarkably, the dynamic energy release rate is an odd function of the electric displacement intensity factor and the magnetic induction intensity factor, which is consistent with experimental observations. The crack driving force and the energy-momentum tensor in electro-thermo-elastodynamic fracture given in Chapter 5 can be taken as a special case. While many efforts have been devoted to the establishment of an advanced fracture mechanics methodology involving multifield analysis, much remains to be done for practical applications. The crack growth problems under combined magnetic, electric, thermal, and mechanical loadings are certainly worthy of more studies, especially involving surface wave phenomena and material hysteresis effects.

Table 6.1 Summary of the elements of dynamic fracture mechanics for nonlinear magneto-electro-thermo-elastic solids

Helmholtz free energy	$\hat{h} = \hat{e} - T\hat{s} = \hat{h}(\mathbf{C}, T, \nabla_R T, \hat{\mathbf{\Pi}}, \hat{\mathbf{B}}; \mathbf{X})$
Poynting vector	$\boldsymbol{S} = \boldsymbol{E} \times \boldsymbol{H}$
Dynamic energy release rate	$\tilde{J}_0 = \lim\limits_{\tilde{\Gamma} \to 0} \frac{1}{\dot{A}} \int_{\tilde{\Gamma}} [\mathbf{n} \cdot (\boldsymbol{\sigma} +_{em}\boldsymbol{\sigma} + \mathbf{v} \otimes \mathbf{G}) \cdot \mathbf{v} - \mathbf{n} \cdot \boldsymbol{S} + (\tilde{\rho}\hat{k} + \tilde{\rho}\hat{h} +_{em} u^f) \mathbf{n} \cdot \boldsymbol{V}_C] d\tilde{\Gamma}$
Invariant integral	$\hat{J} = \frac{1}{\dot{A}} \int_{\tilde{\Gamma}} [\mathbf{n} \cdot (\boldsymbol{\sigma} +_{em}\boldsymbol{\sigma} + \mathbf{v} \otimes \mathbf{G}) \cdot \mathbf{v} - \mathbf{n} \cdot \boldsymbol{S} + (\tilde{\rho}\hat{k} + \tilde{\rho}\hat{h} +_{em} u^f) \mathbf{n} \cdot \boldsymbol{V}_C] d\tilde{\Gamma}$ $- \lim\limits_{\tilde{\Gamma}_0 \to 0} \frac{1}{\dot{A}} \int_{\tilde{V}_{\tilde{\Gamma}} - \tilde{V}_{\tilde{\Gamma}_0}} \frac{\tilde{\partial}}{\tilde{\partial} t} (\tilde{\rho}\hat{k} + \tilde{\rho}\hat{h} +_{em} u^f) d\tilde{V} + \lim\limits_{\tilde{\Gamma}_0 \to 0} \frac{1}{\dot{A}} \int_{\tilde{V}_{\tilde{\Gamma}} - \tilde{V}_{\tilde{\Gamma}_0}} \tilde{\rho}\hat{\mathbf{f}} \cdot \mathbf{v} d\tilde{V}$ $- \lim\limits_{\tilde{\Gamma}_0 \to 0} \frac{1}{\dot{A}} \int_{\tilde{V}_{\tilde{\Gamma}} - \tilde{V}_{\tilde{\Gamma}_0}} \frac{\tilde{\rho}}{\rho_0} \hat{\boldsymbol{E}} \cdot \boldsymbol{J}_e d\tilde{V} - \lim\limits_{\tilde{\Gamma}_0 \to 0} \frac{1}{\dot{A}} \int_{\tilde{V}_{\tilde{\Gamma}} - \tilde{V}_{\tilde{\Gamma}_0}} \tilde{\rho}\hat{s}\dot{T} d\tilde{V}$
Path-independent integral	$\tilde{J} = \tilde{\mathbf{J}} \cdot \overline{\mathbf{E}}_1 = (1/B) \int_{\tilde{\Gamma}} \mathbf{n} \cdot \tilde{\mathbf{b}} d\tilde{\Gamma} \cdot \overline{\mathbf{E}}_1$ (steady-state crack propagation in the absence of mechanical body force, temperature change, and electricity conduction)
Energy-momentum tensor	$\tilde{\mathbf{b}} = -[\boldsymbol{\sigma} +_{em}\boldsymbol{\sigma} + (\mathbf{D} \cdot \mathbf{E})\mathbf{I} - \mathbf{D} \otimes \mathbf{E} + (\mathbf{B} \cdot \mathbf{H})\mathbf{I} - \mathbf{B} \otimes \mathbf{H}$ $- \mathbf{v} \otimes (\mathbf{P} \times \mathbf{B})] \cdot \mathbf{u}\tilde{\nabla} + (\tilde{\rho}\hat{k} + \tilde{\rho}\hat{h} +_{em} u^f)\mathbf{I}$
Irwin-relation	$\tilde{J} = \frac{1}{4} (\tilde{K}_{II}, \tilde{K}_I, \tilde{K}_{III}, 0, 0) \cdot \tilde{\mathbf{H}}^i \cdot (\tilde{K}_{II}, \tilde{K}_I, \tilde{K}_{III}, \tilde{K}_D, \tilde{K}_B)^T$ (linearized theory)

Chapter 7

Dynamic Crack Propagation in Magneto-Electro-Elastic Solids

7.1 Introduction

The transient response of electromagnetic materials in the presence of multifield coupling effects is essentially distinct from those found in purely mechanical problems. For example, shear horizontal (SH) surface waves may occur in a piezoelectric material with hexagonal symmetry (Alshits *et al.*, 1992; Alshits, 2002), whereas there are no antiplane-mode surface waves in a purely elastic material of the same symmetry. Due to the shear horizontal surface wave effects, antiplane dynamic crack propagation in piezoelectric materials (e.g., Li and Mataga, 1996a–b; Ing and Wang, 2004a–b; Melkumyan, 2005; Chen *et al.*, 2007, 2008) exhibits many features only associated with in-plane modes in the elastic case. The magneto-electro-mechanical coupling effects introduce more difficulties to solving transient crack growth problems analytically.

Since the previous chapter demonstrated the solution procedure for steady-state crack propagation in magneto-electro-elastic solids, we now focus on the techniques for analyzing transient crack growth in magneto-electro-elastic solids. Our attention is limited to the illustration of sudden constant-speed extension of a mode-III crack in a magneto-electro-elastic solid, so that the model is mathematically tractable for a closed-form analytical solution following the work by Chen (2009c). For more complex problems, numerical methods are often resorted to because of mathematical difficulties. The treatment of this subject is far from exhaustive and the reader may refer to the literature for further information.

The following section begins with a brief description of the shear horizontal surface wave phenomenon. In Section 7.3, the boundary-initial value problem for a sudden constant-speed extension of a semi-infinite mode-III crack is formulated with a unified treatment of electrically and magnetically permeable, semi-permeable, and impermeable crack-face conditions. In Section 7.4, integral transform, Wiener–Hopf and Cagniard–de Hoop techniques are used to solve the boundary-initial value problem in both the cracked solid region and the interior fluid region. In Section 7.5, the fundamental solutions for traction loading only are attained with the inverse square-root singularity near the crack tip. In Section 7.6, the fundamental solutions are generalized to mixed loads, resulting in self-induced and crossover dynamic field intensity factors. In Section 7.7, the dynamic energy release rate is evaluated based on the near-tip field solutions characterized by the dynamic field intensity factors. In Section 7.8, the surface wave effect on dynamic crack propagation in magneto-electro-elastic solids is discussed.

7.2 Shear Horizontal Surface Waves

In contrast to elastic body waves (P waves or S waves in seismology) that move through the body of an object, Rayleigh waves are a type of commonly known surface waves which travel along a surface and decay exponentially away from the surface. Bleustein (1968) and Gulyaev (1969) independently discovered the propagation of shear horizontal waves in piezoelectric materials with hexagonal symmetry. Lothe and Barnett (1976, 1977) further developed the theory for surface waves in piezoelectric crystals. Alshits *et al.* (1992) studied the existence of surface waves in half-infinite anisotropic elastic media with piezoelectric and piezomagnetic properties. Alshits (2002) also reviewed the role of anisotropy in crystal acoustics. Wang *et al.* (2007c) analyzed a magneto-electro-elastic half-space problem. The surface wave effect is very important for the design and analysis of high-performance devices such as transducers and wave filters. The major solution steps for shear horizontal surface wave problems involving magneto-electro-elastic coupling effects are outlined below.

Consider the propagation of shear horizontal surface waves along the free surface of a magneto-electro-elastic solid poled in the X_3-direction (Fig. 7.1). The field equations as well as the boundary conditions for the out-of-plane displacement component w and the electric and magnetic potentials φ and ψ are independent of X_3 and uncoupled from those for the in-plane displacement components.

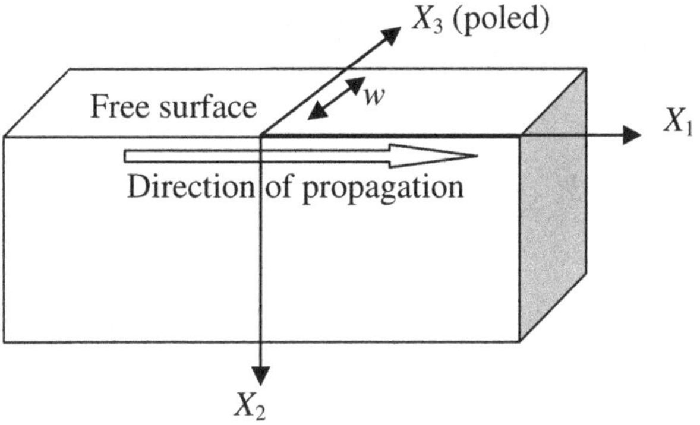

Fig. 7.1 Schematic of shear horizontal wave propagation along the free surface of a magneto-electro-elastic solid occupying the half space.

Based on the quasi-static approximation for the electromagnetic fields, the basic field equations for the half-space solid region $\Omega^{(s)} = \{X_2 \geq 0, -\infty < X_1 < +\infty\}$ and the half-space fluid region $\Omega^{(f)} = \{X_2 \leq 0, -\infty < X_1 < +\infty\}$, in the absence of mechanical body force, electricity conduction, and free electric charge, are expressed as

$$\frac{\partial^2 w}{\partial X_1^2} + \frac{\partial^2 w}{\partial X_2^2} + \frac{2V_C}{c_T^2}\frac{\partial^2 w}{\partial X_1 \partial t} - \frac{1}{c_T^2}\frac{\partial^2 w}{\partial t^2} = 0 \quad \text{in } \Omega^{(s)}, \tag{7.1}$$

$$\frac{\partial^2 \overline{\varphi}^{(s)}}{\partial X_1^2} + \frac{\partial^2 \overline{\varphi}^{(s)}}{\partial X_2^2} = 0 \quad \text{in } \Omega^{(s)}, \tag{7.2}$$

$$\frac{\partial^2 \bar{\psi}^{(s)}}{\partial X_1^2} + \frac{\partial^2 \bar{\psi}^{(s)}}{\partial X_2^2} = 0 \quad \text{in } \Omega^{(s)}, \tag{7.3}$$

$$\frac{\partial^2 \varphi^{(f)}}{\partial X_1^2} + \frac{\partial^2 \varphi^{(f)}}{\partial X_2^2} = 0 \quad \text{in } \Omega^{(f)}, \tag{7.4}$$

$$\frac{\partial^2 \psi^{(f)}}{\partial X_1^2} + \frac{\partial^2 \psi^{(f)}}{\partial X_2^2} = 0 \quad \text{in } \Omega^{(f)}, \tag{7.5}$$

$$\bar{\varphi}^{(s)} = \varphi^{(s)} - \frac{e_{15}\mu_{11} - h_{15}g_{11}}{\kappa_{11}\mu_{11} - g_{11}^2} w \quad \text{in } \Omega^{(s)}, \tag{7.6}$$

$$\bar{\psi}^{(s)} = \psi^{(s)} - \frac{h_{15}\kappa_{11} - e_{15}g_{11}}{\kappa_{11}\mu_{11} - g_{11}^2} w \quad \text{in } \Omega^{(s)}, \tag{7.7}$$

$${}_t\sigma_{k3} = \bar{c}_{44} \frac{\partial w}{\partial X_k} + e_{15} \frac{\partial \bar{\varphi}^{(s)}}{\partial X_k} + h_{15} \frac{\partial \bar{\psi}^{(s)}}{\partial X_k} \quad \text{in } \Omega^{(s)}, \tag{7.8}$$

$$D_k^{(s)} = -\kappa_{11} \frac{\partial \bar{\varphi}^{(s)}}{\partial X_k} - g_{11} \frac{\partial \bar{\psi}^{(s)}}{\partial X_k} \quad \text{in } \Omega^{(s)}, \tag{7.9}$$

$$B_k^{(s)} = -g_{11} \frac{\partial \bar{\varphi}^{(s)}}{\partial X_k} - \mu_{11} \frac{\partial \bar{\psi}^{(s)}}{\partial X_k} \quad \text{in } \Omega^{(s)}, \tag{7.10}$$

$$D_k^{(f)} = -\kappa^f \frac{\partial \varphi^{(f)}}{\partial X_k} \quad \text{in } \Omega^{(f)}, \tag{7.11}$$

$$B_k^{(f)} = -\mu^f \frac{\partial \psi^{(f)}}{\partial X_k} \quad \text{in } \Omega^{(f)}, \tag{7.12}$$

$$E_k^{(s)} = -\frac{\partial \varphi^{(s)}}{\partial X_k} \quad \text{in } \Omega^{(s)}, \tag{7.13}$$

$$H_k^{(s)} = -\frac{\partial \psi^{(s)}}{\partial X_k} \quad \text{in } \Omega^{(s)}, \tag{7.14}$$

$$E_k^{(f)} = -\frac{\partial \varphi^{(f)}}{\partial X_k} \quad \text{in } \Omega^{(f)}, \tag{7.15}$$

$$H_k^{(f)} = -\frac{\partial \psi^{(f)}}{\partial X_k} \quad \text{in } \Omega^{(f)}, \tag{7.16}$$

where k = 1,2, $c_T = (\bar{c}_{44}/\rho)^{1/2}$ is the piezoelectromagnetically stiffened bulk shear wave speed, ρ is the mass density, $\bar{c}_{44} = c_{44} + (e_{15}^2\mu_{11} - 2e_{15}h_{15}g_{11} + h_{15}^2\kappa_{11})/(\kappa_{11}\mu_{11} - g_{11}^2)$ is the piezoelectromagnetically stiffened elastic constant, c_{44}, κ_{11}, μ_{11}, g_{11}, e_{15}, and h_{15} are the elastic, dielectric permittivity, magnetic permeability, magnetoelectric, piezoelectric, and piezomagnetic coefficients for the solid, κ^f and μ^f are the dielectric permittivity and magnetic permeability coefficients for the fluid (vacuum, air, oil, etc.), ${}_t\sigma_{k3}$ are the total stress components, $D_k^{(p)}$ are the electric displacement components, $B_k^{(p)}$ are the magnetic induction components, $E_k^{(p)}$ are the electric field components, $H_k^{(p)}$ are the magnetic field components, where the superscript $p = s$ stands for the solid region and $p = f$ for the fluid region.

The remote conditions may be taken as

$${}_t\sigma_{23} = 0, \quad \bar{\varphi}^{(s)} = 0, \quad \bar{\psi}^{(s)} = 0, \quad \text{as } X_2 \to +\infty, \tag{7.17}$$

$$\varphi^{(f)} = 0, \quad \psi^{(f)} = 0, \quad \text{as } X_2 \to -\infty. \tag{7.18}$$

The boundary conditions at the free surface are given by

$${}_t\sigma_{23} = 0, \quad \text{on } X_2 = 0, \tag{7.19}$$

$$\varphi^{(s)} - \varphi^{(f)} = 0, \quad D_2^{(s)} - D_2^{(f)} = 0, \quad \text{on } X_2 = 0, \tag{7.20}$$

$$\psi^{(s)} - \psi^{(f)} = 0, \; B_2^{(s)} - B_2^{(f)} = 0, \text{ on } X_2 = 0. \tag{7.21}$$

A general solution of Eqs. (7.1)–(7.5) for surface waves propagating along the X_1-direction and decaying in the X_2-direction is represented by

$$w = a_1 \exp(-\xi_2 X_2)\exp[i(\xi_1 X_1 - \omega t)], \tag{7.22}$$

$$\bar{\varphi}^{(s)} = a_2 \exp(-\xi_1 X_2)\exp[i(\xi_1 X_1 - \omega t)], \tag{7.23}$$

$$\bar{\psi}^{(s)} = a_3 \exp(-\xi_1 X_2)\exp[i(\xi_1 X_1 - \omega t)], \tag{7.24}$$

$$\varphi^{(f)} = a_4 \exp(\xi_1 X_2)\exp[i(\xi_1 X_1 - \omega t)], \tag{7.25}$$

$$\psi^{(f)} = a_5 \exp(\xi_1 X_2)\exp[i(\xi_1 X_1 - \omega t)], \tag{7.26}$$

where $i = \sqrt{-1}$, $\omega = c_T \sqrt{\xi_1^2 - \xi_2^2}$ is the frequency of a time-harmonic disturbance, $\xi_1 = \omega / c_{bq}$ is the wave number, and c_{bq} is the shear horizontal surface wave speed.

Application of the boundary conditions (7.19)–(7.21) leads to

$$\bar{c}_{44}\xi_2 a_1 + e_{15}\xi_1 a_2 + h_{15}\xi_1 a_3 = 0, \tag{7.27}$$

$$\frac{e_{15}\mu_{11} - h_{15}g_{11}}{\kappa_{11}\mu_{11} - g_{11}^2} a_1 + a_2 - a_4 = 0, \tag{7.28}$$

$$\frac{h_{15}\kappa_{11} - e_{15}g_{11}}{\kappa_{11}\mu_{11} - g_{11}^2} a_1 + a_3 - a_5 = 0, \tag{7.29}$$

$$\kappa_{11}a_2 + g_{11}a_3 + \kappa^f a_4 = 0, \tag{7.30}$$

$$g_{11}a_2 + \mu_{11}a_3 + \mu^f a_5 = 0. \tag{7.31}$$

For a nontrivial solution, the determinant of the coefficients must be zero, resulting in

$$\xi_2 = k_{em}^2 \xi_1 , \tag{7.32}$$

where k_{em} is the magneto-electro-mechanical coupling factor satisfying

$$k_{em}^2 = (1 - c_{bg}^2 / c_T^2)^{1/2}$$

$$= \frac{\kappa^f [e_{15}(\mu^f + \mu_{11}) - g_{11}h_{15}](e_{15}\mu_{11} - h_{15}g_{11})}{\bar{c}_{44}[(\kappa^f + \kappa_{11})(\mu^f + \mu_{11}) - g_{11}^2](\kappa_{11}\mu_{11} - g_{11}^2)} \tag{7.33}$$

$$+ \frac{\mu^f [h_{15}(\kappa^f + \kappa_{11}) - g_{11}e_{15}](h_{15}\kappa_{11} - e_{15}g_{11})}{\bar{c}_{44}[(\kappa^f + \kappa_{11})(\mu^f + \mu_{11}) - g_{11}^2](\kappa_{11}\mu_{11} - g_{11}^2)} .$$

As $\kappa_{11} / \kappa^f \to 0$ and $\mu_{11} / \mu^f \to 0$ for the electrically and magnetically permeable crack-face condition, the limiting case of (7.33) is

$$k_{em} \to \sqrt{\frac{(e_{15}^2 \mu_{11} + h_{15}^2 \kappa_{11} - 2e_{15}h_{15}g_{11})}{\bar{c}_{44}(\kappa_{11}\mu_{11} - g_{11}^2)}} . \tag{7.34}$$

As $\kappa_{11} / \kappa^f \to \infty$ and $\mu_{11} / \mu^f \to \infty$ for the electrically and magnetically impermeable crack-face condition, the limiting case of (7.33) is

$$k_{em} \to 0 . \tag{7.35}$$

It can be seen that the shear horizontal surface wave speed c_{bg} should only be lower than the piezoelectromagnetically stiffened bulk shear wave speed c_T for the existence of the surface wave-type solution. As the magneto-electro-mechanical coupling factor tends to zero (i.e., $k_{em} \to 0$), the shear horizontal surface wave speed approaches the piezo-electromagnetically stiffened bulk shear wave speed (i.e., $c_{bg} \to c_T$).

Hence, the shear horizontal surface wave may occur under the electromagnetically permeable or semi-permeable crack-face condition, but there is no surface wave of this type under the electromagnetically impermeable crack-face condition. The propagation of Bleustein–Gulyaev surface waves in hexagonally symmetric piezoelectric materials can be taken as a special case.

7.3 Transient Mode-III Crack Growth Problem

Since the class of transient crack growth problems is rather difficult to solve analytically, existing solutions in the literature often involve certain assumptions. For example, Baker (1962) studied constant-speed crack growth under stress wave loading. The imposed constant-speed condition enables extraordinary simplification of the corresponding boundary-initial value problem. A solution procedure for the sudden extension of a pre-existing crack in an elastic body subjected to general time-independent loading is summarized by Freund (1990) in his monograph on dynamic fracture mechanics:

> "The mechanical fields prior to crack growth are equilibrium fields. If the loading is increased to a sufficiently large magnitude, then the crack will begin to extend... The applied loads induce a traction distribution on the crack plane ahead of the crack tip, and the process of crack growth is essentially the negation of this traction distribution. This idea is exploited to obtain a complete solution for general loading by means of superposition... First, the situation of crack growth with a pair of opposed concentrated forces acting on fixed material points on the crack faces is analyzed, giving rise to a very useful result called the fundamental solution for the problem. Then, the corresponding field quantities for any distribution of tractions on the crack faces can be determined directly by superposition over this fundamental solution."

This method can be extended to transient crack growth in the presence of magneto-electro-elastic coupling effects. Consider a semi-infinite crack propagating at constant speed V_C in a magneto-electro-elastic solid (Fig. 7.2) under the assumption that there is vacuum, air, or other fluid of negligible mechanical influence inside the crack occupying the region $\Omega^{(f)} = \{(\tilde{X}_1, \tilde{X}_2) | -\infty < \tilde{X}_1 < 0, -\delta < \tilde{X}_2 < \delta\}$. A reference Cartesian coordinate system $\{\tilde{X}_K, K = 1,2,3\}$ attached to the moving crack tip is chosen, which coincides at time $t=0$ with the fixed Cartesian coordinate system $\{X_K, K=1,2,3\}$. It is assumed that, for time $t<0$, the crack tip is at $X_1=0$ and the magneto-electro-elastic solid is load-free and at rest everywhere. At time $t=0$, the crack tip begins to move at speed V_C in the positive X_1-direction and leaves behind a pair of mixed concentrated loads. Then, the crack tip at any time $t>0$ is at $X_1=V_C t$. Analyzing the situation of crack growth with a pair of mixed concentrated loads acting upon fixed material points on the crack faces gives rise to the fundamental solutions, which can be used to determine the corresponding field quantities for general mixed loading by means of superposition.

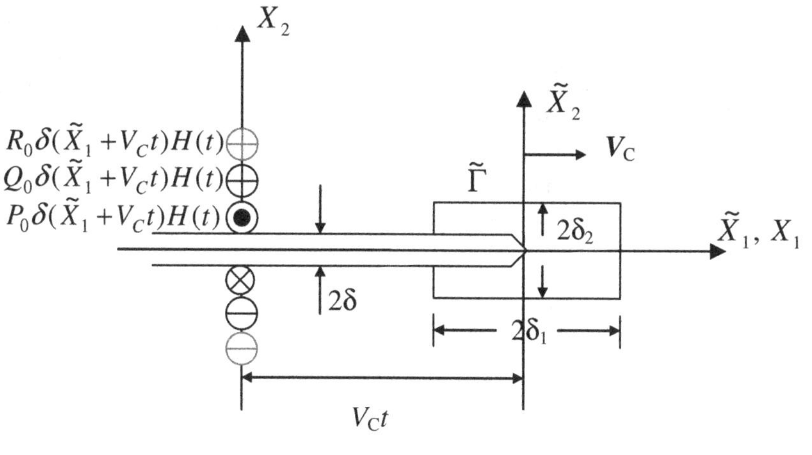

Fig. 7.2 Kernel crack growth problem with a pair of concentrated loads equal in magnitude and opposite in sign applied to the upper and lower surfaces of a semi-infinite mode-III crack propagating at constant speed V_C in a magneto-electro-elastic solid. (After Chen, 2009c, with permission from Elsevier.)

The Galilean transformation can be introduced as

$$\tilde{X}_1=X_1-V_C t\,,\ \tilde{X}_2=X_2\,,\ \tilde{X}_3=X_3\,,\ \tilde{t}=t\,. \tag{7.36}$$

For a transient mode-III crack growth problem in a transversely isotropic magneto-electro-elastic solid with the $\tilde{X}_3$-axis along the poling direction, the out-of-plane displacement and the electric and magnetic potentials are independent of $\tilde{X}_3$ in the reference frame moving with the crack tip, that is, $w = w(\tilde{X}_1, \tilde{X}_2, \tilde{t})$, $\varphi^{(p)} = \varphi^{(p)}(\tilde{X}_1, \tilde{X}_2, \tilde{t})$, $\psi^{(p)} = \psi^{(p)}(\tilde{X}_1, \tilde{X}_2, \tilde{t})$. Here, the superscript $p = s$ stands for the cracked solid region and $p = f$ stands for the interior fluid region. Based on the quasi-static approximation for the electromagnetic fields, the governing equations in the cracked solid region $\Omega^{(s)}$ and the interior fluid region $\Omega^{(f)}$ in the absence of mechanical body force, electricity conduction, and free electric charge are expressed in the reference frame moving with the crack tip as

$$s^2 \frac{\partial^2 w}{\partial \tilde{X}_1^2} + \frac{\partial^2 w}{\partial \tilde{X}_2^2} + \frac{2V_C}{c_T^2} \frac{\partial^2 w}{\partial \tilde{X}_1 \partial \tilde{t}} - \frac{1}{c_T^2} \frac{\partial^2 w}{\partial \tilde{t}^2} = 0 \quad \text{in } \Omega^{(s)}, \tag{7.37}$$

$$\frac{\partial^2 \overline{\varphi}^{(s)}}{\partial \tilde{X}_1^2} + \frac{\partial^2 \overline{\varphi}^{(s)}}{\partial \tilde{X}_2^2} = 0 \quad \text{in } \Omega^{(s)}, \tag{7.38}$$

$$\frac{\partial^2 \overline{\psi}^{(s)}}{\partial \tilde{X}_1^2} + \frac{\partial^2 \overline{\psi}^{(s)}}{\partial \tilde{X}_2^2} = 0 \quad \text{in } \Omega^{(s)}, \tag{7.39}$$

$$\frac{\partial^2 \varphi^{(f)}}{\partial \tilde{X}_1^2} + \frac{\partial^2 \varphi^{(f)}}{\partial \tilde{X}_2^2} = 0 \quad \text{in } \Omega^{(f)}, \tag{7.40}$$

$$\frac{\partial^2 \psi^{(f)}}{\partial \tilde{X}_1^2} + \frac{\partial^2 \psi^{(f)}}{\partial \tilde{X}_2^2} = 0 \quad \text{in } \Omega^{(f)}, \tag{7.41}$$

where $s = (1 - V_C^2/c_T^2)^{1/2}$, $V_C = |\boldsymbol{V}_C|$ is the crack tip velocity, and the relations (7.6)–(7.16) are still valid.

Under the assumption that the pre-existing state is quiescent and can be removed by superposition, the remote conditions are taken as

$$ {}_t\sigma_{23}(\tilde{X}_1, \tilde{X}_2, \tilde{t}) = 0, \; E_2^{(s)}(\tilde{X}_1, \tilde{X}_2, \tilde{t}) = 0, \; H_2^{(s)}(\tilde{X}_1, \tilde{X}_2, \tilde{t}) = 0, $$

$$\text{as } |\tilde{X}_2| \to \infty . \tag{7.42}$$

To obtain the fundamental solutions, a pair of mixed concentrated loads equal in magnitude and opposite in sign are suddenly applied on the upper and lower surfaces of the crack at time $t = 0^+$. In the reference frame moving with the crack tip, the corresponding boundary conditions are imposed

$$ {}_t\sigma_{23}(\tilde{X}_1, \delta, \tilde{t}) = -P_0 \delta(\tilde{X}_1 + V_C \tilde{t}) H(\tilde{t}), \tag{7.43} $$

$$ {}_t\sigma_{23}(\tilde{X}_1, -\delta, \tilde{t}) = -P_0 \delta(\tilde{X}_1 + V_C \tilde{t}) H(\tilde{t}), \tag{7.44} $$

$$ D_2^{(s)}(\tilde{X}_1, \delta, \tilde{t}) - D_2^{(f)}(\tilde{X}_1, \delta, \tilde{t}) = -Q_0 \delta(\tilde{X}_1 + V_C \tilde{t}) H(\tilde{t}), \tag{7.45} $$

$$ D_2^{(s)}(\tilde{X}_1, -\delta, \tilde{t}) - D_2^{(f)}(\tilde{X}_1, -\delta, \tilde{t}) = -Q_0 \delta(\tilde{X}_1 + V_C \tilde{t}) H(\tilde{t}), \tag{7.46} $$

$$ B_2^{(s)}(\tilde{X}_1, \delta, \tilde{t}) - B_2^{(f)}(\tilde{X}_1, \delta, \tilde{t}) = -R_0 \delta(\tilde{X}_1 + V_C \tilde{t}) H(\tilde{t}), \tag{7.47} $$

$$ B_2^{(s)}(\tilde{X}_1, -\delta, \tilde{t}) - B_2^{(f)}(\tilde{X}_1, -\delta, \tilde{t}) = -R_0 \delta(\tilde{X}_1 + V_C \tilde{t}) H(\tilde{t}), \tag{7.48} $$

$$ \varphi^{(s)}(\tilde{X}_1, \delta, \tilde{t}) - \varphi^{(f)}(\tilde{X}_1, \delta, \tilde{t}) = 0, \tag{7.49} $$

$$ \varphi^{(s)}(\tilde{X}_1, -\delta, \tilde{t}) - \varphi^{(f)}(\tilde{X}_1, -\delta, \tilde{t}) = 0, \tag{7.50} $$

$$ \psi^{(s)}(\tilde{X}_1, \delta, \tilde{t}) - \psi^{(f)}(\tilde{X}_1, \delta, \tilde{t}) = 0, \tag{7.51} $$

$$ \psi^{(s)}(\tilde{X}_1, -\delta, \tilde{t}) - \psi^{(f)}(\tilde{X}_1, -\delta, \tilde{t}) = 0. \tag{7.52} $$

The initial conditions are

$$ w(\tilde{X}_1, \tilde{X}_2, 0) = 0, \tag{7.53} $$

$$ \dot{w}(\tilde{X}_1, \tilde{X}_2, 0) = 0, \tag{7.54} $$

$$ \varphi^{(p)}(\tilde{X}_1, \tilde{X}_2, 0) = 0, \tag{7.55} $$

$$\psi^{(p)}(\tilde{X}_1, \tilde{X}_2, 0) = 0. \tag{7.56}$$

7.4 Integral Transform, Wiener–Hopf Technique, and Cagniard–de Hoop Method

The Wiener–Hopf technique was originally developed to solve a particular type of integral equation and then extended to a variety of applications (Noble, 1958). The essence of the solution process is to determine two unknown analytical functions from one single equation based on the theory of complex variables using the integral transforms such as Laplace, Fourier, or Mellin transforms. This technique was applied to the analysis of half-plane diffraction problems by de Hoop (1958) through suppression of time dependence by the Laplace transform. Once a solution in the transformed domain was attained, the Cagniard–de Hoop method was used to invert the transforms to obtain a solution in the physical domain. The reader may refer to the books by Freund (1990) and Broberg (1999) for further details.

The one-sided Laplace transform with respect to the time variable $\tilde{t}$ and the two-sided Laplace transform with respect to the spatial variable $\tilde{X}_1$ are applied as follows:

$$f^*(\tilde{X}_1, \tilde{X}_2, p) = \int_0^\infty f(\tilde{X}_1, \tilde{X}_2, \tilde{t}) \exp(-p\tilde{t}) d\tilde{t}, \tag{7.57}$$

$$f(\tilde{X}_1, \tilde{X}_2, \tilde{t}) = \frac{1}{2\pi i} \int_{Br_1} f^*(\tilde{X}_1, \tilde{X}_2, p) \exp(p\tilde{t}) dp, \tag{7.58}$$

$$\hat{f}^*(\zeta, \tilde{X}_2, p) = \int_{-\infty}^{\infty} f^*(\tilde{X}_1, \tilde{X}_2, p) \exp(-p\zeta\tilde{X}_1) d\tilde{X}_1, \tag{7.59}$$

$$f^*(\tilde{X}_1, \tilde{X}_2, p) = \frac{p}{2\pi i} \int_{Br_2} \hat{f}^*(\zeta, \tilde{X}_2, p) \exp(p\zeta\tilde{X}_1) d\zeta, \tag{7.60}$$

where the inversion integration is taken over the Bromwich path.

Application of the transforms to Eqs. (7.37)–(7.41) yields a set of ordinary differential equations:

$$\left[\frac{d^2}{d\tilde{X}_2^2} - p^2\left(\frac{1}{c_T^2} - 2\frac{V_C}{c_T^2}\zeta - s^2\zeta^2\right)\right]\hat{w}^*(\zeta, \tilde{X}_2, p) = 0, \quad \forall(\zeta, \tilde{X}_2, p) \in \Omega_p^{(s)}, \tag{7.61}$$

$$\left[\frac{d^2}{d\tilde{X}_2^2} - p^2\left(\varepsilon^2 - \zeta^2\right)\right]\hat{\bar{\varphi}}^{(s)*}(\zeta, \tilde{X}_2, p) = 0, \ \forall(\zeta, \tilde{X}_2, p) \in \Omega_p^{(s)}, \tag{7.62}$$

$$\left[\frac{d^2}{d\tilde{X}_2^2} - p^2\left(\varepsilon^2 - \zeta^2\right)\right]\hat{\bar{\psi}}^{(s)*}(\zeta, \tilde{X}_2, p) = 0, \ \forall(\zeta, \tilde{X}_2, p) \in \Omega_p^{(s)}, \tag{7.63}$$

$$\left[\frac{d^2}{d\tilde{X}_2^2} - p^2\left(\varepsilon^2 - \zeta^2\right)\right]\hat{\varphi}^{(f)*}(\zeta, \tilde{X}_2, p) = 0, \ \forall(\zeta, \tilde{X}_2, p) \in \Omega_p^{(f)}, \tag{7.64}$$

$$\left[\frac{d^2}{d\tilde{X}_2^2} - p^2\left(\varepsilon^2 - \zeta^2\right)\right]\hat{\psi}^{(f)*}(\zeta, \tilde{X}_2, p) = 0, \ \forall(\zeta, \tilde{X}_2, p) \in \Omega_p^{(f)}, \tag{7.65}$$

where $\varepsilon \to 0^+$ is an auxiliary (positive real) perturbation parameter. The technique of introducing an auxiliary perturbation parameter may be viewed as the quasi-static approximation for the electromagnetic fields since the crack speed is much lower than light speed.

Consideration of the remote conditions (7.42) leads to general solutions of the form

$$\hat{w}^*(\zeta, \tilde{X}_2, p) = \operatorname{sgn}(\tilde{X}_2)\frac{1}{p^2}A(\zeta)\exp(-p\alpha|\tilde{X}_2|), \tag{7.66}$$

$$\hat{\bar{\varphi}}^{(s)*}(\zeta, \tilde{X}_2, p) = \operatorname{sgn}(\tilde{X}_2)\frac{1}{p^2}B(\zeta)\exp(-p\beta|\tilde{X}_2|), \tag{7.67}$$

$$\hat{\bar{\psi}}^{(s)*}(\zeta, \tilde{X}_2, p) = \operatorname{sgn}(\tilde{X}_2)\frac{1}{p^2}C(\zeta)\exp(-p\beta|\tilde{X}_2|), \tag{7.68}$$

where $\alpha(\zeta)=\sqrt{(1/c_T^2-2V_C\zeta/c_T^2-s^2\zeta^2)}$ and $\beta(\zeta)=\lim_{\varepsilon\to 0^+}\sqrt{(\varepsilon^2-\zeta^2)}$.

Since the solution in the interior fluid region is not subjected to any remote conditions at $\left|\tilde{X}_2\right|\to\infty$, the complete form of the solution should be used:

$$\hat{\varphi}^{(f)*}(\zeta,\tilde{X}_2,p)=\frac{1}{p^2}[D_+(\zeta)\exp(p\beta\tilde{X}_2)+D_-(\zeta)\exp(-p\beta\tilde{X}_2)], \quad (7.69)$$

$$\hat{\psi}^{(f)*}(\zeta,\tilde{X}_2,p)=\frac{1}{p^2}[E_+(\zeta)\exp(p\beta\tilde{X}_2)+E_-(\zeta)\exp(-p\beta\tilde{X}_2)]. \quad (7.70)$$

In the complex ζ-plane, the function α has branch points at $\zeta=-1/(c_T-V_C)$ and $\zeta=1/(c_T+V_C)$, and the function β has branch points at $\zeta=-\varepsilon$ and $\zeta=+\varepsilon$. The branch of α with a positive real part $\mathrm{Re}(\alpha)>0$ and the branch of β with a positive real part $\mathrm{Re}(\beta)>0$ should be chosen, where the branch cuts run from the branch points outwards along the real axis.

To apply the Wiener–Hopf technique, the traction and displacement boundary conditions are expanded over the full range of the $\tilde{X}_1$-axis, that is,

$${}_t\sigma_{23}(\tilde{X}_1,0^+,\tilde{t})=\sigma_+(\tilde{X}_1,\tilde{t})-P_0\delta(\tilde{X}_1+V_C\tilde{t})H(\tilde{t}), \quad -\infty<\tilde{X}_1<\infty, \quad (7.71)$$

$$w(\tilde{X}_1,0^+,\tilde{t})=w_-(\tilde{X}_1,\tilde{t})+0, \quad -\infty<\tilde{X}_1<\infty, \quad (7.72)$$

with

$$\sigma_+(\tilde{X}_1,\tilde{t})=\begin{cases}{}_t\sigma_{23}(\tilde{X}_1,0^+,\tilde{t}) & \tilde{X}_1\geq 0\\ 0 & \tilde{X}_1<0\end{cases}, \quad (7.73)$$

$$w_-(\tilde{X}_1,\tilde{t})=\begin{cases}0 & \tilde{X}_1\geq 0\\ w(\tilde{X}_1,0^+,\tilde{t}) & \tilde{X}_1<0\end{cases}. \quad (7.74)$$

We first solve the case of traction loading only. To satisfy the transformed boundary conditions (7.45)–(7.52) with $Q_0 = 0$ and $R_0 = 0$, it follows that

$$A(\zeta) = U_-(\zeta) = p^2 \int_{-\infty}^{0} w_-^*(\tilde{X}_1, p)\exp(-p\zeta\tilde{X}_1)d\tilde{X}_1, \tag{7.75}$$

$$\begin{aligned} B(\zeta) = &\frac{\exp(p\beta\delta - p\alpha\delta)\cosh(p\beta\delta)}{f_\mu f_\kappa - f_g^2} \\ &\times (c_1 f_\mu \kappa_{11}^f - c_2 f_g \mu_{11}^f)A(\zeta), \end{aligned} \tag{7.76}$$

$$\begin{aligned} C(\zeta) = &\frac{\exp(p\beta\delta - p\alpha\delta)\cosh(p\beta\delta)}{f_\mu f_\kappa - f_g^2} \\ &\times (c_2 f_\kappa \mu_{11}^f - c_1 f_g \kappa_{11}^f)A(\zeta), \end{aligned} \tag{7.77}$$

$$D_+(\zeta) = -D_-(\zeta) = \frac{\exp(-p\beta\delta)}{2\sinh(p\beta\delta)} B(\zeta) + c_1 \frac{\exp(-p\alpha\delta)}{2\sinh(p\beta\delta)} A(\zeta), \tag{7.78}$$

$$E_+(\zeta) = -E_-(\zeta) = \frac{\exp(-p\beta\delta)}{2\sinh(p\beta\delta)} C(\zeta) + c_2 \frac{\exp(-p\alpha\delta)}{2\sinh(p\beta\delta)} A(\zeta), \tag{7.79}$$

where $f_\kappa = \sinh(p\beta\delta)\kappa_{11} - \cosh(p\beta\delta)\kappa_{11}^f$, $f_g = \sinh(p\beta\delta)g_{11}$,

$f_\mu = \sinh(p\beta\delta)\mu_{11} - \cosh(p\beta\delta)\mu_{11}^f$, $c_1 = (e_{15}\mu_{11} - h_{15}g_{11})/(\kappa_{11}\mu_{11} - g_{11}^2)$,

and $c_2 = (h_{15}\kappa_{11} - e_{15}g_{11})/(\kappa_{11}\mu_{11} - g_{11}^2)$.

By letting $\delta \to 0$ while keeping $\sinh(p\beta\delta)\kappa_{11}/\kappa_{11}^f \to \lambda_e$ and $\sinh(p\beta\delta)\mu_{11}/\mu_{11}^f \to \lambda_m$, Eqs. (7.76) and (7.77) become

$$B(\zeta) = -\bar{c}_1 A(\zeta), \tag{7.80}$$

$$C(\zeta) = -\bar{c}_2 A(\zeta), \tag{7.81}$$

where λ_e and λ_m are mutually dependent with $\lambda_e/\lambda_m = (\kappa_{11}\mu_{11}^f)/(\kappa_{11}^f\mu_{11})$,

$$\bar{c}_1 = [\kappa_{11}\mu_{11}(1-\lambda_m)c_1 + \mu_{11}g_{11}\lambda_e c_2]/[\kappa_{11}\mu_{11}(1-\lambda_e)(1-\lambda_m) - \lambda_e\lambda_m g_{11}^2],$$

and

$$\bar{c}_2 = [\kappa_{11}\mu_{11}(1-\lambda_e)c_2 + \kappa_{11}g_{11}\lambda_m c_1]/[\kappa_{11}\mu_{11}(1-\lambda_e)(1-\lambda_m) - \lambda_e\lambda_m g_{11}^2].$$

It is noted that $\bar{c}_1 \to c_1$ as $\lambda_e \to 0$ and $\bar{c}_2 \to c_2$ as $\lambda_m \to 0$.

There are four limiting conditions: (i) electrically and magnetically permeable crack-face condition as $\lambda_e \to 0$ and $\lambda_m \to 0$, (ii) electrically and magnetically impermeable crack-face condition as $\lambda_e \to \infty$ and $\lambda_m \to \infty$, (iii) electrically permeable and magnetically impermeable crack-face condition as $\lambda_e \to 0$ and $\lambda_m \to \infty$, and (iv) electrically impermeable and magnetically permeable crack-face condition as $\lambda_e \to \infty$ and $\lambda_m \to 0$. The electromagnetically semi-permeable crack-face condition may be approximated if λ_e and λ_m are considered as finite nonzero parameters. For simplicity, λ_e and λ_m are assumed to be constant in the following analysis.

The transformed total stress, electric displacement, and magnetic induction are expressed in terms of the single unknown function $U_-(\zeta)$ as

$$ {}_t\sigma_{13}^*(\tilde{X}_1,\tilde{X}_2,p) = \frac{\bar{c}_{44}}{2\pi i}\left\{\int_{\zeta_\alpha - i\infty}^{\zeta_\alpha + i\infty} \zeta U_-(\zeta)\exp[-p(\alpha\tilde{X}_2 - \zeta\tilde{X}_1)]d\zeta \right. \tag{7.82}$$
$$\left. -\int_{\zeta_\beta - i\infty}^{\zeta_\beta + i\infty} (k_{em}^\lambda)^2 \zeta U_-(\zeta)\exp[-p(\beta\tilde{X}_2 - \zeta\tilde{X}_1)]d\zeta \right\},$$

$$ {}_t\sigma_{23}^*(\tilde{X}_1,\tilde{X}_2,p) = -\frac{\bar{c}_{44}}{2\pi i}\left\{\int_{\zeta_\alpha - i\infty}^{\zeta_\alpha + i\infty} \alpha(\zeta) U_-(\zeta)\exp[-p(\alpha\tilde{X}_2 - \zeta\tilde{X}_1)]d\zeta \right. \tag{7.83}$$
$$\left. -\int_{\zeta_\beta - i\infty}^{\zeta_\beta + i\infty} (k_{em}^\lambda)^2 \beta(\zeta) U_-(\zeta)\exp[-p(\beta\tilde{X}_2 - \zeta\tilde{X}_1)]d\zeta \right\},$$

$$D_1^{(s)*}(\tilde{X}_1,\tilde{X}_2,p)=\frac{\bar{c}_1\kappa_{11}+\bar{c}_2g_{11}}{2\pi i}\times\int_{\zeta_\beta-i\infty}^{\zeta_\beta+i\infty}\zeta U_-(\zeta)\exp[-p(\beta\tilde{X}_2-\zeta\tilde{X}_1)]d\zeta, \tag{7.84}$$

$$D_2^{(s)*}(\tilde{X}_1,\tilde{X}_2,p)=-\frac{\bar{c}_1\kappa_{11}+\bar{c}_2g_{11}}{2\pi i}\times\int_{\zeta_\beta-i\infty}^{\zeta_\beta+i\infty}\beta(\zeta)U_-(\zeta)\exp[-p(\beta\tilde{X}_2-\zeta\tilde{X}_1)]d\zeta, \tag{7.85}$$

$$B_1^{(s)*}(\tilde{X}_1,\tilde{X}_2,p)=\frac{\bar{c}_1g_{11}+\bar{c}_2\mu_{11}}{2\pi i}\times\int_{\zeta_\beta-i\infty}^{\zeta_\beta+i\infty}\zeta U_-(\zeta)\exp[-p(\beta\tilde{X}_2-\zeta\tilde{X}_1)]d\zeta, \tag{7.86}$$

$$B_2^{(s)*}(\tilde{X}_1,\tilde{X}_2,p)=-\frac{\bar{c}_1g_{11}+\bar{c}_2\mu_{11}}{2\pi i}\times\int_{\zeta_\beta-i\infty}^{\zeta_\beta+i\infty}\beta(\zeta)U_-(\zeta)\exp[-p(\beta\tilde{X}_2-\zeta\tilde{X}_1)]d\zeta, \tag{7.87}$$

where $k_{em}^{\lambda}=\sqrt{(\bar{c}_1e_{15}+\bar{c}_2h_{15})/\bar{c}_{44}}$ is the magneto-electro-mechanical coupling factor which depends on the permeability parameters λ_e and λ_m through $\bar{c}_1$ and $\bar{c}_2$, $-1/(c_T-V_C)<\zeta_\alpha<1/(c_T+V_C)$, $-\varepsilon<\zeta_\beta<\varepsilon$.

Substituting Eq. (7.83) into the transformed traction boundary condition (7.71) leads to the following Wiener–Hopf equation:

$$\Sigma_+(\zeta)+\frac{P_0}{V_C(\zeta-1/V_C)}=K(\zeta)U_-(\zeta), \tag{7.88}$$

where

$$K(\zeta)=-\bar{c}_{44}[\alpha(\zeta)-(k_{em}^{\lambda})^2\beta(\zeta)], \tag{7.89}$$

$$\Sigma_+(\zeta)=p\int_0^{+\infty}\sigma_+^*(\tilde{X}_1,p)\exp(-p\zeta\tilde{X}_1)d\tilde{X}_1. \tag{7.90}$$

The modified form of the Bleustein–Gulyaev wave function

$$BG(\zeta) = \alpha(\zeta) - (k_{em}^{\lambda})^2 \beta(\zeta), \tag{7.91}$$

has a simple structure with roots at $\zeta = -1/(c_{bg}^{\lambda} - V_C)$ and $\zeta = 1/(c_{bg}^{\lambda} + V_C)$, where the shear horizontal surface wave speed is defined as

$$c_{bg}^{\lambda} = \sqrt{\frac{\overline{c}_{44}[1 - (k_{em}^{\lambda})^4]}{\rho}} = c_T \sqrt{[1 - (k_{em}^{\lambda})^4]}. \tag{7.92}$$

For the electrically and magnetically permeable crack-face condition ($\lambda_e \to 0$ and $\lambda_m \to 0$), we have $c_{bg}^{\lambda} \to c_{bg}^{0} = c_T \sqrt{[1 - (k_{em}^{0})^4]}$ and $k_{em}^{\lambda} \to k_{em}^{0} = \sqrt{(c_1 e_{15} + c_2 h_{15})/\overline{c}_{44}}$, whereas for the electrically and magnetically impermeable crack-face condition ($\lambda_e \to \infty$ and $\lambda_m \to \infty$), we have $c_{bg}^{\lambda} \to c_T$ and $k_{em}^{\lambda} \to 0$. In particular, for an electrically permeable mode-III crack propagating in a hexagonally symmetric piezoelectric medium, we retrieve the electromechanical coupling factor and the Bleustein–Gulyaev surface wave speed (Li and Mataga, 1996a; Ing and Wang, 2004b). The equality of the shear horizontal surface wave speed to the bulk shear wave speed in the limit of electromagnetic impermeability indicates that there is no shear horizontal surface wave mode under the electrically and magnetically impermeable crack-face condition.

It is convenient to rewrite Eq. (7.89) in the following form:

$$K(\zeta) = -\overline{c}_{44}[s - (k_{em}^{\lambda})^2] \sqrt{[1/(c_{bg}^{\lambda} + V_C) - \zeta][1/(c_{bg}^{\lambda} - V_C) + \zeta]}\, S(\zeta), \tag{7.93}$$

where an auxiliary function $S(\zeta)$ is introduced by

$$S(\zeta) = \frac{\alpha(\zeta) - (k_{em}^{\lambda})^2 \beta(\zeta)}{[s - (k_{em}^{\lambda})^2] \sqrt{[1/(c_{bg}^{\lambda} + V_C) - \zeta][1/(c_{bg}^{\lambda} - V_C) + \zeta]}}. \tag{7.94}$$

It should be noted that $S(\zeta) \to 1$ as $|\zeta| \to \infty$. The essence of deriving the solution of the Wiener–Hopf equation (7.88) is to decompose $S(\zeta)$ such that

$$S(\zeta) = S_+(\zeta) S_-(\zeta), \tag{7.95}$$

where $S_+(\zeta)$ and $S_-(\zeta)$ are analytical in their respective half planes with an overlapping strip.

The primary features of the complex ζ-plane pertinent to the solution of the Wiener–Hopf equation (7.88) are depicted in Fig. 7.3 with the branch points at $\zeta = -1/(c_T - V_C)$, $\zeta = 1/(c_T + V_C)$, $\zeta = -\varepsilon$, and $\zeta = +\varepsilon$, and the roots at $\zeta = -1/(c_{bg}^{\lambda} - V_C)$ and $\zeta = 1/(c_{bg}^{\lambda} + V_C)$. The common strip of analyticity is between the two dashed lines, and the overlapping half planes in which the functions labeled with subscripts (+) and (−) are analytical and are indicated with arrows.

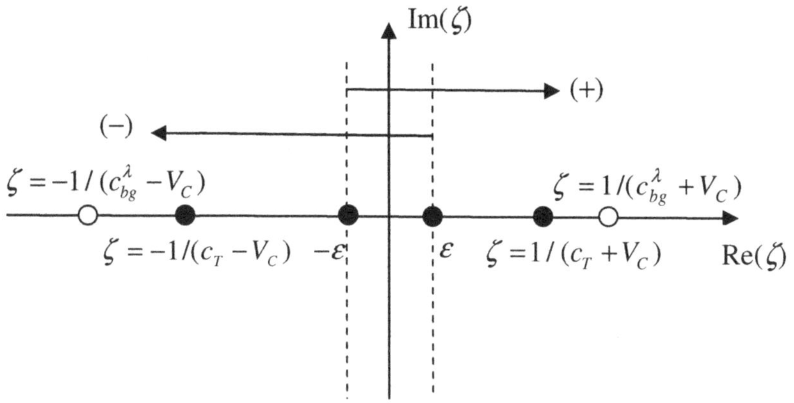

Fig. 7.3 The complex ζ-plane showing the primary features pertinent to the solution of the Wiener–Hopf equation (7.88).

The final factorization of $S(\zeta)$ into products of two sectional analytical functions leads to

$$S_+(\zeta) = \sqrt{\frac{1/(c_{bg}^{\lambda} - V_C) + \zeta}{1/(c_T - V_C) + \zeta}}\,\phi_+(\zeta), \tag{7.96}$$

$$S_{-}(\zeta)=\sqrt{\frac{1/(c_{bg}^{\lambda}+V_C)-\zeta}{1/(c_T+V_C)-\zeta}}\phi_{-}(\zeta), \tag{7.97}$$

$$\phi_{\pm}(\zeta)=\exp\left\{\frac{1}{\pi}\int_{\varepsilon}^{1/(c_T\mp V_C)}\arctan[\Theta(\mp\eta)]\frac{d\eta}{\eta\pm\zeta}\right\}, \tag{7.98}$$

$$\Theta(\eta)=\frac{(k_{em}^{\lambda})^2\sqrt{\eta^2-\varepsilon^2}}{s\sqrt{[1/(c_T-V_C)+\eta][1/(c_T+V_C)-\eta]}}. \tag{7.99}$$

Hence, the two unknown functions in the Wiener–Hopf equation (7.88) are determined to be

$$U_{-}(\zeta)=-\frac{P_0}{\bar{c}_{44}\sqrt{V_C}}\frac{[s+(k_{em}^{\lambda})^2]}{[1-(k_{em}^{\lambda})^4]}\frac{1}{\sqrt{1-V_C/c_T}}$$

$$\times\frac{\mathcal{D}_{+}(1/V_C)}{(1+V_C/c_{bg}^{\lambda})}\frac{\mathcal{D}_{-}(\zeta)}{(\zeta-1/V_C)}\frac{\sqrt{1/(c_T+V_C)-\zeta}}{[1/(c_{bg}^{\lambda}+V_C)-\zeta]}, \tag{7.100}$$

$$\Sigma_{+}(\zeta)=\frac{P_0}{V_C(\zeta-1/V_C)}$$

$$\times\left[\frac{\sqrt{V_C}(1-V_C/c_{bg}^{\lambda})}{\sqrt{1-V_C/c_T}}\frac{1/(c_{bg}^{\lambda}-V_C)+\zeta}{\sqrt{1/(c_T-V_C)+\zeta}}\frac{\mathcal{D}_{+}(1/V_C)}{\mathcal{D}_{+}(\zeta)}-1\right], \tag{7.101}$$

where

$$\mathcal{D}_{\pm}(\zeta)=\frac{1}{\phi_{\pm}(\zeta)}. \tag{7.102}$$

Once $U_{-}(\zeta)$ and $\Sigma_{+}(\zeta)$ are obtained by the Wiener–Hopf technique, the Cagniard–de Hoop method can be used to find the closed-form solutions in the physical domain by employing the following inversion paths in the complex ζ-plane:

$$\zeta_{1+}(\tilde{X}_1,\tilde{t})=\frac{1}{\tilde{X}_1^2+s^2\tilde{X}_2^2}\left[-\left(\tilde{X}_1\tilde{t}+\frac{V_C}{c_T^2}\tilde{X}_2^2\right)\right.$$
$$\left.+i\tilde{X}_2\sqrt{s^2\tilde{t}^2-\frac{2V_C}{c_T^2}\tilde{X}_1\tilde{t}-\frac{1}{c_T^2}(\tilde{X}_2^2+\tilde{X}_1^2)}\right], \tag{7.103}$$

$$\zeta_{2+}(\tilde{X}_1,\tilde{t})=\frac{1}{\tilde{X}_1^2+\tilde{X}_2^2}\left(-\tilde{X}_1\tilde{t}+i\tilde{X}_2\tilde{t}\sqrt{1-\varepsilon^2\frac{\tilde{X}_2^2+\tilde{X}_1^2}{\tilde{t}^2}}\right). \tag{7.104}$$

7.5 Fundamental Solutions for Traction Loading Only

The closed-form fundamental solutions for traction loading only are obtained as

$$w(\tilde{X}_1,\tilde{X}_2,\tilde{t})=\frac{1}{\pi}\int_{\tilde{t}_1}^{\tilde{t}}\mathrm{Im}\left[U_-(\zeta_{1+}(\tau))\frac{\partial\zeta_{1+}(\tau)}{\partial\tau}\right]d\tau, \tag{7.105}$$

$$\varphi^{(s)}(\tilde{X}_1,\tilde{X}_2,\tilde{t})=\frac{1}{\pi}\left\{c_1\int_{\tilde{t}_1}^{\tilde{t}}\mathrm{Im}\left[U_-(\zeta_{1+}(\tau))\frac{\partial\zeta_{1+}(\tau)}{\partial\tau}\right]d\tau\right.$$
$$\left.-\bar{c}_1\int_{\tilde{t}_2}^{\tilde{t}}\mathrm{Im}\left[U_-(\zeta_{2+}(\tau))\frac{\partial\zeta_{2+}(\tau)}{\partial\tau}\right]d\tau\right\}, \tag{7.106}$$

$$\psi^{(s)}(\tilde{X}_1,\tilde{X}_2,\tilde{t})=\frac{1}{\pi}\left\{c_2\int_{\tilde{t}_1}^{\tilde{t}}\mathrm{Im}\left[U_-(\zeta_{1+}(\tau))\frac{\partial\zeta_{1+}(\tau)}{\partial\tau}\right]d\tau\right.$$
$$\left.-\bar{c}_2\int_{\tilde{t}_2}^{\tilde{t}}\mathrm{Im}\left[U_-(\zeta_{2+}(\tau))\frac{\partial\zeta_{2+}(\tau)}{\partial\tau}\right]d\tau\right\}, \tag{7.107}$$

$${}_t\sigma_{13}(\tilde{X}_1,\tilde{X}_2,\tilde{t})=\frac{\bar{c}_{44}}{\pi}\left\{\mathrm{Im}\left[\zeta_{1+}U_-(\zeta_{1+})\frac{\partial\zeta_{1+}}{\partial\tilde{t}}\right]H(\tilde{t}-\tilde{t}_1)\right.$$
$$\left.-(k_{em}^{\lambda})^2\,\mathrm{Im}\left[\zeta_{2+}U_-(\zeta_{2+})\frac{\partial\zeta_{2+}}{\partial\tilde{t}}\right]H(\tilde{t}-\tilde{t}_2)\right\}, \tag{7.108}$$

$$
{}_t\sigma_{23}(\tilde{X}_1,\tilde{X}_2,\tilde{t}) = -\frac{\bar{c}_{44}}{\pi}\left\{ \mathrm{Im}\left[\alpha(\zeta_{1+})U_-(\zeta_{1+})\frac{\partial \zeta_{1+}}{\partial \tilde{t}} \right] H(\tilde{t}-\tilde{t}_1) \right.
$$
$$
\left. -(k_{em}^{\lambda})^2\,\mathrm{Im}\left[\beta(\zeta_{2+})U_-(\zeta_{2+})\frac{\partial \zeta_{2+}}{\partial \tilde{t}} \right] H(\tilde{t}-\tilde{t}_2) \right\}, \tag{7.109}
$$

$$
D_1^{(s)}(\tilde{X}_1,\tilde{X}_2,\tilde{t}) = \frac{\bar{c}_1\kappa_{11}+\bar{c}_2 g_{11}}{\pi} \times \mathrm{Im}\left[\zeta_{2+}U_-(\zeta_{2+})\frac{\partial \zeta_{2+}}{\partial \tilde{t}} \right] H(\tilde{t}-\tilde{t}_2), \tag{7.110}
$$

$$
D_2^{(s)}(\tilde{X}_1,\tilde{X}_2,\tilde{t}) = -\frac{\bar{c}_1\kappa_{11}+\bar{c}_2 g_{11}}{\pi} \times \mathrm{Im}\left[\beta(\zeta_{2+})U_-(\zeta_{2+})\frac{\partial \zeta_{2+}}{\partial \tilde{t}} \right] H(\tilde{t}-\tilde{t}_2), \tag{7.111}
$$

$$
B_1^{(s)}(\tilde{X}_1,\tilde{X}_2,\tilde{t}) = \frac{\bar{c}_1 g_{11}+\bar{c}_2 \mu_{11}}{\pi} \times \mathrm{Im}\left[\zeta_{2+}U_-(\zeta_{2+})\frac{\partial \zeta_{2+}}{\partial \tilde{t}} \right] H(\tilde{t}-\tilde{t}_2), \tag{7.112}
$$

$$
B_2^{(s)}(\tilde{X}_1,\tilde{X}_2,\tilde{t}) = -\frac{\bar{c}_1 g_{11}+\bar{c}_2 \mu_{11}}{\pi} \times \mathrm{Im}\left[\beta(\zeta_{2+})U_-(\zeta_{2+})\frac{\partial \zeta_{2+}}{\partial \tilde{t}} \right] H(\tilde{t}-\tilde{t}_2), \tag{7.113}
$$

where $\tilde{t}_1 = [V_C\tilde{X}_1/c_T^2 + \sqrt{(\tilde{X}_1^2 + s^2\tilde{X}_2^2)}/c_T]/s^2$ and $\tilde{t}_2 = \varepsilon\sqrt{(\tilde{X}_1^2+\tilde{X}_2^2)}$.

The asymptotic behavior of the solutions near the moving crack tip will be examined below. As $\tilde{X}_2 \to 0$, both inversion contours take the same path $\zeta_{1+} = \zeta_{2+} = \zeta_+$, i.e.,

$$
\zeta_+ = -\frac{\tilde{t}}{\tilde{X}_1}, \tag{7.114}
$$

$$\frac{\partial \zeta_+}{\partial \tilde{t}} = -\frac{1}{\tilde{X}_1}. \tag{7.115}$$

Hence, the total stress, electric displacement, and magnetic induction fields ahead of the moving crack tip are represented by

$${}_t\sigma_{23}(\tilde{X}_1, 0^+, \tilde{t}) = \frac{(1 - V_C / c_{bg}^\lambda)}{\sqrt{1 - V_C / c_T}} \mathscr{D}_+(1/V_C) \frac{P_0}{\pi\sqrt{\tilde{X}_1}} \frac{1}{\sqrt{V_C \tilde{t}}}, \tag{7.116}$$

$$\begin{aligned} D_2(\tilde{X}_1, 0^+, \tilde{t}) = {} & \frac{(\bar{c}_1 \kappa_{11} + \bar{c}_2 g_{11})}{\bar{c}_{44}} \frac{1}{[1 - (k_{em}^\lambda)^4]} \frac{[s + (k_{em}^\lambda)^2]}{(1 + V_C / c_{bg}^\lambda)} \\ & \times \frac{\mathscr{D}_+(1/V_C)}{\sqrt{1 - V_C / c_T}} \frac{P_0}{\pi\sqrt{\tilde{X}_1}} \frac{1}{\sqrt{V_C \tilde{t}}}, \end{aligned} \tag{7.117}$$

$$\begin{aligned} B_2(\tilde{X}_1, 0^+, \tilde{t}) = {} & \frac{(\bar{c}_1 g_{11} + \bar{c}_2 \mu_{11})}{\bar{c}_{44}} \frac{1}{[1 - (k_{em}^\lambda)^4]} \frac{[s + (k_{em}^\lambda)^2]}{(1 + V_C / c_{bg}^\lambda)} \\ & \times \frac{\mathscr{D}_+(1/V_C)}{\sqrt{1 - V_C / c_T}} \frac{P_0}{\pi\sqrt{\tilde{X}_1}} \frac{1}{\sqrt{V_C \tilde{t}}}. \end{aligned} \tag{7.118}$$

It is evident that the near-tip total stress, electric displacement, and magnetic induction fields possess an inverse square-root singularity, similar to the near-tip stress field in classical elastodynamic fracture mechanics. The dynamic total stress, electric displacement, and magnetic induction intensity factors for mode-III crack propagation are defined as

$$\tilde{K}_{III} = \lim_{\tilde{X}_1 \to 0^+} \sqrt{2\pi \tilde{X}_1}\, {}_t\sigma_{23}(\tilde{X}_1, 0^+, \tilde{t}), \tag{7.119}$$

$$\tilde{K}_D = \lim_{\tilde{X}_1 \to 0^+} \sqrt{2\pi \tilde{X}_1} D_2(\tilde{X}_1, 0^+, \tilde{t}), \tag{7.120}$$

$$\tilde{K}_B = \lim_{\tilde{X}_1 \to 0^+} \sqrt{2\pi \tilde{X}_1} B_2(\tilde{X}_1, 0^+, \tilde{t}). \tag{7.121}$$

With normalization based on the corresponding quasi-static value, the dimensionless dynamic field intensity factors are obtained as

$$\begin{aligned} \frac{\tilde{K}_{III}^{(T)}(V_C\tilde{t}, V_C)}{\tilde{K}_{III}^{(T)}(V_C\tilde{t}, 0)} &= \frac{(1 - V_C / c_{bg}^{\lambda})}{\sqrt{1 - V_C / c_T}} \mathscr{D}_+ (1/V_C) \\ &= f_1^{\lambda}(V_C), \end{aligned} \tag{7.122}$$

$$\begin{aligned} \frac{\tilde{K}_{D}^{(T)}(V_C\tilde{t}, V_C)}{\tilde{K}_{D}^{(T)}(V_C\tilde{t}, 0)} &= \frac{[s + (k_{em}^{\lambda})^2]}{[1 + (k_{em}^{\lambda})^2](1 + V_C / c_{bg}^{\lambda})} \frac{\mathscr{D}_+ (1/V_C)}{\sqrt{1 - V_C / c_T}} \\ &= f_2^{\lambda}(V_C), \end{aligned} \tag{7.123}$$

$$\begin{aligned} \frac{\tilde{K}_{B}^{(T)}(V_C\tilde{t}, V_C)}{\tilde{K}_{B}^{(T)}(V_C\tilde{t}, 0)} &= \frac{[s + (k_{em}^{\lambda})^2]}{[1 + (k_{em}^{\lambda})^2](1 + V_C / c_{bg}^{\lambda})} \frac{\mathscr{D}_+ (1/V_C)}{\sqrt{1 - V_C / c_T}} \\ &= f_2^{\lambda}(V_C), \end{aligned} \tag{7.124}$$

where the superscript (T) indicates traction loading.

The functions f_1^{λ} and f_2^{λ} are universal functions of the crack tip velocity. As shown in Figs. 7.4 and 7.5, the functions $f_1^{\lambda}(V_C)$ and $f_2^{\lambda}(V_C)$ decrease monotonically with increasing crack tip velocity V_C. The larger the magneto-electro-mechanical coupling factor k_{em}^{λ}, the lower the values of f_1^{λ} and f_2^{λ}. It is noted that the function $f_1^{\lambda}(V_C)$ approaches zero but the function $f_2^{\lambda}(V_C)$ does not tend to zero as $V_C / c_{bg}^{\lambda} \to 1$. Consequently, the dynamic total stress intensity factor tends to zero but the dynamic electric displacement and magnetic induction intensity factors do not vanish as the crack tip velocity V_C approaches the shear horizontal surface wave speed c_{bg}^{λ}. The dynamic total stress, electric displacement, and magnetic induction intensity factors are reduced to those for the electromagnetically impermeable crack-face condition as $k_{em}^{\lambda} \to 0$.

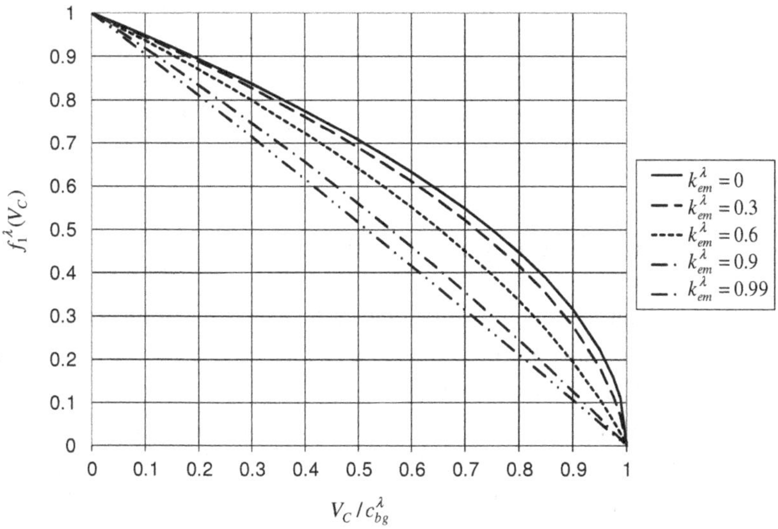

Fig. 7.4 Universal function f_1^λ versus dimensionless crack tip velocity V_C / c_{bg}^λ for a broad range of magneto-electro-mechanical coupling factors. (After Chen, 2009c, with permission from Elsevier.)

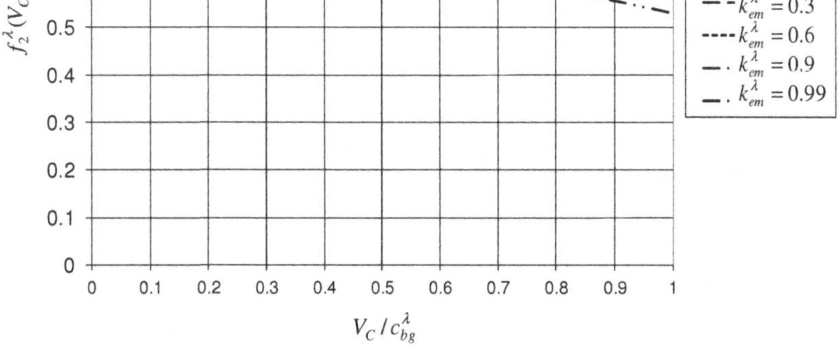

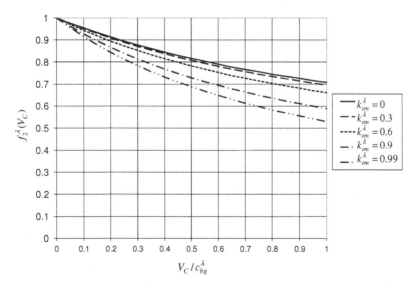

Fig. 7.5 Universal function f_2^λ versus dimensionless crack tip velocity V_C / c_{bg}^λ for a broad range of magneto-electro-mechanical coupling factors. (After Chen, 2009c, with permission from Elsevier.)

The solution for dynamic antiplane crack propagation in a purely elastic medium (Freund, 1990; Ma and Chen, 1992) and the electrode solution for dynamic mode-III crack propagation in a hexagonally symmetric piezoelectric material (Li and Mataga, 1996a; Melkumyan, 2005) can be taken as special cases.

7.6 Fundamental Solutions for Mixed Loads

For a pair of mixed concentrated loads, the general solutions given by (7.66)–(7.70) still hold but the unknown functions need to be determined under the boundary conditions with nonzero P_0, Q_0, and R_0. Since the continuity conditions (7.49)–(7.52) for the electric and magnetic potentials are kept the same, the functions $D_+(\zeta)$, $D_-(\zeta)$, $E_+(\zeta)$, and $E_-(\zeta)$ can be calculated from (7.78)–(7.79) so long as the functions $A(\zeta)$, $B(\zeta)$, and $C(\zeta)$ are attained.

Substituting the unknown functions $A(\zeta)$, $B(\zeta)$ and $C(\zeta)$ back into the transformed version of the boundary conditions (7.43)–(7.48) yields the following integral equations:

$$\frac{1}{2\pi i}\int_{\zeta_c-i\infty}^{\zeta_c+i\infty}[\alpha(\zeta)-(k_{em}^{\lambda})^2\beta(\zeta)]A(\zeta)\exp(p\zeta\tilde{X}_1)d\zeta$$

$$=\frac{\tilde{L}(P_0,Q_0,R_0)}{\bar{c}_{44}V_C}\exp\left(\frac{p\tilde{X}_1}{V_C}\right),\quad \text{for } \tilde{X}_1<0,$$

$$\frac{1}{2\pi ip}\int_{\zeta_c-i\infty}^{\zeta_c+i\infty}A(\zeta)\exp(p\zeta\tilde{X}_1)d\zeta=0,\ \text{for } \tilde{X}_1>0, \tag{7.125}$$

$$\frac{1}{2\pi i}\int_{\zeta_c-i\infty}^{\zeta_c+i\infty}\beta(\zeta)\bar{B}(\zeta)\exp(p\zeta\tilde{X}_1)d\zeta=-\frac{\tilde{L}_1(Q_0,R_0)}{V_C}\exp\left(\frac{p\tilde{X}_1}{V_C}\right),\quad \text{for } \tilde{X}_1<0,$$

$$\frac{1}{2\pi ip}\int_{\zeta_c-i\infty}^{\zeta_c+i\infty}\bar{B}(\zeta)\exp(p\zeta\tilde{X}_1)d\zeta=0,\ \text{for } \tilde{X}_1>0, \tag{7.126}$$

$$\frac{1}{2\pi i}\int_{\zeta_c-i\infty}^{\zeta_c+i\infty}\beta(\zeta)\bar{C}(\zeta)\exp(p\zeta\tilde{X}_1)d\zeta=-\frac{\tilde{L}_2(Q_0,R_0)}{V_C}\exp\left(\frac{p\tilde{X}_1}{V_C}\right),\quad \text{for } \tilde{X}_1<0,$$

$$\frac{1}{2\pi ip}\int_{\zeta_c-i\infty}^{\zeta_c+i\infty}\bar{C}(\zeta)\exp(p\zeta\tilde{X}_1)d\zeta=0,\ \text{for}\ \tilde{X}_1>0, \tag{7.127}$$

where $-\varepsilon<\mathrm{Re}(\zeta_c)<\varepsilon$ and the functions $\bar{B}(\zeta)$, $\bar{C}(\zeta)$, $\tilde{L}(P_0,Q_0,R_0)$, $\tilde{L}_1(Q_0,R_0)$, and $\tilde{L}_2(Q_0,R_0)$ are defined by

$$\bar{B}(\zeta)=B(\zeta)+\bar{c}_1A(\zeta), \tag{7.128}$$

$$\bar{C}(\zeta)=C(\zeta)+\bar{c}_2A(\zeta), \tag{7.129}$$

$$\tilde{L}(P_0,Q_0,R_0)=P_0+e_{15}\tilde{L}_1(Q_0,R_0)+h_{15}\tilde{L}_2(Q_0,R_0), \tag{7.130}$$

$$\tilde{L}_1(Q_0,R_0)=[Q_0\mu_{11}(1-1/\lambda_m)-R_0g_{11}]/[(1-1/\lambda_e)(1-1/\lambda_m)\kappa_{11}\mu_{11}-g_{11}^2], \tag{7.131}$$

$$\tilde{L}_2(Q_0,R_0)=[R_0\kappa_{11}(1-1/\lambda_e)-Q_0g_{11}]/[(1-1/\lambda_e)(1-1/\lambda_m)\kappa_{11}\mu_{11}-g_{11}^2]. \tag{7.132}$$

It is noted that the load function $\tilde{L}_1(Q_0,R_0)$ goes to zero for the electrically permeable crack-face condition ($\lambda_e\to 0$) and the load function $\tilde{L}_2(Q_0,R_0)$ goes to zero for the magnetically permeable crack-face condition ($\lambda_m\to 0$). Thus, the total traction is the only contribution to the load function $\tilde{L}(P_0,Q_0,R_0)$ for the electromagnetically permeable crack-face condition ($\lambda_e\to 0$ and $\lambda_m\to 0$).

The dual integral equation (7.125) has the same structure as the Wiener–Hopf equation encountered for the traction loading-only case. Thus, the solution is obtained as

$$A(\zeta)=-\frac{\tilde{L}(P_0,Q_0,R_0)}{\bar{c}_{44}\sqrt{V_C}}\frac{[s+(k_{em}^{\lambda})^2]}{[1-(k_{em}^{\lambda})^4]}\frac{1}{\sqrt{1-V_C/c_T}}$$
$$\times\frac{\mathscr{D}_+(1/V_C)}{(1+V_C/c_{bg}^{\lambda})}\frac{\mathscr{D}_-(\zeta)}{(\zeta-1/V_C)}\frac{\sqrt{1/(c_T+V_C)-\zeta}}{[1/(c_{bg}^{\lambda}+V_C)-\zeta]}. \tag{7.133}$$

Following the procedure outlined by Sih and Chen (1977) and Li and Mataga (1996b), the dual integral equations (7.126) and (7.127) are satisfied if

$$\bar{B}_{-}(\zeta) = \tilde{L}_1(Q_0, R_0)\frac{1}{\sqrt{V_C}}\frac{1}{(\zeta - 1/V_C)}\frac{1}{\sqrt{\varepsilon - \zeta}}, \tag{7.134}$$

$$\bar{C}_{-}(\zeta) = \tilde{L}_2(Q_0, R_0)\frac{1}{\sqrt{V_C}}\frac{1}{(\zeta - 1/V_C)}\frac{1}{\sqrt{\varepsilon - \zeta}}. \tag{7.135}$$

Similar to the traction loading-only case, the closed-form fundamental solutions for the mixed loading case can be achieved by the Cagniard–de Hoop inversion scheme, that is,

$$w(\tilde{X}_1, \tilde{X}_2, \tilde{t}) = \frac{1}{\pi}\int_{\tilde{t}_1}^{\tilde{t}} \mathrm{Im}\left[A(\zeta_{1+}(\tau))\frac{\partial \zeta_{1+}(\tau)}{\partial \tau}\right]d\tau, \tag{7.136}$$

$$\begin{aligned}\varphi^{(s)}(\tilde{X}_1, \tilde{X}_2, \tilde{t}) = \frac{1}{\pi}\Bigg\{ & c_1\int_{\tilde{t}_1}^{\tilde{t}} \mathrm{Im}\left[A(\zeta_{1+}(\tau))\frac{\partial \zeta_{1+}(\tau)}{\partial \tau}\right]d\tau \\ & -\bar{c}_1\int_{\tilde{t}_2}^{\tilde{t}} \mathrm{Im}\left[A(\zeta_{2+}(\tau))\frac{\partial \zeta_{2+}(\tau)}{\partial \tau}\right]d\tau + \tilde{L}_1(Q_0, R_0) \\ & \times\int_{\tilde{t}_2}^{\tilde{t}} \mathrm{Im}\left[\frac{1}{(\zeta_{2+}(\tau) - 1/V_C)}\frac{1}{\sqrt{\varepsilon - \zeta_{2+}(\tau)}}\frac{\partial \zeta_{2+}(\tau)}{\partial \tau}\right]d\tau\Bigg\},\end{aligned} \tag{7.137}$$

$$\begin{aligned}\psi^{(s)}(\tilde{X}_1, \tilde{X}_2, \tilde{t}) = \frac{1}{\pi}\Bigg\{ & c_2\int_{\tilde{t}_1}^{\tilde{t}} \mathrm{Im}\left[A(\zeta_{1+}(\tau))\frac{\partial \zeta_{1+}(\tau)}{\partial \tau}\right]d\tau \\ & -\bar{c}_2\int_{\tilde{t}_2}^{\tilde{t}} \mathrm{Im}\left[A(\zeta_{2+}(\tau))\frac{\partial \zeta_{2+}(\tau)}{\partial \tau}\right]d\tau + \tilde{L}_2(Q_0, R_0) \\ & \times\int_{\tilde{t}_2}^{\tilde{t}} \mathrm{Im}\left[\frac{1}{(\zeta_{2+}(\tau) - 1/V_C)}\frac{1}{\sqrt{\varepsilon - \zeta_{2+}(\tau)}}\frac{\partial \zeta_{2+}(\tau)}{\partial \tau}\right]d\tau\Bigg\},\end{aligned} \tag{7.138}$$

$$
\begin{aligned}
{}_t\sigma_{13}(\tilde{X}_1,\tilde{X}_2,\tilde{t}) &= \frac{\bar{c}_{44}}{\pi}\left\{\operatorname{Im}\left[\zeta_{1+}A(\zeta_{1+})\frac{\partial\zeta_{1+}}{\partial\tilde{t}}\right]H(\tilde{t}-\tilde{t}_1)\right. \\
&-(k_{em}^{\lambda})^2\operatorname{Im}\left[\zeta_{2+}A(\zeta_{2+})\frac{\partial\zeta_{2+}}{\partial\tilde{t}}\right]H(\tilde{t}-\tilde{t}_2) \\
&+\frac{e_{15}\tilde{L}_1(Q_0,R_0)+h_{15}\tilde{L}_2(Q_0,R_0)}{\bar{c}_{44}\sqrt{V_C}} \\
&\left.\times\operatorname{Im}\left[\frac{\zeta_{2+}}{(\zeta_{2+}-1/V_C)}\frac{1}{\sqrt{\varepsilon-\zeta_{2+}}}\frac{\partial\zeta_{2+}}{\partial\tilde{t}}\right]H(\tilde{t}-\tilde{t}_2)\right\},
\end{aligned}
\tag{7.139}
$$

$$
\begin{aligned}
{}_t\sigma_{23}(\tilde{X}_1,\tilde{X}_2,\tilde{t}) &= -\frac{\bar{c}_{44}}{\pi}\left\{\operatorname{Im}\left[\alpha(\zeta_{1+})A(\zeta_{1+})\frac{\partial\zeta_{1+}}{\partial\tilde{t}}\right]H(\tilde{t}-\tilde{t}_1)\right. \\
&-(k_{em}^{\lambda})^2\operatorname{Im}\left[\beta(\zeta_{2+})A(\zeta_{2+})\frac{\partial\zeta_{2+}}{\partial\tilde{t}}\right]H(\tilde{t}-\tilde{t}_2) \\
&+\frac{e_{15}\tilde{L}_1(Q_0,R_0)+h_{15}\tilde{L}_2(Q_0,R_0)}{\bar{c}_{44}\sqrt{V_C}} \\
&\left.\times\operatorname{Im}\left[\frac{\sqrt{\varepsilon+\zeta_{2+}}}{(\zeta_{2+}-1/V_C)}\frac{\partial\zeta_{2+}}{\partial\tilde{t}}\right]H(\tilde{t}-\tilde{t}_2)\right\},
\end{aligned}
\tag{7.140}
$$

$$
\begin{aligned}
D_1^{(s)}(\tilde{X}_1,\tilde{X}_2,\tilde{t}) &= \frac{1}{\pi}\left\{(\bar{c}_1\kappa_{11}+\bar{c}_2 g_{11})\operatorname{Im}\left[\zeta_{2+}A(\zeta_{2+})\frac{\partial\zeta_{2+}}{\partial\tilde{t}}\right]\right. \\
&\times H(\tilde{t}-\tilde{t}_2)-\frac{\kappa_{11}\tilde{L}_1(Q_0,R_0)+g_{11}\tilde{L}_2(Q_0,R_0)}{\sqrt{V_C}} \\
&\left.\times\operatorname{Im}\left[\frac{\zeta_{2+}}{(\zeta_{2+}-1/V_C)}\frac{1}{\sqrt{\varepsilon-\zeta_{2+}}}\frac{\partial\zeta_{2+}}{\partial\tilde{t}}\right]H(\tilde{t}-\tilde{t}_2)\right\},
\end{aligned}
\tag{7.141}
$$

$$D_2^{(s)}(\tilde{X}_1,\tilde{X}_2,\tilde{t}) = -\frac{1}{\pi}\left\{(\overline{c}_1\kappa_{11}+\overline{c}_2 g_{11})\,\mathrm{Im}\left[\beta(\zeta_{2+})A(\zeta_{2+})\frac{\partial\zeta_{2+}}{\partial\tilde{t}}\right]\right.$$
$$\times H(\tilde{t}-\tilde{t}_2) - \frac{\kappa_{11}\tilde{L}_1(Q_0,R_0)+g_{11}\tilde{L}_2(Q_0,R_0)}{\sqrt{V_C}} \tag{7.142}$$
$$\left.\times\mathrm{Im}\left[\frac{\sqrt{\varepsilon+\zeta_{2+}}}{(\zeta_{2+}-1/V_C)}\frac{\partial\zeta_{2+}}{\partial\tilde{t}}\right]H(\tilde{t}-\tilde{t}_2)\right\},$$

$$B_1^{(s)}(\tilde{X}_1,\tilde{X}_2,\tilde{t}) = \frac{1}{\pi}\left\{(\overline{c}_1 g_{11}+\overline{c}_2\mu_{11})\,\mathrm{Im}\left[\zeta_{2+}A(\zeta_{2+})\frac{\partial\zeta_{2+}}{\partial\tilde{t}}\right]\right.$$
$$\times H(\tilde{t}-\tilde{t}_2) - \frac{g_{11}\tilde{L}_1(Q_0,R_0)+\mu_{11}\tilde{L}_2(Q_0,R_0)}{\sqrt{V_C}} \tag{7.143}$$
$$\left.\times\mathrm{Im}\left[\frac{\zeta_{2+}}{(\zeta_{2+}-1/V_C)}\frac{1}{\sqrt{\varepsilon-\zeta_{2+}}}\frac{\partial\zeta_{2+}}{\partial\tilde{t}}\right]H(\tilde{t}-\tilde{t}_2)\right\},$$

$$B_2^{(s)}(\tilde{X}_1,\tilde{X}_2,\tilde{t}) = -\frac{1}{\pi}\left\{(\overline{c}_1 g_{11}+\overline{c}_2\mu_{11})\,\mathrm{Im}\left[\beta(\zeta_{2+})A(\zeta_{2+})\frac{\partial\zeta_{2+}}{\partial\tilde{t}}\right]\right.$$
$$\times H(\tilde{t}-\tilde{t}_2) - \frac{g_{11}\tilde{L}_1(Q_0,R_0)+\mu_{11}\tilde{L}_2(Q_0,R_0)}{\sqrt{V_C}} \tag{7.144}$$
$$\left.\times\mathrm{Im}\left[\frac{\sqrt{\varepsilon+\zeta_{2+}}}{(\zeta_{2+}-1/V_C)}\frac{\partial\zeta_{2+}}{\partial\tilde{t}}\right]H(\tilde{t}-\tilde{t}_2)\right\}.$$

The results may be generalized to mixed load distributions following the procedure outlined by Freund (1990) for elastodynamic crack growth. Let $f_P(X_1-X_1^{'},X_2,t-X_1^{'}/V_C)$, $f_Q(X_1-X_1^{'},X_2,t-X_1^{'}/V_C)$, and $f_R(X_1-X_1^{'},X_2,t-X_1^{'}/V_C)$ denote, respectively, a field quantity in the fundamental solutions for unit concentrated shear loads, unit concentrated electric loads, and unit concentrated magnetic loads appearing on the crack faces at $X_1=X_1^{'}$ as the crack tip passes the point $X_1^{'}$ at time $t=X_1^{'}/V_C$. The field quantity for the case of distributed shear loading $p(X_1^{'})$, distributed electric loading $q(X_1^{'})$, and distributed magnetic loading $r(X_1^{'})$ appearing through the crack tip is thus given by the superposition integral

$$f(X_1,X_2,t)=\int_0^{V_Ct} f_P(X_1-X_1^{'},X_2,t-X_1^{'}/V_C)p(X_1^{'})dX_1^{'} \\ +\int_0^{V_Ct} f_Q(X_1-X_1^{'},X_2,t-X_1^{'}/V_C)q(X_1^{'})dX_1^{'} \\ +\int_0^{V_Ct} f_R(X_1-X_1^{'},X_2,t-X_1^{'}/V_C)r(X_1^{'})dX_1^{'}. \tag{7.145}$$

Consequently, the near-tip fields for mixed load distributions are expressed as

$$\begin{aligned} {}_t\sigma_{23}(\tilde{X}_1,0^+,\tilde{t}) &= \frac{H(\tilde{X}_1)}{\pi\sqrt{\tilde{X}_1}} \\ &\times\left\{\frac{(1-V_C/c_{bg}^{\lambda})\mathcal{D}_+(1/V_C)}{\sqrt{1-V_C/c_T}}\tilde{L}(1,0,0)\int_0^{V_C\tilde{t}}\frac{p(X_1^{'})}{\sqrt{V_C\tilde{t}-X_1^{'}}}dX_1^{'}\right. \\ &+\left[\frac{(1-V_C/c_{bg}^{\lambda})\mathcal{D}_+(1/V_C)}{\sqrt{1-V_C/c_T}}-1\right]\tilde{L}(0,1,0)\int_0^{V_C\tilde{t}}\frac{q(X_1^{'})}{\sqrt{V_C\tilde{t}-X_1^{'}}}dX_1^{'} \\ &\left.+\left[\frac{(1-V_C/c_{bg}^{\lambda})\mathcal{D}_+(1/V_C)}{\sqrt{1-V_C/c_T}}-1\right]\tilde{L}(0,0,1)\int_0^{V_C\tilde{t}}\frac{r(X_1^{'})}{\sqrt{V_C\tilde{t}-X_1^{'}}}dX_1^{'}\right\}, \end{aligned} \tag{7.146}$$

$$\begin{aligned} D_2(\tilde{X}_1,0^+,\tilde{t}) &= \frac{H(\tilde{X}_1)}{\pi\sqrt{\tilde{X}_1}}\times\left\{\frac{(\bar{c}_1\kappa_{11}+\bar{c}_2 g_{11})}{\bar{c}_{44}}\frac{1}{[1-(k_{em}^{\lambda})^4]}\right. \\ &\times\frac{[s+(k_{em}^{\lambda})^2]}{(1+V_C/c_{bg}^{\lambda})}\frac{\mathcal{D}_+(1/V_C)}{\sqrt{1-V_C/c_T}}\tilde{L}(1,0,0)\int_0^{V_C\tilde{t}}\frac{p(X_1^{'})}{\sqrt{V_C\tilde{t}-X_1^{'}}}dX_1^{'} \\ &+\frac{(\bar{c}_1\kappa_{11}+\bar{c}_2 g_{11})}{\bar{c}_{44}}\frac{1}{[1-(k_{em}^{\lambda})^4]}\frac{[s+(k_{em}^{\lambda})^2]}{(1+V_C/c_{bg}^{\lambda})}\frac{\mathcal{D}_+(1/V_C)}{\sqrt{1-V_C/c_T}} \\ &\times\tilde{L}(0,1,0)\int_0^{V_C\tilde{t}}\frac{q(X_1^{'})}{\sqrt{V_C\tilde{t}-X_1^{'}}}dX_1^{'}+[\kappa_{11}\tilde{L}_1(1,0)+g_{11}\tilde{L}_2(1,0)] \end{aligned}$$

$$\times\int_0^{V_C\tilde{t}} \frac{q(X_1')}{\sqrt{V_C\tilde{t}-X_1'}}dX_1' + \frac{(\bar{c}_1\kappa_{11}+\bar{c}_2 g_{11})}{\bar{c}_{44}} \frac{1}{[1-(k_{em}^\lambda)^4]}$$

$$\times\frac{[s+(k_{em}^\lambda)^2]}{(1+V_C/c_{bg}^\lambda)} \frac{\mathcal{D}_+(1/V_C)}{\sqrt{1-V_C/c_T}} \tilde{L}(0,0,1)\int_0^{V_C\tilde{t}} \frac{r(X_1')}{\sqrt{V_C\tilde{t}-X_1'}}dX_1' \quad (7.147)$$

$$\left.+[\kappa_{11}\tilde{L}_1(0,1)+g_{11}\tilde{L}_2(0,1)]\int_0^{V_C\tilde{t}} \frac{r(X_1')}{\sqrt{V_C\tilde{t}-X_1'}}dX_1'\right\},$$

$$B_2(\tilde{X}_1,0^+,\tilde{t}) = \frac{H(\tilde{X}_1)}{\pi\sqrt{\tilde{X}_1}}\left\{\frac{(\bar{c}_1 g_{11}+\bar{c}_2\mu_{11})}{\bar{c}_{44}} \frac{1}{[1-(k_{em}^\lambda)^4]}\right.$$

$$\times\frac{[s+(k_{em}^\lambda)^2]}{(1+V_C/c_{bg}^\lambda)} \frac{\mathcal{D}_+(1/V_C)}{\sqrt{1-V_C/c_T}} \tilde{L}(1,0,0)\int_0^{V_C\tilde{t}} \frac{p(X_1')}{\sqrt{V_C\tilde{t}-X_1'}}dX_1'$$

$$+\frac{(\bar{c}_1 g_{11}+\bar{c}_2\mu_{11})}{\bar{c}_{44}} \frac{1}{[1-(k_{em}^\lambda)^4]} \frac{[s+(k_{em}^\lambda)^2]}{(1+V_C/c_{bg}^\lambda)} \frac{\mathcal{D}_+(1/V_C)}{\sqrt{1-V_C/c_T}}$$

$$\times\tilde{L}(0,1,0)\int_0^{V_C\tilde{t}} \frac{q(X_1')}{\sqrt{V_C\tilde{t}-X_1'}}dX_1' + [g_{11}\tilde{L}_1(1,0)+\mu_{11}\tilde{L}_2(1,0)]$$

$$\times\int_0^{V_C\tilde{t}} \frac{q(X_1')}{\sqrt{V_C\tilde{t}-X_1'}}dX_1' + \frac{(\bar{c}_1 g_{11}+\bar{c}_2\mu_{11})}{\bar{c}_{44}} \frac{1}{[1-(k_{em}^\lambda)^4]}$$

$$\times\frac{[s+(k_{em}^\lambda)^2]}{(1+V_C/c_{bg}^\lambda)} \frac{\mathcal{D}_+(1/V_C)}{\sqrt{1-V_C/c_T}} \tilde{L}(0,0,1)\int_0^{V_C\tilde{t}} \frac{r(X_1')}{\sqrt{V_C\tilde{t}-X_1'}}dX_1'$$

$$\left.+[g_{11}\tilde{L}_1(0,1)+\mu_{11}\tilde{L}_2(0,1)]\int_0^{V_C\tilde{t}} \frac{r(X_1')}{\sqrt{V_C\tilde{t}-X_1'}}dX_1'\right\}. \quad (7.148)$$

It can be seen that the near-tip fields still exhibit the inverse square-root singularity in the local coordinate affixed to the moving crack tip. The self-induced and cross-over dynamic total stress, electric displacement, and magnetic induction intensity factors can be expressed in the form of a universal function of the crack tip velocity times the corresponding quasi-static value, that is,

$$\tilde{K}_{III}^{(T)}(V_C\tilde{t},V_C)=k_{III}^{(T)}(V_C)\tilde{K}_{III}^{(T)}(V_C\tilde{t},0)$$
$$=\sqrt{\frac{2}{\pi}}\tilde{L}(1,0,0)\frac{(1-V_C/c_{bg}^{\lambda})}{\sqrt{1-V_C/c_T}}\mathscr{D}_+(1/V_C)\int_0^{V_C\tilde{t}}\frac{p(X_1^{'})}{\sqrt{V_C\tilde{t}-X_1^{'}}}dX_1^{'}, \tag{7.149}$$

$$\tilde{K}_{III}^{(D)}(V_C\tilde{t},V_C)=k_{III}^{(D)}(V_C)\tilde{K}_{III}^{(D)}(V_C\tilde{t},0)$$
$$=\sqrt{\frac{2}{\pi}}\tilde{L}(0,1,0)\left[\frac{(1-V_C/c_{bg}^{\lambda})}{\sqrt{1-V_C/c_T}}\mathscr{D}_+(1/V_C)-1\right]\int_0^{V_C\tilde{t}}\frac{q(X_1^{'})}{\sqrt{V_C\tilde{t}-X_1^{'}}}dX_1^{'}, \tag{7.150}$$

$$\tilde{K}_{III}^{(B)}(V_C\tilde{t},V_C)=k_{III}^{(B)}(V_C)\tilde{K}_{III}^{(B)}(V_C\tilde{t},0)$$
$$=\sqrt{\frac{2}{\pi}}\tilde{L}(0,0,1)\left[\frac{(1-V_C/c_{bg}^{\lambda})}{\sqrt{1-V_C/c_T}}\mathscr{D}_+(1/V_C)-1\right]\int_0^{V_C\tilde{t}}\frac{r(X_1^{'})}{\sqrt{V_C\tilde{t}-X_1^{'}}}dX_1^{'}, \tag{7.151}$$

$$\tilde{K}_{D}^{(T)}(V_C\tilde{t},V_C)=k_{D}^{(T)}(V_C)\tilde{K}_{D}^{(T)}(V_C\tilde{t},0)$$
$$=\sqrt{\frac{2}{\pi}}\tilde{L}(1,0,0)\frac{(\bar{c}_1\kappa_{11}+\bar{c}_2g_{11})}{\bar{c}_{44}}\frac{1}{[1-(k_{em}^{\lambda})^4]}\frac{[s+(k_{em}^{\lambda})^2]}{(1+V_C/c_{bg}^{\lambda})} \tag{7.152}$$
$$\times\frac{\mathscr{D}_+(1/V_C)}{\sqrt{1-V_C/c_T}}\int_0^{V_C\tilde{t}}\frac{p(X_1^{'})}{\sqrt{V_C\tilde{t}-X_1^{'}}}dX_1^{'},$$

$$\tilde{K}_{D}^{(D)}(V_C\tilde{t},V_C)=k_{D}^{(D)}(V_C)\tilde{K}_{D}^{(D)}(V_C\tilde{t},0)$$
$$=\sqrt{\frac{2}{\pi}}\tilde{L}(0,1,0)\frac{(\bar{c}_1\kappa_{11}+\bar{c}_2g_{11})}{\bar{c}_{44}}\frac{1}{[1-(k_{em}^{\lambda})^4]}\frac{[s+(k_{em}^{\lambda})^2]}{(1+V_C/c_{bg}^{\lambda})}$$
$$\times\frac{\mathscr{D}_+(1/V_C)}{\sqrt{1-V_C/c_T}}\int_0^{V_C\tilde{t}}\frac{q(X_1^{'})}{\sqrt{V_C\tilde{t}-X_1^{'}}}dX_1^{'} \tag{7.153}$$
$$+\sqrt{\frac{2}{\pi}}[\kappa_{11}\tilde{L}_1(1,0)+g_{11}\tilde{L}_2(1,0)]\int_0^{V_C\tilde{t}}\frac{q(X_1^{'})}{\sqrt{V_C\tilde{t}-X_1^{'}}}dX_1^{'},$$

$$\tilde{K}_D^{(B)}(V_C\tilde{t},V_C) = k_D^{(B)}(V_C)\tilde{K}_D^{(B)}(V_C\tilde{t},0)$$

$$= \sqrt{\frac{2}{\pi}}\tilde{L}(0,0,1)\frac{(\bar{c}_1\kappa_{11}+\bar{c}_2 g_{11})}{\bar{c}_{44}}\frac{1}{[1-(k_{em}^{\lambda})^4]}\frac{[s+(k_{em}^{\lambda})^2]}{(1+V_C/c_{bg}^{\lambda})}$$

$$\times\frac{\mathcal{D}_+(1/V_C)}{\sqrt{1-V_C/c_T}}\int_0^{V_C\tilde{t}}\frac{r(X_1^{'})}{\sqrt{V_C\tilde{t}-X_1^{'}}}dX_1^{'} \tag{7.154}$$

$$+\sqrt{\frac{2}{\pi}}[\kappa_{11}\tilde{L}_1(0,1)+g_{11}\tilde{L}_2(0,1)]\int_0^{V_C\tilde{t}}\frac{r(X_1^{'})}{\sqrt{V_C\tilde{t}-X_1^{'}}}dX_1^{'},$$

$$\tilde{K}_B^{(T)}(V_C\tilde{t},V_C) = k_B^{(T)}(V_C)\tilde{K}_B^{(T)}(V_C\tilde{t},0)$$

$$= \sqrt{\frac{2}{\pi}}\tilde{L}(1,0,0)\frac{(\bar{c}_1 g_{11}+\bar{c}_2\mu_{11})}{\bar{c}_{44}}\frac{1}{[1-(k_{em}^{\lambda})^4]}\frac{[s+(k_{em}^{\lambda})^2]}{(1+V_C/c_{bg}^{\lambda})} \tag{7.155}$$

$$\times\frac{\mathcal{D}_+(1/V_C)}{\sqrt{1-V_C/c_T}}\int_0^{V_C\tilde{t}}\frac{p(X_1^{'})}{\sqrt{V_C\tilde{t}-X_1^{'}}}dX_1^{'},$$

$$\tilde{K}_B^{(D)}(V_C\tilde{t},V_C) = k_B^{(D)}(V_C)\tilde{K}_B^{(D)}(V_C\tilde{t},0)$$

$$= \sqrt{\frac{2}{\pi}}\tilde{L}(0,1,0)\frac{(\bar{c}_1 g_{11}+\bar{c}_2\mu_{11})}{\bar{c}_{44}}\frac{1}{[1-(k_{em}^{\lambda})^4]}\frac{[s+(k_{em}^{\lambda})^2]}{(1+V_C/c_{bg}^{\lambda})}$$

$$\times\frac{\mathcal{D}_+(1/V_C)}{\sqrt{1-V_C/c_T}}\int_0^{V_C\tilde{t}}\frac{q(X_1^{'})}{\sqrt{V_C\tilde{t}-X_1^{'}}}dX_1^{'} \tag{7.156}$$

$$+\sqrt{\frac{2}{\pi}}[g_{11}\tilde{L}_1(1,0)+\mu_{11}\tilde{L}_2(1,0)]\int_0^{V_C\tilde{t}}\frac{q(X_1^{'})}{\sqrt{V_C\tilde{t}-X_1^{'}}}dX_1^{'},$$

$$\tilde{K}_B^{(B)}(V_C\tilde{t},V_C)=k_B^{(B)}(V_C)\tilde{K}_B^{(B)}(V_C\tilde{t},0)$$
$$=\sqrt{\frac{2}{\pi}}\tilde{L}(0,0,1)\frac{(\bar{c}_1g_{11}+\bar{c}_2\mu_{11})}{\bar{c}_{44}}\frac{1}{[1-(k_{em}^{\lambda})^4]}\frac{[s+(k_{em}^{\lambda})^2]}{(1+V_C/c_{bg}^{\lambda})}$$
$$\times\frac{\mathscr{D}_+(1/V_C)}{\sqrt{1-V_C/c_T}}\int_0^{V_C\tilde{t}}\frac{r(X_1^{'})}{\sqrt{V_C\tilde{t}-X_1^{'}}}dX_1^{'} \tag{7.157}$$
$$+\sqrt{\frac{2}{\pi}}[g_{11}\tilde{L}_1(0,1)+\mu_{11}\tilde{L}_2(0,1)]\int_0^{V_C\tilde{t}}\frac{r(X_1^{'})}{\sqrt{V_C\tilde{t}-X_1^{'}}}dX_1^{'},$$

where the superscript (*T*) stands for traction loading, superscript (*D*) for electric loading, and superscript (*B*) for magnetic loading.

In general, the dynamic total stress intensity factor does not tend to zero as $V_C \to c_{bg}^{\lambda}$ under mixed loading due to the existence of the cross-over terms. The self-induced and cross-over dynamic total stress, electric displacement, and magnetic induction intensity factors are reduced to those for the electrically and magnetically permeable crack-face condition as $\lambda_e \to 0$ and $\lambda_m \to 0$, the electrically and magnetically impermeable crack-face condition as $\lambda_e \to \infty$ and $\lambda_m \to \infty$, the electrically permeable and magnetically impermeable crack-face condition as $\lambda_e \to 0$ and $\lambda_m \to \infty$, and the electrically impermeable and magnetically permeable crack-face condition as $\lambda_e \to \infty$ and $\lambda_m \to 0$. In particular, it emerges that the dynamic field intensity factors are not altered by electric displacement and magnetic induction loads on the surfaces of an electromagnetically permeable crack because there is no gap assumed between the top and bottom surfaces of the crack and the electric displacement and magnetic induction loads on the upper surface effectively cancel out those on the lower surface. This outcome is analogous to the finding by Haug and McMeeking (2006) on a permeable crack with surface charge in poled ferroelectrics. As the crack propagation velocity approaches zero, the quasi-static limits of (7.149)–(7.157) are consistent with the existing static crack solutions (e.g., Liu *et al.*, 2001; Gao *et al.*, 2004; Wang and Mai, 2003, 2007a) with the replacement of the Cauchy stress tensor with the total stress tensor. In particular, the cross-over terms due to electric and magnetic loadings in (7.150) and (7.151) become negligible for quasi-static crack propagation.

7.7 Evaluation of Dynamic Energy Release Rate

The dynamic energy release rate, which is defined as the rate of energy flow out of the body and into the crack front per unit crack advance, can be evaluated by the definition (6.25) given in Section 6.2.2. By choosing the contour shown in Fig. 7.2 and allowing the contour to shrink onto the crack tip by first letting $\delta_2 \to 0$ and then $\delta_1 \to 0$, there is no contribution to $\tilde{J}_0$ from the segments parallel to the $\tilde{X}_2$-axis and the segments along the crack faces. Furthermore, this is a convenient choice because $n_1 = 0$ along the segments parallel to the $\tilde{X}_1$-axis. Consequently, the dynamic energy release rate for mode-III crack propagation is calculated from the near-tip field solutions as

$$\begin{aligned}\tilde{J}_0 &= 2\lim_{\delta_1 \to 0}\left\{\lim_{\delta_2 \to 0}\frac{1}{V_C}\int_{-\delta_1}^{\delta_1} t\sigma_{23}(\tilde{X}_1,\delta_2,\tilde{t})\dot{w}(\tilde{X}_1,\delta_2,\tilde{t})d\tilde{X}_1\right\} \\ &= \frac{1}{4}\tilde{K}_{III}\tilde{K}_{III}^{COD},\end{aligned} \tag{7.158}$$

where the mode-III dynamic crack opening displacement intensity factor is defined as

$$\tilde{K}_{III}^{COD} = \lim_{\tilde{X}_1 \to 0^-}\sqrt{\frac{\pi}{-2\tilde{X}_1}}\Delta w(\tilde{X}_1,0^{\pm},\tilde{t}). \tag{7.159}$$

Based on the near-tip field solutions, the mode-III dynamic total stress and crack opening displacement intensity factors are given by

$$\begin{aligned}\tilde{K}_{III}(V_C\tilde{t},V_C) &= \sqrt{\frac{2}{\pi}}\tilde{L}(1,0,0)\frac{(1-V_C/c_{bg}^{\lambda})}{\sqrt{1-V_C/c_T}}\mathscr{D}_+(1/V_C)\int_0^{V_C\tilde{t}}\frac{p(X_1')}{\sqrt{V_C\tilde{t}-X_1'}}dX_1' \\ &+\sqrt{\frac{2}{\pi}}\tilde{L}(0,1,0)\left[\frac{(1-V_C/c_{bg}^{\lambda})}{\sqrt{1-V_C/c_T}}\mathscr{D}_+(1/V_C)-1\right]\int_0^{V_C\tilde{t}}\frac{q(X_1')}{\sqrt{V_C\tilde{t}-X_1'}}dX_1' \\ &+\sqrt{\frac{2}{\pi}}\tilde{L}(0,0,1)\left[\frac{(1-V_C/c_{bg}^{\lambda})}{\sqrt{1-V_C/c_T}}\mathscr{D}_+(1/V_C)-1\right]\int_0^{V_C\tilde{t}}\frac{r(X_1')}{\sqrt{V_C\tilde{t}-X_1'}}dX_1',\end{aligned} \tag{7.160}$$

$$\tilde{K}_{III}^{COD}(V_C\tilde{t},V_C) = 2\sqrt{\frac{2}{\pi}}\frac{\tilde{L}(1,0,0)}{\bar{c}_{44}[1-(k_{em}^{\lambda})^4]}\frac{[s+(k_{em}^{\lambda})^2]}{(1+V_C/c_{bg}^{\lambda})}\frac{\mathscr{D}_+(1/V_C)}{\sqrt{1-V_C/c_T}}$$

$$\times\int_0^{V_C\tilde{t}}\frac{p(X_1')}{\sqrt{V_C\tilde{t}-X_1'}}dX_1' + 2\sqrt{\frac{2}{\pi}}\frac{\tilde{L}(0,1,0)}{\bar{c}_{44}[1-(k_{em}^{\lambda})^4]}\frac{[s+(k_{em}^{\lambda})^2]}{(1+V_C/c_{bg}^{\lambda})}$$

$$\times\frac{\mathscr{D}_+(1/V_C)}{\sqrt{1-V_C/c_T}}\int_0^{V_C\tilde{t}}\frac{q(X_1')}{\sqrt{V_C\tilde{t}-X_1'}}dX_1' + 2\sqrt{\frac{2}{\pi}}\frac{\tilde{L}(0,0,1)}{\bar{c}_{44}[1-(k_{em}^{\lambda})^4]}$$

$$\times\frac{[s+(k_{em}^{\lambda})^2]}{(1+V_C/c_{bg}^{\lambda})}\frac{\mathscr{D}_+(1/V_C)}{\sqrt{1-V_C/c_T}}\int_0^{V_C\tilde{t}}\frac{r(X_1')}{\sqrt{V_C\tilde{t}-X_1'}}dX_1'$$

$$= \frac{2}{\bar{c}_{44}\sqrt{1-V_C^2/c_T^2}}\left[\tilde{K}_{III}(V_C\tilde{t},V_C) + \frac{e_{15}\mu_{11}-h_{15}g_{11}}{\kappa_{11}\mu_{11}-g_{11}^2}\tilde{K}_D(V_C\tilde{t},V_C)\right.$$

$$\left.+\frac{h_{15}\kappa_{11}-e_{15}g_{11}}{\kappa_{11}\mu_{11}-g_{11}^2}\tilde{K}_B(V_C\tilde{t},V_C)\right]. \tag{7.161}$$

From Eqs. (7.158) and (7.161), the dynamic energy release rate for mode-III crack propagation in the presence of magneto-electro-mechanical coupling effects has an odd dependence on the dynamic electric displacement intensity factor and the dynamic magnetic induction intensity factor, that is,

$$\tilde{J}_0(V_C\tilde{t},V_C) = \frac{\tilde{K}_{III}(V_C\tilde{t},V_C)}{2\bar{c}_{44}\sqrt{1-V_C^2/c_T^2}}[\tilde{K}_{III}(V_C\tilde{t},V_C)$$

$$+\frac{e_{15}\mu_{11}-h_{15}g_{11}}{\kappa_{11}\mu_{11}-g_{11}^2}\tilde{K}_D(V_C\tilde{t},V_C) \tag{7.162}$$

$$+\frac{h_{15}\kappa_{11}-e_{15}g_{11}}{\kappa_{11}\mu_{11}-g_{11}^2}\tilde{K}_B(V_C\tilde{t},V_C)].$$

The dynamic energy release rate is reduced to that for the electrically and magnetically permeable crack-face condition as $\lambda_e \to 0$ and

$\lambda_m \to 0$, the electrically and magnetically impermeable crack-face condition as $\lambda_e \to \infty$ and $\lambda_m \to \infty$, the electrically permeable and magnetically impermeable crack-face condition as $\lambda_e \to 0$ and $\lambda_m \to \infty$, and the electrically impermeable and magnetically permeable crack-face condition as $\lambda_e \to \infty$ and $\lambda_m \to 0$.

7.8 Influence of Shear Horizontal Surface Wave Speed and Crack Tip Velocity

As the crack tip velocity V_C tends to zero, the quasi-static case is recovered, that is,

$$\tilde{K}_{III}(V_C\tilde{t},0) = \sqrt{\frac{2}{\pi}}\tilde{L}(1,0,0)\int_0^{V_C\tilde{t}} \frac{p(X_1^{'})}{\sqrt{V_C\tilde{t} - X_1^{'}}} dX_1^{'}, \tag{7.163}$$

$$\begin{aligned} \tilde{K}_{III}^{COD}(V_C\tilde{t},0) &= 2\sqrt{\frac{2}{\pi}} \frac{\tilde{L}(1,0,0)}{\bar{c}_{44}[1-(k_{em}^{\lambda})^2]} \int_0^{V_C\tilde{t}} \frac{p(X_1^{'})}{\sqrt{V_C\tilde{t} - X_1^{'}}} dX_1^{'} \\ &+ 2\sqrt{\frac{2}{\pi}} \frac{\tilde{L}(0,1,0)}{\bar{c}_{44}[1-(k_{em}^{\lambda})^2]} \int_0^{V_C\tilde{t}} \frac{q(X_1^{'})}{\sqrt{V_C\tilde{t} - X_1^{'}}} dX_1^{'} \\ &+ 2\sqrt{\frac{2}{\pi}} \frac{\tilde{L}(0,0,1)}{\bar{c}_{44}[1-(k_{em}^{\lambda})^2]} \int_0^{V_C\tilde{t}} \frac{r(X_1^{'})}{\sqrt{V_C\tilde{t} - X_1^{'}}} dX_1^{'}, \end{aligned} \tag{7.164}$$

$$\tilde{J}_0(V_C\tilde{t},0) = \frac{1}{4}\tilde{K}_{III}(V_C\tilde{t},0)\tilde{K}_{III}^{COD}(V_C\tilde{t},0). \tag{7.165}$$

Next, we will examine the special case of the electromagnetically permeable crack-face condition. As $\lambda_e \to 0$ and $\lambda_m \to 0$, we have $c_{bg}^0 = c_T\sqrt{[1-(k_{em}^0)^4]}$ and $k_{em}^0 = \sqrt{(c_1 e_{15}^0 + c_2 h_{15}^0)/\bar{c}_{44}^0}$. The dynamic total stress intensity factor, dynamic crack opening displacement intensity factor, and dynamic energy release rate normalized by the corresponding quasi-static value become

$$\frac{\tilde{K}_{III}(V_C\tilde{t}, V_C)}{\tilde{K}_{III}(V_C\tilde{t}, 0)} = \frac{(1 - V_C / c_{bg}^0)}{\sqrt{1 - V_C / c_T}} \mathscr{D}_+ (1/V_C) = f_1^0(V_C), \tag{7.166}$$

$$\frac{\tilde{K}_{III}^{COD}(V_C\tilde{t}, V_C)}{\tilde{K}_{III}^{COD}(V_C\tilde{t}, 0)} = \frac{[s + (k_{em}^0)^2]}{[1 + (k_{em}^0)^2](1 + V_C / c_{bg}^0)} \frac{\mathscr{D}_+ (1/V_C)}{\sqrt{1 - V_C / c_T}} = f_2^0(V_C), \tag{7.167}$$

$$\frac{\tilde{J}_0(V_C\tilde{t}, V_C)}{\tilde{J}_0(V_C\tilde{t}, 0)} = \frac{[s + (k_{em}^0)^2](1 - V_C / c_{bg}^0)[\mathscr{D}_+ (1/V_C)]^2}{[1 + (k_{em}^0)^2](1 + V_C / c_{bg}^0)(1 - V_C / c_T)} = f_3^0(V_C). \tag{7.168}$$

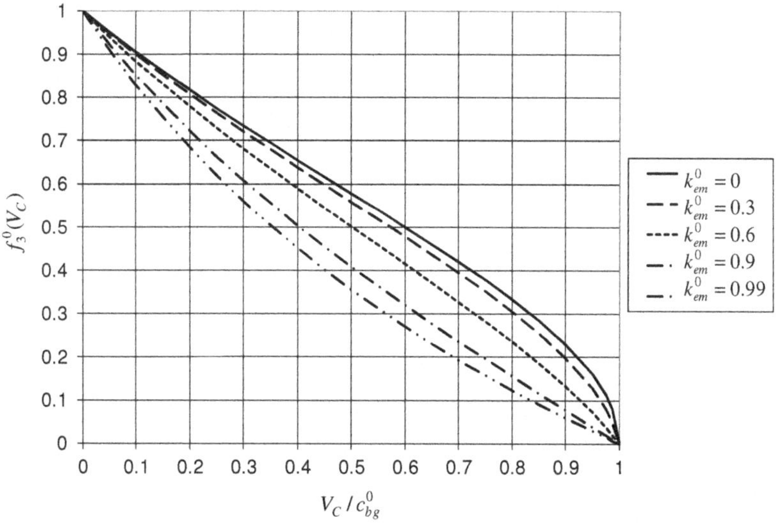

Fig. 7.6 Universal function f_3^0 versus dimensionless crack tip velocity V_C / c_{bg}^0 for a broad range of magneto-electro-mechanical coupling factor. (After Chen, 2009c, with permission from Elsevier).

Like the functions $f_1^{\lambda}(V_C)$ and $f_2^{\lambda}(V_C)$, the function $f_3^0(V_C)$ is a universal function of the crack tip velocity. The dimensionless function $f_3^0(V_C)$ is plotted against the dimensionless crack tip velocity V_C / c_{bg}^0 in Fig. 7.6. Similar to the universal function $f_1^{\lambda}(V_C)$ (see Fig. 7.4), the universal function $f_3^0(V_C)$ has the feature that $f_3^0(V_C)=1$ for $V_C / c_{bg}^0 = 0$ and $f_3^0(V_C)=0$ for $V_C / c_{bg}^0 = 1$. It also decreases monotonically with increasing dimensionless crack tip velocity and increasing magneto-electro-mechanical coupling factor k_{em}^0. Hence, as the crack tip velocity V_C approaches the shear horizontal surface wave speed c_{bg}^0, the dynamic energy release rate tends to zero so that the shear horizontal surface wave speed c_{bg}^0 serves as a speed barrier for the propagation of an electromagnetically permeable mode-III crack.

Chapter 8

Fracture of Functionally Graded Materials

8.1 Introduction

Functionally graded materials (FGMs) are nonhomogeneous materials whose properties vary continuously along one or more directions. Bones and wood may be taken as FGMs in nature. The idea of FGMs originated from high-temperature applications of thermal barrier coatings for aircraft and aerospace industries in the mid-1980s. Due to the gradual spatial variation in properties instead of a sharp jump across interfaces, FGMs have potential advantages in reducing stress concentration and increasing fracture toughness. Eischen (1987a–b) developed a path-domain independent J_k^*-integral for fracture of nonhomogeneous materials. Honein and Herrmann (1997) studied the conservation laws in nonhomogeneous elastostatics by means of a special version of Noether's theorem and proposed a path-independent J_e-integral. The near-tip stress field in a FGM possesses a classical inverse square-root singularity like that in a homogeneous material, so that existing crack-tip finite element modeling codes can be used to analyze structural components made of FGMs (Erdogan, 1995). Suresh and Mortensen (1998) and Jin (2003) provided comprehensive reviews on the fundamentals of FGMs and the progress in fracture mechanics of FGMs. Wang and Mai (2005, 2006) investigated a periodic array of cracks in FGMs subjected to thermomechanical loading and transient loading.

This new concept of tailoring materials can also be extended to piezoelectric/piezomagnetic materials to improve reliability and achieve optimized performance in aerospace, transportation, communication, biomedical, and other applications (see for example Wu *et al.*, 1996; Zhu *et al.*, 2000; Hudnut *et al.*, 2000; Li and Weng, 2002a–b; Takagi *et al.*, 2002, 2003; Kwon, 2004; Chakraborty *et al.*, 2005; Chue and Ou, 2005; Zhou *et al.*, 2005; Feng and Su, 2006, 2007; Ma *et al.*, 2007; Wang and Mai, 2007b; Sladek *et al.*, 2007a–b; Ueda, 2005–2008; Singh *et al.*, 2009; Rao and Kuna, 2008, 2010; Shin and Lee, 2010; Dineva *et al.*, 2010). Successful applications of FGMs rely on a thorough understanding of the fracture behavior of such materials under various aggressive operational conditions.

This chapter is focused on the extension of fracture mechanics methodologies to this emerging class of FGMs subjected to combined magnetic, electric, thermal, and mechanical loadings, covering boundary-initial value problems, typical solution methods, and fracture characterizing parameters. While this subject is far from mature, the formulation presented in this chapter will likely form the basis for further advances.

8.2 Formulation of Boundary-Initial Value Problems

The boundary-initial value problems under combined magnetic, electric, thermal, and mechanical loadings can be mathematically formulated by the basic field equations with appropriate boundary and initial conditions. The fundamental difference between functionally graded materials and homogeneous materials is whether material properties vary spatially or not. Since the quasi-static approximation for the electromagnetic fields may be adopted in many practical engineering applications without loss of solution accuracy, it is employed in this formulation for simplicity.

Consider a FGM occupying the region V in the absence of mechanical body forces, free electric charges, and electricity conduction. The boundary of V is denoted by S. The equations to be satisfied in the region V are listed as follows:

Constitutive relations (linearized theory):

$$ {}_t\sigma_{kl} = c_{klmn}\varepsilon_{mn} - e_{mkl}E_m - h_{mkl}H_m - \beta_{kl}\theta , \tag{8.1} $$

$$D_k = e_{klm}\varepsilon_{lm} + \kappa_{kl}E_l + g_{kl}H_l + \omega_k\theta\,, \tag{8.2}$$

$$B_k = h_{klm}\varepsilon_{lm} + g_{lk}E_l + \mu_{kl}H_l + \mu_0\gamma_k\theta\,, \tag{8.3}$$

$$\rho\hat{s} = \beta_{kl}\varepsilon_{kl} + \omega_k E_k + \mu_0\gamma_k H_k + \frac{1}{T_0}C_v\theta\,, \tag{8.4}$$

$$\mathbf{j}_q = -\mathbf{k}^{qq}\cdot\nabla\theta\,, \tag{8.5}$$

where material properties and mass density are functions of the coordinates X_K (K=1,2,3), i.e.,

$$c_{klnm} = c_{lkmn} = c_{mnkl} = c_{klmn} = c_{klmn}(X_1, X_2, X_3)\,, \tag{8.6}$$

$$e_{mlk} = e_{mkl} = e_{mkl}(X_1, X_2, X_3)\,, \tag{8.7}$$

$$h_{mlk} = h_{mkl} = h_{mkl}(X_1, X_2, X_3)\,, \tag{8.8}$$

$$\beta_{lk} = \beta_{kl} = \beta_{kl}(X_1, X_2, X_3)\,, \tag{8.9}$$

$$\kappa_{lk} = \kappa_{kl} = \kappa_{kl}(X_1, X_2, X_3)\,, \tag{8.10}$$

$$\mu_{lk} = \mu_{kl} = \mu_{kl}(X_1, X_2, X_3)\,, \tag{8.11}$$

$$g_{kl} = g_{kl}(X_1, X_2, X_3)\,, \tag{8.12}$$

$$\omega_k = \omega_k(X_1, X_2, X_3)\,, \tag{8.13}$$

$$\gamma_k = \gamma_k(X_1, X_2, X_3)\,, \tag{8.14}$$

$$C_v = C_v(X_1, X_2, X_3)\,, \tag{8.15}$$

$$\mathbf{k}^{qq} = \mathbf{k}^{qq}(X_1, X_2, X_3)\,, \tag{8.16}$$

$$\rho = \rho(X_1, X_2, X_3)\,. \tag{8.17}$$

Kinematic relations:

$$\varepsilon_{mn} = (u_{m,n} + u_{n,m})/2\,, \tag{8.18}$$

$$E_m = -\phi_{,m}\,, \tag{8.19}$$

$$H_m = -\psi_{,m}\,. \tag{8.20}$$

Balance equations:

$$\nabla \cdot \mathbf{D} = 0\,, \tag{8.21}$$

$$\nabla \cdot \mathbf{B} = 0\,, \tag{8.22}$$

$$\frac{d\rho}{dt} = -\rho \nabla \cdot \mathbf{v}\,, \tag{8.23}$$

$$\rho \frac{d\mathbf{v}}{dt} = \nabla \cdot_t \boldsymbol{\sigma}\,, \tag{8.24}$$

$$\rho T \frac{d\hat{s}}{dt} = -\nabla \cdot \mathbf{j}_q\,. \tag{8.25}$$

The boundary conditions are

$$\mathbf{n} \cdot [[\mathbf{D}]] = 0 \text{ across } S\,, \tag{8.26}$$

$$\mathbf{n} \times [[\mathbf{E}]] = 0 \text{ across } S\,, \tag{8.27}$$

$$\mathbf{n} \cdot [[\mathbf{B}]] = 0 \text{ across } S\,, \tag{8.28}$$

$$\mathbf{n} \times [[\mathbf{H}]] = 0 \text{ across } S\,, \tag{8.29}$$

$$\mathbf{n} \cdot_t \boldsymbol{\sigma} = \mathbf{t}_B \text{ on } S_\sigma\,, \tag{8.30}$$

$$\mathbf{u} = \mathbf{u}_B \text{ on } S_u\,, \tag{8.31}$$

$$\mathbf{n} \cdot \mathbf{j}_q = q_B \text{ on } S_q\,, \tag{8.32}$$

$$T = T_B \text{ on } S_T\,, \tag{8.33}$$

where $\mathbf{n}$ is the unit outer normal vector of the boundary S and $S = S_\sigma \cup S_u = S_q \cup S_T$. Other mixed boundary conditions may also be employed.

The initial conditions are

$$\mathbf{u}\big|_{t=t_0} = \mathbf{u}_0 , \tag{8.34}$$

$$\dot{\mathbf{u}}\big|_{t=t_0} = \mathbf{v}_0 , \tag{8.35}$$

$$T\big|_{t=t_0} = T_0 . \tag{8.36}$$

8.3 Basic Solution Techniques

Due to the difficulties and complexities of the spatial dependence of graded properties plus multifield coupling effects involved in this class of boundary-initial value problems, numerical methods such as finite element method (FEM), boundary element method (BEM), or meshless local Petrov–Galerkin method (MLPG) are often used. Analytical or semi-analytical solutions may be obtained only for some limited variations of graded properties such as exponential or power-law functions of spatial coordinates. For example, Li and Weng (2002a) were among the first to study a stationary crack problem in a strip of functionally graded piezoelectric material (FGPM) subjected to antiplane mechanical and in-plane electric loadings with variations of the material properties one-dimensionally perpendicular to the crack plane, by using the Fourier transform to reduce the problem to two pairs of dual integral equations and then into Fredholm integral equations of the second kind. Their results showed that the near-tip stress and electric displacement fields in a FGPM exhibit the same inverse square-root singularity as those in a homogeneous piezoelectric material, but the magnitudes of the field intensity factors depend significantly on the gradient of the graded properties. Li and Weng (2002b) further investigated the Yoffe-type moving crack problem in a strip of FGPM subjected to antiplane mechanical loading and in-plane electric loading using the Galilean transformation and the Fourier transform. They found that the increase in the gradient of the material properties can reduce the magnitudes of the stress and electric displacement intensity factors, which has the same effect as the electromechanical coupling factor. Zhou *et al.* (2005) studied the behavior of a crack in functionally graded piezoelectric/piezomagnetic materials subjected to an antiplane shear loading with the variations of the material properties one-dimensionally parallel to the crack, by using the Fourier transform to reduce the problem to a pair of dual integral equations which are solved by the Schmidt method. Feng and Su (2006, 2007) and Ma *et al.* (2007) considered dynamic and static mode-III embedded or edge-crack problems in a functionally graded magneto-electro-elastic strip/plate with variations of material properties one-dimensionally parallel to the crack, by using integral transforms and dislocation density functions to reduce the problem to a system of singular integral equations. Wang and Mai (2007b) analyzed a mode-III crack problem in functionally graded magneto-electro-elastic materials with the variations of the material properties one-dimensionally perpendicular to the crack plane by using the Fourier transform to reduce the problem by means of the singular integral equation technique.

The integral transform/integral equation method is illustrated below for the Yoffe-type moving crack problem in a transversely isotropic functionally graded magneto-electro-elastic strip subjected to antiplane shear loading and in-plane electric and magnetic loadings (Fig. 8.1). Following the treatment by Yoffe (1951) and Li and Weng (2002b), consider a crack of length $2a$ moving at constant velocity V_C while keeping its length unchanged. A reference Cartesian coordinate system $\{\tilde{X}_K, K=1,2,3\}$ attached to the moving crack tip is chosen, which coincides at time $t=0$ with the fixed Cartesian coordinate system $\{X_K, K=1,2,3\}$. The principal material axes are taken to coincide with the reference axes with the $\tilde{X}_3$-axis in the poling direction, where the $\tilde{X}_3$-axis is parallel to the crack front.

Thus, the Galilean transformation can be introduced as

$$\tilde{X}_1 = X_1 - V_C t, \quad \tilde{X}_2 = X_2, \quad \tilde{X}_3 = X_3, \quad \tilde{t} = t. \qquad (8.37)$$

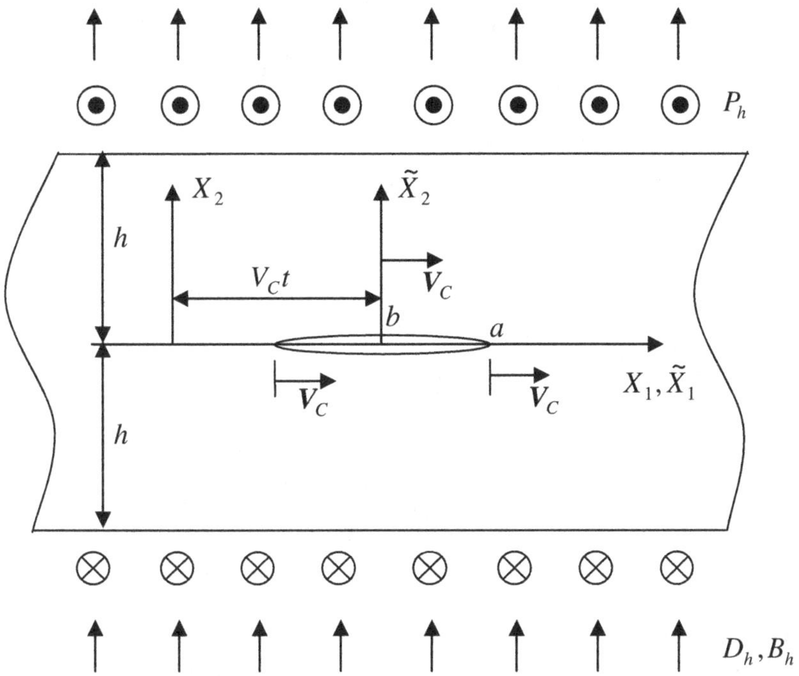

Fig. 8.1 A Yoffe-type mode-III moving crack problem in a functionally graded magneto-electro-elastic strip.

For the Yoffe-type mode-III moving crack problem, only the out-of-plane displacement w, electric potential ϕ, and magnetic potential ψ are non-vanishing, which are independent of $\tilde{X}_3$ and $\tilde{t}$ in the reference frame moving with the crack tip, that is,

$$w = w(\tilde{X}_1, \tilde{X}_2), \; \phi^{(p)} = \phi^{(p)}(\tilde{X}_1, \tilde{X}_2), \; \psi^{(p)} = \psi^{(p)}(\tilde{X}_1, \tilde{X}_2), \quad (8.38)$$

where the superscript $p = s$ stands for the cracked solid region and $p = f$ stands for the interior fluid region filled with vacuum, air, or oil of negligible mechanical influence.

The linearized constitutive equations are given in the Voigt notation by

$$ {}_t\sigma_{i3} = c_{44}\frac{\partial w}{\partial \tilde{X}_i} + e_{15}\frac{\partial \phi^{(s)}}{\partial \tilde{X}_i} + h_{15}\frac{\partial \psi^{(s)}}{\partial \tilde{X}_i} , \tag{8.39} $$

$$ D_i^{(s)} = e_{15}\frac{\partial w}{\partial \tilde{X}_i} - \kappa_{11}\frac{\partial \phi^{(s)}}{\partial \tilde{X}_i} - g_{11}\frac{\partial \psi^{(s)}}{\partial \tilde{X}_i} , \tag{8.40} $$

$$ B_i^{(s)} = h_{15}\frac{\partial w}{\partial \tilde{X}_i} - g_{11}\frac{\partial \phi^{(s)}}{\partial \tilde{X}_i} - \mu_{11}\frac{\partial \psi^{(s)}}{\partial \tilde{X}_i} , \tag{8.41} $$

$$ D_i^{(f)} = -\kappa^f \frac{\partial \phi^{(f)}}{\partial \tilde{X}_i} , \tag{8.42} $$

$$ B_i^{(f)} = -\mu^f \frac{\partial \psi^{(f)}}{\partial \tilde{X}_i} , \tag{8.43} $$

where $i = 1, 2$.

The graded properties are taken to vary continuously along the $\tilde{X}_2$-axis inside the strip in the same proportion with the following distribution:

$$ c_{44}(\tilde{X}_2) = c_{44}^0(1+\alpha|\tilde{X}_2|)^k , \tag{8.44} $$

$$ \kappa_{11}(\tilde{X}_2) = \kappa_{11}^0(1+\alpha|\tilde{X}_2|)^k , \tag{8.45} $$

$$ \mu_{11}(\tilde{X}_2) = \mu_{11}^0(1+\alpha|\tilde{X}_2|)^k , \tag{8.46} $$

$$ e_{15}(\tilde{X}_2) = e_{15}^0(1+\alpha|\tilde{X}_2|)^k , \tag{8.47} $$

$$ h_{15}(\tilde{X}_2) = h_{15}^0(1+\alpha|\tilde{X}_2|)^k , \tag{8.48} $$

$$ g_{11}(\tilde{X}_2) = g_{11}^0(1+\alpha|\tilde{X}_2|)^k , \tag{8.49} $$

$$ \rho(\tilde{X}_2) = \rho^0(1+\alpha|\tilde{X}_2|)^k , \tag{8.50} $$

where k is a constant and the parameter α can be determined by the values of the material properties at the $\tilde{X}_2 = 0$ and $\tilde{X}_2 = \pm h$ planes, i.e.,

$$\begin{aligned}\alpha &= (\sqrt[k]{c_{44}^h / c_{44}^0} - 1)/h = (\sqrt[k]{\kappa_{11}^h / \kappa_{11}^0} - 1)/h = (\sqrt[k]{\mu_{11}^h / \mu_{11}^0} - 1)/h \\ &= (\sqrt[k]{e_{15}^h / e_{15}^0} - 1)/h = (\sqrt[k]{f_{15}^h / f_{15}^0} - 1)/h = (\sqrt[k]{g_{11}^h / g_{11}^0} - 1)/h \\ &= (\sqrt[k]{\rho^h / \rho^0} - 1)/h.\end{aligned} \tag{8.51}$$

Due to the symmetry in geometry and loading, it is sufficient to consider the problem for $\tilde{X}_1 \geq 0, \tilde{X}_2 \geq 0$ only. Since the Yoffe-type moving crack problem is in a steady state, the governing equations with respect to the reference frame moving with the crack tip may be rewritten as

$$s^2 \frac{\partial^2 w}{\partial \tilde{X}_1^2} + \frac{\partial^2 w}{\partial \tilde{X}_2^2} + \frac{k\alpha}{1 + \alpha \tilde{X}_2} \frac{\partial w}{\partial \tilde{X}_2} = 0, \tag{8.52}$$

$$\frac{\partial^2 \bar{\phi}^{(s)}}{\partial \tilde{X}_1^2} + \frac{\partial^2 \bar{\phi}^{(s)}}{\partial \tilde{X}_2^2} + \frac{k\alpha}{1 + \alpha \tilde{X}_2} \frac{\partial \bar{\phi}^{(s)}}{\partial \tilde{X}_2} = 0 , \tag{8.53}$$

$$\frac{\partial^2 \bar{\psi}^{(s)}}{\partial \tilde{X}_1^2} + \frac{\partial^2 \bar{\psi}^{(s)}}{\partial \tilde{X}_2^2} + \frac{k\alpha}{1 + \alpha \tilde{X}_2} \frac{\partial \bar{\psi}^{(s)}}{\partial \tilde{X}_2} = 0 , \tag{8.54}$$

$$\frac{\partial^2 \phi^{(f)}}{\partial \tilde{X}_1^2} + \frac{\partial^2 \phi^{(f)}}{\partial \tilde{X}_2^2} = 0, \tag{8.55}$$

$$\frac{\partial^2 \psi^{(f)}}{\partial \tilde{X}_1^2} + \frac{\partial^2 \psi^{(f)}}{\partial \tilde{X}_2^2} = 0, \tag{8.56}$$

$$\bar{\phi}^{(s)} = \phi^{(s)} - c_1^0 w, \tag{8.57}$$

$$\bar{\psi}^{(s)} = \psi^{(s)} - c_2^0 w, \tag{8.58}$$

where $s^2 = 1 - (V_C / c_T^0)^2$, $c_T^0 = (\bar{c}_{44}^0 / \rho^0)^{1/2}$ is the piezoelectro-magnetically stiffened bulk shear wave speed at the $\tilde{X}_2 = 0$ plane,

$\bar{c}_{44}^0 = c_{44}^0 + [(e_{15}^0)^2 \mu_{11}^0 - 2e_{15}^0 h_{15}^0 g_{11}^0 + (h_{15}^0)^2 \kappa_{11}^0] / [\kappa_{11}^0 \mu_{11}^0 - (g_{11}^0)^2]$ is the piezoelectromagnetically stiffened elastic constant at the $\tilde{X}_2 = 0$ plane, $c_1^0 = (e_{15}^0 \mu_{11}^0 - h_{15}^0 g_{11}^0) / [\kappa_{11}^0 \mu_{11}^0 - (g_{11}^0)^2]$, and

$c_2^0 = (h_{15}^0 \kappa_{11}^0 - e_{15}^0 g_{11}^0) / [\kappa_{11}^0 \mu_{11}^0 - (g_{11}^0)^2]$.

For an elliptical cavity-like crack, the following exact boundary conditions are imposed:

$$ {}_t\sigma_{23}(\tilde{X}_1, h) = P_h, \tag{8.59}$$

$$ D_2^{(s)}(\tilde{X}_1, h) = D_h, \tag{8.60}$$

$$ B_2^{(s)}(\tilde{X}_1, h) = B_h, \tag{8.61}$$

$$ {}_t\sigma_{n3}(\tilde{X}_1, \tilde{X}_2) = 0, \ (\tilde{X}_1 / a)^2 + (\tilde{X}_2 / b)^2 = 1, \tag{8.62}$$

$$ D_n^{(s)}(\tilde{X}_1, \tilde{X}_2) - D_n^{(f)}(\tilde{X}_1, \tilde{X}_2) = 0, \ (\tilde{X}_1 / a)^2 + (\tilde{X}_2 / b)^2 = 1, \tag{8.63}$$

$$ B_n^{(s)}(\tilde{X}_1, \tilde{X}_2) - B_n^{(f)}(\tilde{X}_1, \tilde{X}_2) = 0, \ (\tilde{X}_1 / a)^2 + (\tilde{X}_2 / b)^2 = 1, \tag{8.64}$$

$$ E_t^{(s)}(\tilde{X}_1, \tilde{X}_2) - E_t^{(f)}(\tilde{X}_1, \tilde{X}_2) = 0, \ (\tilde{X}_1 / a)^2 + (\tilde{X}_2 / b)^2 = 1, \tag{8.65}$$

$$ H_t^{(s)}(\tilde{X}_1, \tilde{X}_2) - H_t^{(f)}(\tilde{X}_1, \tilde{X}_2) = 0, \ (\tilde{X}_1 / a)^2 + (\tilde{X}_2 / b)^2 = 1, \tag{8.66}$$

$$ w(\tilde{X}_1, 0) = 0, \ |\tilde{X}_1| \geq a, \tag{8.67}$$

$$ \phi^{(s)}(\tilde{X}_1, 0) = 0, \ |\tilde{X}_1| \geq a, \tag{8.68}$$

$$ \psi^{(s)}(\tilde{X}_1, 0) = 0, \ |\tilde{X}_1| \geq a, \tag{8.69}$$

where the subscript "n" stands for the normal component and the subscript "t" for the tangential component on the crack surface.

With the introduction of the Fourier cosine transform to Eqs. (8.52)–(8.56), the general solutions can be found as

$$w(\tilde{X}_1,\tilde{X}_2)=\frac{2}{\pi}\int_0^\infty(1+\alpha\tilde{X}_2)^{-\beta}\{A_1(\zeta)I_\beta[(1+\alpha\tilde{X}_2)s\zeta/\alpha] \\ +A_2(\zeta)K_\beta[(1+\alpha\tilde{X}_2)s\zeta/\alpha]\}\cos(\zeta\tilde{X}_1)d\zeta+a_0\tilde{X}_2, \tag{8.70}$$

$$\bar{\phi}^{(s)}(\tilde{X}_1,\tilde{X}_2)=\frac{2}{\pi}\int_0^\infty(1+\alpha\tilde{X}_2)^{-\beta}\{B_1(\zeta)I_\beta[(1+\alpha\tilde{X}_2)\zeta/\alpha] \\ +B_2(\zeta)K_\beta[(1+\alpha\tilde{X}_2)\zeta/\alpha]\}\cos(\zeta\tilde{X}_1)d\zeta-b_0\tilde{X}_2, \tag{8.71}$$

$$\bar{\psi}^{(s)}(\tilde{X}_1,\tilde{X}_2)=\frac{2}{\pi}\int_0^\infty(1+\alpha\tilde{X}_2)^{-\beta}\{C_1(\zeta)I_\beta[(1+\alpha\tilde{X}_2)\zeta/\alpha] \\ +C_2(\zeta)K_\beta[(1+\alpha\tilde{X}_2)\zeta/\alpha]\}\cos(\zeta\tilde{X}_1)d\zeta-c_0\tilde{X}_2, \tag{8.72}$$

$$\phi^{(f)}(\tilde{X}_1,\tilde{X}_2)=\frac{2}{\pi}\int_0^\infty D_1(\zeta)\sinh(\zeta\tilde{X}_2)\cos(\zeta\tilde{X}_1)d\zeta, \tag{8.73}$$

$$\psi^{(f)}(\tilde{X}_1,\tilde{X}_2)=\frac{2}{\pi}\int_0^\infty E_1(\zeta)\sinh(\zeta\tilde{X}_2)\cos(\zeta\tilde{X}_1)d\zeta, \tag{8.74}$$

where $\beta=(k-1)/2$, I_β and K_β are the first- and second-kind modified Bessel functions, a_0, b_0, and c_0 are real constants, and $A_1(\zeta)$, $A_2(\zeta)$, $B_1(\zeta)$, $B_2(\zeta)$, $C_1(\zeta)$, $C_2(\zeta)$, $D_1(\zeta)$, and $E_1(\zeta)$ are unknown functions to be determined.

Hence, the expressions for the total stress, electric displacement, and magnetic induction are obtained as

$$\begin{aligned}{}_t\sigma_{13}(\tilde{X}_1,\tilde{X}_2)&=\tilde{c}_{44}(\tilde{X}_2)w_{,1}+e_{15}(\tilde{X}_2)\bar{\phi}_{,1}+h_{15}(\tilde{X}_2)\bar{\psi}_{,1} \\ &=-\frac{2}{\pi}\int_0^\infty\zeta(1+\alpha\tilde{X}_2)^{-\beta}\{\tilde{c}_{44}(\tilde{X}_2)A_1(\zeta)I_\beta[(1+\alpha\tilde{X}_2)s\zeta/\alpha] \\ &\quad+e_{15}(\tilde{X}_2)B_1(\zeta)I_\beta[(1+\alpha\tilde{X}_2)\zeta/\alpha] \\ &\quad+h_{15}(\tilde{X}_2)C_1(\zeta)I_\beta[(1+\alpha\tilde{X}_2)\zeta/\alpha] \\ &\quad+\tilde{c}_{44}(\tilde{X}_2)A_2(\zeta)K_\beta[(1+\alpha\tilde{X}_2)s\zeta/\alpha] \\ &\quad+e_{15}(\tilde{X}_2)B_2(\zeta)K_\beta[(1+\alpha\tilde{X}_2)\zeta/\alpha] \\ &\quad+h_{15}(\tilde{X}_2)C_2(\zeta)K_\beta[(1+\alpha\tilde{X}_2)\zeta/\alpha]\}\sin(\zeta\tilde{X}_1)d\zeta,\end{aligned} \tag{8.75}$$

$$
\begin{aligned}
{}_t\sigma_{23}(\tilde{X}_1,\tilde{X}_2) &= \tilde{c}_{44}(\tilde{X}_2)w_{,2} + e_{15}(\tilde{X}_2)\overline{\varphi}_{,2} + h_{15}(\tilde{X}_2)\overline{\psi}_{,2} \\
&= -\frac{2}{\pi}\int_0^\infty \{[\tilde{c}_{44}(\tilde{X}_2)A_1(\zeta)P_I(\zeta,\tilde{X}_2) \\
&\quad + e_{15}(\tilde{X}_2)B_1(\zeta)Q_I(\zeta,\tilde{X}_2) + h_{15}(\tilde{X}_2)C_1(\zeta)Q_I(\zeta,\tilde{X}_2)] \\
&\quad + [\tilde{c}_{44}(\tilde{X}_2)A_2(\zeta)P_K(\zeta,\tilde{X}_2) + e_{15}(\tilde{X}_2)B_2(\zeta)Q_K(\zeta,\tilde{X}_2) \\
&\quad + h_{15}(\tilde{X}_2)C_2(\zeta)Q_K(\zeta,\tilde{X}_2)]\}\cos(\zeta\tilde{X}_1)d\zeta \\
&\quad + \tilde{c}_{44}(\tilde{X}_2)a_0 - e_{15}(\tilde{X}_2)b_0 - h_{15}(\tilde{X}_2)c_0,
\end{aligned} \tag{8.76}
$$

$$
\begin{aligned}
D_1(\tilde{X}_1,\tilde{X}_2) &= \tilde{e}_{15}(\tilde{X}_2)w_{,1} - \kappa_{11}(\tilde{X}_2)\overline{\varphi}_{,1} - g_{11}(\tilde{X}_2)\overline{\psi}_{,1} \\
&= -\frac{2}{\pi}\int_0^\infty \zeta(1+\alpha\tilde{X}_2)^{-\beta}\{\tilde{e}_{15}(\tilde{X}_2)A_1(\zeta)I_\beta[(1+\alpha\tilde{X}_2)s\zeta/\alpha] \\
&\quad - \kappa_{11}(\tilde{X}_2)B_1(\zeta)I_\beta[(1+\alpha\tilde{X}_2)\zeta/\alpha] \\
&\quad - g_{11}(\tilde{X}_2)C_1(\zeta)I_\beta[(1+\alpha\tilde{X}_2)\zeta/\alpha] \\
&\quad + \tilde{e}_{15}(\tilde{X}_2)A_2(\zeta)K_\beta[(1+\alpha\tilde{X}_2)s\zeta/\alpha] \\
&\quad - \kappa_{11}(\tilde{X}_2)B_2(\zeta)K_\beta[(1+\alpha\tilde{X}_2)\zeta/\alpha] \\
&\quad - g_{11}(\tilde{X}_2)C_2(\zeta)K_\beta[(1+\alpha\tilde{X}_2)\zeta/\alpha]\}\sin(\zeta\tilde{X}_1)d\zeta,
\end{aligned} \tag{8.77}
$$

$$
\begin{aligned}
D_2(\tilde{X}_1,\tilde{X}_2) &= \tilde{e}_{15}(\tilde{X}_2)w_{,2} - \kappa_{11}(\tilde{X}_2)\overline{\varphi}_{,2} - g_{11}(\tilde{X}_2)\overline{\psi}_{,2} \\
&= -\frac{2}{\pi}\int_0^\infty \{[\tilde{e}_{15}(\tilde{X}_2)A_1(\zeta)P_I(\zeta,\tilde{X}_2) \\
&\quad - \kappa_{11}(\tilde{X}_2)B_1(\zeta)Q_I(\zeta,\tilde{X}_2) - g_{11}(\tilde{X}_2)C_1(\zeta)Q_I(\zeta,\tilde{X}_2)] \\
&\quad + [\tilde{e}_{15}(\tilde{X}_2)A_2(\zeta)P_K(\zeta,\tilde{X}_2) - \kappa_{11}(\tilde{X}_2)B_2(\zeta)Q_K(\zeta,\tilde{X}_2) \\
&\quad - g_{11}(\tilde{X}_2)C_2(\zeta)Q_K(\zeta,\tilde{X}_2)]\}\cos(\zeta\tilde{X}_1)d\zeta \\
&\quad + \tilde{e}_{15}(\tilde{X}_2)a_0 + \kappa_{11}(\tilde{X}_2)b_0 + g_{11}(\tilde{X}_2)c_0,
\end{aligned} \tag{8.78}
$$

$$
\begin{aligned}
B_1(\tilde{X}_1,\tilde{X}_2) &= \tilde{h}_{15}(\tilde{X}_2)w_{,1} - g_{11}(\tilde{X}_2)\overline{\phi}_{,1} - \mu_{11}(\tilde{X}_2)\overline{\psi}_{,1} \\
&= -\frac{2}{\pi}\int_0^\infty \zeta(1+\alpha\tilde{X}_2)^{-\beta}\{\tilde{h}_{15}(\tilde{X}_2)A_1(\zeta)I_\beta[(1+\alpha\tilde{X}_2)s\zeta/\alpha] \\
&- g_{11}(\tilde{X}_2)B_1(\zeta)I_\beta[(1+\alpha\tilde{X}_2)\zeta/\alpha] \\
&- \mu_{11}(\tilde{X}_2)C_1(\zeta)I_\beta[(1+\alpha\tilde{X}_2)\zeta/\alpha] \\
&+ \tilde{h}_{15}(\tilde{X}_2)A_2(\zeta)K_\beta[(1+\alpha\tilde{X}_2)s\zeta/\alpha] \\
&- g_{11}(\tilde{X}_2)B_2(\zeta)K_\beta[(1+\alpha\tilde{X}_2)\zeta/\alpha] \\
&- \mu_{11}(\tilde{X}_2)C_2(\zeta)K_\beta[(1+\alpha\tilde{X}_2)\zeta/\alpha]\}\sin(\zeta\tilde{X}_1)d\zeta,
\end{aligned} \tag{8.79}
$$

$$
\begin{aligned}
B_2(\tilde{X}_1,\tilde{X}_2) &= \tilde{h}_{15}(\tilde{X}_2)w_{,2} - g_{11}(\tilde{X}_2)\overline{\varphi}_{,2} - \mu_{11}(\tilde{X}_2)\overline{\psi}_{,2} \\
&= -\frac{2}{\pi}\int_0^\infty \{[\tilde{h}_{15}(\tilde{X}_2)A_1(\zeta)P_I(\zeta,\tilde{X}_2) \\
&-g_{11}(\tilde{X}_2)B_1(\zeta)Q_I(\zeta,\tilde{X}_2) - \mu_{11}(\tilde{X}_2)C_1(\zeta)Q_I(\zeta,\tilde{X}_2)] \\
&+[\tilde{h}_{15}(\tilde{X}_2)A_2(\zeta)P_K(\zeta,\tilde{X}_2) - g_{11}(\tilde{X}_2)B_2(\zeta)Q_K(\zeta,\tilde{X}_2) \\
&-\mu_{11}(\tilde{X}_2)C_2(\zeta)Q_K(\zeta,\tilde{X}_2)]\}\cos(\zeta\tilde{X}_1)d\zeta \\
&+\tilde{h}_{15}(\tilde{X}_2)a_0 + g_{11}(\tilde{X}_2)b_0 + \mu_{11}(\tilde{X}_2)c_0,
\end{aligned} \tag{8.80}
$$

where

$$\tilde{c}_{44}(\tilde{X}_2) = c_{44}(\tilde{X}_2) + c_1^0 e_{15}(\tilde{X}_2) + c_2^0 h_{15}(\tilde{X}_2), \tag{8.81}$$

$$\tilde{e}_{15}(\tilde{X}_2) = e_{15}(\tilde{X}_2) - c_1^0\kappa_{11}(\tilde{X}_2) - c_2^0 g_{11}(\tilde{X}_2), \tag{8.82}$$

$$\tilde{h}_{15}(\tilde{X}_2) = h_{15}(\tilde{X}_2) - c_1^0 g_{11}(\tilde{X}_2) - c_2^0\mu_{11}(\tilde{X}_2), \tag{8.83}$$

$$
\begin{aligned}
P_I(\zeta,\tilde{X}_2) &= \beta\alpha(1+\alpha\tilde{X}_2)^{-\beta-1}I_\beta[(1+\alpha\tilde{X}_2)s\zeta/\alpha] \\
&\quad - s\zeta(1+\alpha\tilde{X}_2)^{-\beta}I'_\beta[(1+\alpha\tilde{X}_2)s\zeta/\alpha],
\end{aligned} \tag{8.84}
$$

$$P_K(\zeta,\tilde{X}_2)=\beta\alpha(1+\alpha\tilde{X}_2)^{-\beta-1}K_\beta[(1+\alpha\tilde{X}_2)s\zeta/\alpha] \\ -s\zeta(1+\alpha\tilde{X}_2)^{-\beta}K_\beta^{'}[(1+\alpha\tilde{X}_2)s\zeta/\alpha], \tag{8.85}$$

$$Q_I(\zeta,\tilde{X}_2)=\beta\alpha(1+\alpha\tilde{X}_2)^{-\beta-1}I_\beta[(1+\alpha\tilde{X}_2)\zeta/\alpha] \\ -\zeta(1+\alpha\tilde{X}_2)^{-\beta}I_\beta^{'}[(1+\alpha\tilde{X}_2)\zeta/\alpha], \tag{8.86}$$

$$Q_K(\zeta,\tilde{X}_2)=\beta\alpha(1+\alpha\tilde{X}_2)^{-\beta-1}K_\beta[(1+\alpha\tilde{X}_2)\zeta/\alpha] \\ -\zeta(1+\alpha\tilde{X}_2)^{-\beta}K_\beta^{'}[(1+\alpha\tilde{X}_2)\zeta/\alpha]. \tag{8.87}$$

Application of the edge-loading conditions (8.59)–(8.61) results in the following relations:

$$P_h=\tilde{c}_{44}^h a_1-e_{15}^h b_1-h_{15}^h c_1, \tag{8.88}$$

$$D_h=\tilde{e}_{15}^h a_1+\kappa_{11}^h b_1+g_{11}^h c_1, \tag{8.89}$$

$$B_h=\tilde{h}_{15}^h a_1+g_{11}^h b_1+\mu_{11}^h c_1, \tag{8.90}$$

$$A_2(\zeta)=P_{21}A_1(\zeta), \tag{8.91}$$

$$B_2(\zeta)=Q_{21}B_1(\zeta), \tag{8.92}$$

$$C_2(\zeta)=Q_{21}C_1(\zeta), \tag{8.93}$$

where $P_{21}=-P_I(\zeta,h)/P_K(\zeta,h)$ and $Q_{21}=-Q_I(\zeta,h)/Q_K(\zeta,h)$.

By analogy with the treatment in Chapter 7, letting $b/a\to 0$ while keeping $(b\kappa_{11}^0)/(a\kappa_{11}^f)\to\lambda_e$ and $(b\mu_{11}^0)/(a\mu_{11}^f)\to\lambda_m$, satisfaction of the crack-face boundary conditions (8.62)–(8.69) leads to three pairs of dual integral equations:

$$\frac{2}{\pi}\int_0^\infty \zeta F_1(\zeta)A(\zeta)\cos(\zeta\tilde{X}_1)d\zeta=\frac{\tilde{L}(P_0,D_0,B_0)}{c_{44}^0},\ \left|\tilde{X}_1\right|<a, \tag{8.94}$$

$$\int_0^\infty A(\zeta)\cos(\zeta\tilde{X}_1)d\zeta=0,\ \left|\tilde{X}_1\right|\ge a, \tag{8.95}$$

$$\frac{2}{\pi}\int_0^\infty \zeta F_2(\zeta)\overline{B}(\zeta)\cos(\zeta\tilde{X}_1)d\zeta = -\tilde{L}_1(D_0, B_0), \quad \left|\tilde{X}_1\right| < a, \tag{8.96}$$

$$\int_0^\infty \overline{B}(\zeta)\cos(\zeta\tilde{X}_1)d\zeta = 0\,, \quad \left|\tilde{X}_1\right| \geq a, \tag{8.97}$$

$$\frac{2}{\pi}\int_0^\infty \zeta F_2(\zeta)\overline{C}(\zeta)\cos(\zeta\tilde{X}_1)d\zeta = -\tilde{L}_2(D_0, B_0), \quad \left|\tilde{X}_1\right| < a, \tag{8.98}$$

$$\int_0^\infty \overline{C}(\zeta)\cos(\zeta\tilde{X}_1)d\zeta = 0\,, \quad \left|\tilde{X}_1\right| \geq a, \tag{8.99}$$

where

$$A(\zeta) = A_1(\zeta)[I_\beta(s\zeta/\alpha) + P_{21}K_\beta(s\zeta/\alpha)]\,, \tag{8.100}$$

$$\overline{B}(\zeta) = B_1(\zeta)[I_\beta(\zeta/\alpha) + Q_{21}K_\beta(\zeta/\alpha)] + \overline{c}_1^0 A(\zeta)\,, \tag{8.101}$$

$$\overline{C}(\zeta) = C_1(\zeta)[I_\beta(\zeta/\alpha) + Q_{21}K_\beta(\zeta/\alpha)] + \overline{c}_2^0 A(\zeta)\,, \tag{8.102}$$

$$\begin{aligned} F_1(\zeta) = \frac{\overline{c}_{44}^0}{c_{44}^0\zeta}\Bigg[& \frac{P_I(\zeta,0) + P_{21}P_K(\zeta,0)}{I_\beta(s\zeta/\alpha) + P_{21}K_\beta(s\zeta/\alpha)} \\ & - (k_{em}^\lambda)^2 \frac{Q_I(\zeta,0) + Q_{21}Q_K(\zeta,0)}{I_\beta(\zeta/\alpha) + Q_{21}K_\beta(\zeta/\alpha)}\Bigg], \end{aligned} \tag{8.103}$$

$$F_2(\zeta) = \frac{1}{\zeta}\left[\frac{Q_I(\zeta,0) + Q_{21}Q_K(\zeta,0)}{I_\beta(\zeta/\alpha) + Q_{21}K_\beta(\zeta/\alpha)}\right], \tag{8.104}$$

$$\tilde{L}(P_0, D_0, B_0) = P_0 + e_{15}^0\tilde{L}_1(D_0, B_0) + h_{15}^0\tilde{L}_2(D_0, B_0), \tag{8.105}$$

$$\tilde{L}_1(D_0, B_0) = \frac{D_0\mu_{11}^0(1 - 1/\lambda_m) - B_0 g_{11}^0}{(1 - 1/\lambda_e)(1 - 1/\lambda_m)\kappa_{11}^0\mu_{11}^0 - (g_{11}^0)^2}, \tag{8.106}$$

$$\tilde{L}_2(D_0, B_0) = \frac{B_0\kappa_{11}^0(1 - 1/\lambda_e) - D_0 g_{11}^0}{(1 - 1/\lambda_e)(1 - 1/\lambda_m)\kappa_{11}^0\mu_{11}^0 - (g_{11}^0)^2}, \tag{8.107}$$

$$P_0 = \bar{c}_{44}^0 a_1 - e_{15}^0 b_1 - h_{15}^0 c_1, \tag{8.108}$$

$$D_0 = \kappa_{11}^0 b_1 + g_{11}^0 c_1, \tag{8.109}$$

$$B_0 = g_{11}^0 b_1 + \mu_{11}^0 c_1, \tag{8.110}$$

$$\bar{c}_1^0 = \frac{\kappa_{11}^0 \mu_{11}^0 (1-\lambda_m) c_1^0 + \mu_{11}^0 g_{11}^0 \lambda_e c_2^0}{\kappa_{11}^0 \mu_{11}^0 (1-\lambda_e)(1-\lambda_m) - \lambda_e \lambda_m (g_{11}^0)^2}, \tag{8.111}$$

$$\bar{c}_2^0 = \frac{\kappa_{11}^0 \mu_{11}^0 (1-\lambda_e) c_2^0 + \kappa_{11}^0 g_{11}^0 \lambda_m c_1^0}{\kappa_{11}^0 \mu_{11}^0 (1-\lambda_e)(1-\lambda_m) - \lambda_e \lambda_m (g_{11}^0)^2}, \tag{8.112}$$

$$(k_{em}^{\lambda})^2 = \frac{e_{15}^0 \bar{c}_1^0 + h_{15}^0 \bar{c}_2^0}{\bar{c}_{44}^0} = \sqrt{1 - \left(\frac{c_{bg}^{\lambda}}{c_T^0} \right)^2}. \tag{8.113}$$

It is noted that the magneto-electro-mechanical coupling factor k_{em}^{λ} and the shear horizontal surface wave speed c_{bg}^{λ} defined at the $\tilde{X}_2 = 0$ plane depend on the permeability parameters λ_e and λ_m through $\bar{c}_1^0$ and $\bar{c}_2^0$. Like the homogeneous materials discussed in Chapter 7, there are four limiting conditions for FGMs: (i) electrically and magnetically permeable crack-face condition as $\lambda_e \to 0$ and $\lambda_m \to 0$, (ii) electrically and magnetically impermeable crack-face condition as $\lambda_e \to \infty$ and $\lambda_m \to \infty$, (iii) electrically permeable and magnetically impermeable crack-face condition as $\lambda_e \to 0$ and $\lambda_m \to \infty$, and (iv) electrically impermeable and magnetically permeable crack-face condition as $\lambda_e \to \infty$ and $\lambda_m \to 0$. The electromagnetically semi-permeable crack-face condition may be approximated if λ_e and λ_m are considered as finite nonzero parameters.

The dual equations have the following solution (Copson, 1961):

$$A(\zeta) = \frac{\pi a^2}{2} \frac{\tilde{L}(P_0, D_0, B_0)}{c_{44}^0} \int_0^1 \sqrt{\xi} \Phi_1(\xi) J_0(\zeta a \xi) d\xi, \tag{8.114}$$

$$\bar{B}(\zeta) = -\frac{\pi a^2}{2}\tilde{L}_1(D_0, B_0)\int_0^1 \sqrt{\xi}\Phi_2(\xi)J_0(\zeta a\xi)d\xi, \tag{8.115}$$

$$\bar{C}(\zeta) = -\frac{\pi a^2}{2}\tilde{L}_2(D_0, B_0)\int_0^1 \sqrt{\xi}\Phi_2(\xi)J_0(\zeta a\xi)d\xi, \tag{8.116}$$

where $J_0(\zeta a\xi)$ is the zero-order Bessel function of the first kind and the auxiliary functions $\Phi_1(\xi)$ and $\Phi_2(\xi)$ should be governed by the standard Fredholm integral equations of the second kind:

$$\Phi_1(\xi) + \int_0^1 \Phi_1(\xi)G_1(\xi,\eta)d\eta = \sqrt{\xi}, \tag{8.117}$$

$$\Phi_2(\xi) + \int_0^1 \Phi_2(\xi)G_2(\xi,\eta)d\eta = \sqrt{\xi}, \tag{8.118}$$

with the kernel function in the form

$$G_1(\xi,\eta) = \sqrt{\xi\eta}\int_0^\infty \bar{s}[F_1(\bar{s}/a) - 1]J_0(\bar{s}\xi)J_0(\bar{s}\eta)d\bar{s}, \tag{8.119}$$

$$G_2(\xi,\eta) = \sqrt{\xi\eta}\int_0^\infty \bar{s}[F_2(\bar{s}/a) - 1]J_0(\bar{s}\xi)J_0(\bar{s}\eta)d\bar{s}. \tag{8.120}$$

As the crack velocity V_C tends to zero (i.e., $s \to 1$), the quasi-static solution is retrieved.

8.4 Fracture Characterizing Parameters

Next, we discuss the extension of classical fracture mechanics concepts such as intensity factors, energy release rate, and path-independent integrals to FGMs under combined magnetic, electric, thermal, and mechanical loadings.

8.4.1 *Field intensity factors*

The near-tip field solutions can be derived from the asymptotic analysis when $\zeta \to \infty$. The singular parts of the total stress, electric displacement, and magnetic induction near the right crack tip are given by

$$
\begin{aligned}
{}_t\sigma_{13} = -\sqrt{\pi a}\,\frac{\bar{c}_{44}^{0}}{c_{44}^{0}}[&\tilde{L}(P_0,D_0,B_0)\Phi_1(1)\frac{\sin(\tilde{\theta}_1/2)}{\sqrt{2\pi\tilde{r}_1}} \\
&-\tilde{L}(P_0,D_0,B_0)(k_{em}^{\lambda})^2\Phi_1(1)\frac{\sin(\theta_1/2)}{\sqrt{2\pi r_1}} \\
&-\tilde{L}(0,D_0,B_0)\Phi_2(1)\frac{\sin(\theta_1/2)}{\sqrt{2\pi r_1}}],
\end{aligned} \tag{8.121}
$$

$$
\begin{aligned}
{}_t\sigma_{23} = \sqrt{\pi a}\,\frac{\bar{c}_{44}^{0}}{c_{44}^{0}}[&\tilde{L}(P_0,D_0,B_0)s\Phi_1(1)\frac{\cos(\tilde{\theta}_1/2)}{\sqrt{2\pi\tilde{r}_1}} \\
&-\tilde{L}(P_0,D_0,B_0)(k_{em}^{\lambda})^2\Phi_1(1)\frac{\cos(\theta_1/2)}{\sqrt{2\pi r_1}} \\
&-\tilde{L}(0,D_0,B_0)\Phi_2(1)\frac{\cos(\theta_1/2)}{\sqrt{2\pi r_1}}],
\end{aligned} \tag{8.122}
$$

$$
\begin{aligned}
D_1 = -\sqrt{\pi a}\{&\tilde{L}(P_0,D_0,B_0)\frac{(\bar{c}_1\kappa_{11}^0+\bar{c}_2 g_{11}^0)}{c_{44}^0}\Phi_1(1) \\
&+[\kappa_{11}^0\tilde{L}_1(D_0,B_0)+g_{11}^0\tilde{L}_2(D_0,B_0)]\Phi_2(1)\}\frac{\sin(\theta_1/2)}{\sqrt{2\pi r_1}},
\end{aligned} \tag{8.123}
$$

$$
\begin{aligned}
D_2 = \sqrt{\pi a}\{&\tilde{L}(P_0,D_0,B_0)\frac{(\bar{c}_1\kappa_{11}^0+\bar{c}_2 g_{11}^0)}{c_{44}^0}\Phi_1(1) \\
&+[\kappa_{11}^0\tilde{L}_1(D_0,B_0)+g_{11}^0\tilde{L}_2(D_0,B_0)]\Phi_2(1)\}\frac{\cos(\theta_1/2)}{\sqrt{2\pi r_1}},
\end{aligned} \tag{8.124}
$$

$$
\begin{aligned}
B_1 = -\sqrt{\pi a}\{&\tilde{L}(P_0,D_0,B_0)\frac{(\bar{c}_1 g_{11}^0+\bar{c}_2\mu_{11}^0)}{c_{44}^0}\Phi_1(1) \\
&+[g_{11}^0\tilde{L}_1(D_0,B_0)+\mu_{11}^0\tilde{L}_2(D_0,B_0)]\Phi_2(1)\}\frac{\sin(\theta_1/2)}{\sqrt{2\pi r_1}},
\end{aligned} \tag{8.125}
$$

$$B_2 = \sqrt{\pi a}[\tilde{L}(P_0, D_0, B_0)\frac{(\bar{c}_1 g_{11}^0 + \bar{c}_2 \mu_{11}^0)}{c_{44}^0}\Phi_1(1)$$
$$+ [g_{11}^0 \tilde{L}_1(D_0, B_0) + \mu_{11}^0 \tilde{L}_2(D_0, B_0)]\Phi_2(1)]\frac{\cos(\theta_1/2)}{\sqrt{2\pi r_1}}, \qquad (8.126)$$

where

$$r_1 = \sqrt{(\tilde{X}_1 - a)^2 + \tilde{X}_2^2}, \quad \theta_1 = \arctan[\tilde{X}_2/(\tilde{X}_1 - a)], \qquad (8.127)$$

$$\tilde{r}_1 = \sqrt{(\tilde{X}_1 - a)^2 + (s\tilde{X}_2)^2}, \quad \tilde{\theta}_1 = \arctan[s\tilde{X}_2/(\tilde{X}_1 - a)]. \qquad (8.128)$$

It can be seen that the near-tip field solutions for FGMs exhibit the inverse square-root singularity in the local coordinate system affixed to the moving crack tip, like those for homogeneous materials. Hence, the definition of the dynamic field intensity factors introduced in Chapter 6 can be extended to FGMs, that is,

$$\begin{pmatrix} \tilde{K}_{III}(V_C) \\ \tilde{K}_D(V_C) \\ \tilde{K}_B(V_C) \end{pmatrix} = \lim_{\tilde{X}_1 \to a^+} \sqrt{2\pi(\tilde{X}_1 - a)} \begin{pmatrix} {}_t\sigma_{23}(\tilde{X}_1, 0) \\ D_{22}(\tilde{X}_1, 0) \\ B_{22}(\tilde{X}_1, 0) \end{pmatrix}. \qquad (8.129)$$

The self-induced and cross-over dynamic total stress, electric displacement, and magnetic induction intensity factors can be expressed in the form of a universal function of the crack velocity times the corresponding quasi-static value, that is,

$$\tilde{K}_{III}^{(T)}(V_C) = k_{III}^{(T)}(V_C)\tilde{K}_{III}^{(T)}(0)$$
$$= \sqrt{\pi a}\frac{\bar{c}_{44}^0}{c_{44}^0}\tilde{L}(P_0, 0, 0)[s - (k_{em}^\lambda)^2]\Phi_1(1), \qquad (8.130)$$

$$\tilde{K}_{III}^{(D)}(V_C) = k_{III}^{(D)}(V_C)\tilde{K}_{III}^{(D)}(0)$$
$$= \sqrt{\pi a}\frac{\bar{c}_{44}^0}{c_{44}^0}\tilde{L}(0, D_0, 0)\{[s - (k_{em}^\lambda)^2]\Phi_1(1) - \Phi_2(1)\}, \qquad (8.131)$$

$$\widetilde{K}_{III}^{(B)}(V_C) = k_{III}^{(B)}(V_C)\widetilde{K}_{III}^{(B)}(0)$$
$$= \sqrt{\pi a}\,\frac{\overline{c}_{44}^{0}}{c_{44}^{0}}\widetilde{L}(0,0,B_0)\{[s-(k_{em}^{\lambda})^2]\Phi_1(1)-\Phi_2(1)\}, \tag{8.132}$$

$$\widetilde{K}_{D}^{(T)}(V_C) = k_{D}^{(T)}(V_C)\widetilde{K}_{D}^{(T)}(0)$$
$$= \sqrt{\pi a}\,\frac{(\overline{c}_1^{0}\kappa_{11}^{0}+\overline{c}_2^{0}g_{11}^{0})}{c_{44}^{0}}\widetilde{L}(P_0,0,0)\Phi_1(1), \tag{8.133}$$

$$\widetilde{K}_{D}^{(D)}(V_C) = k_{D}^{(D)}(V_C)\widetilde{K}_{D}^{(D)}(0)$$
$$= \sqrt{\pi a}\,\frac{(\overline{c}_1^{0}\kappa_{11}^{0}+\overline{c}_2^{0}g_{11}^{0})}{c_{44}^{0}}\widetilde{L}(0,D_0,0)\Phi_1(1) \tag{8.134}$$
$$+\sqrt{\pi a}[\kappa_{11}^{0}\widetilde{L}_1(D_0,0)+g_{11}^{0}\widetilde{L}_2(D_0,0)]\Phi_2(1),$$

$$\widetilde{K}_{D}^{(B)}(V_C) = k_{D}^{(B)}(V_C)\widetilde{K}_{D}^{(B)}(0)$$
$$= \sqrt{\pi a}\,\frac{(\overline{c}_1^{0}\kappa_{11}^{0}+\overline{c}_2^{0}g_{11}^{0})}{c_{44}^{0}}\widetilde{L}(0,0,B_0)\Phi_1(1) \tag{8.135}$$
$$+\sqrt{\pi a}[\kappa_{11}^{0}\widetilde{L}_1(0,B_0)+g_{11}^{0}\widetilde{L}_2(0,B_0)]\Phi_2(1),$$

$$\widetilde{K}_{B}^{(T)}(V_C) = k_{B}^{(T)}(V_C)\widetilde{K}_{B}^{(T)}(0)$$
$$= \sqrt{\pi a}\,\frac{(\overline{c}_1^{0}g_{11}^{0}+\overline{c}_2^{0}\mu_{11}^{0})}{c_{44}^{0}}\widetilde{L}(P_0,0,0)\Phi_1(1), \tag{8.136}$$

$$\widetilde{K}_{B}^{(D)}(V_C) = k_{B}^{(D)}(V_C)\widetilde{K}_{B}^{(D)}(0)$$
$$= \sqrt{\pi a}\,\frac{(\overline{c}_1^{0}g_{11}^{0}+\overline{c}_2^{0}\mu_{11}^{0})}{c_{44}^{0}}\widetilde{L}(0,D_0,0)\Phi_1(1) \tag{8.137}$$
$$+\sqrt{\pi a}[g_{11}^{0}\widetilde{L}_1(D_0,0)+\mu_{11}^{0}\widetilde{L}_2(D_0,0)]\Phi_2(1),$$

$$\widetilde{K}_{B}^{(B)}(V_C) = k_{B}^{(B)}(V_C)\widetilde{K}_{B}^{(B)}(0)$$
$$= \sqrt{\pi a}\,\frac{(\overline{c}_1^{0}g_{11}^{0}+\overline{c}_2^{0}\mu_{11}^{0})}{c_{44}^{0}}\widetilde{L}(0,0,B_0)\Phi_1(1) \tag{8.138}$$
$$+\sqrt{\pi a}[g_{11}^{0}\widetilde{L}_1(0,B_0)+\mu_{11}^{0}\widetilde{L}_2(0,B_0)]\Phi_2(1),$$

where superscripts (T), (D) and (B) indicate traction loading, electric loading and magnetic loading, respectively.

For the special case of an electromagnetically permeable crack ($\lambda_e \to 0$ and $\lambda_m \to 0$), the near-tip fields are expressed as

$$ {}_t\sigma_{13} = -\frac{1}{s-(k_{em}^0)^2}\frac{\tilde{K}_{III}(V_C)}{\sqrt{2\pi\tilde{r}_1}}\sin(\tilde{\theta}_1/2) + \frac{(k_{em}^0)^2}{s-(k_{em}^0)^2}\frac{\tilde{K}_{III}(V_C)}{\sqrt{2\pi r_1}}\sin(\theta_1/2), \tag{8.139} $$

$$ {}_t\sigma_{23} = \frac{s}{s-(k_{em}^0)^2}\frac{\tilde{K}_{III}(V_C)}{\sqrt{2\pi\tilde{r}_1}}\cos(\tilde{\theta}_1/2) - \frac{(k_{em}^0)^2}{s-(k_{em}^0)^2}\frac{\tilde{K}_{III}(V_C)}{\sqrt{2\pi r_1}}\cos(\theta_1/2), \tag{8.140} $$

$$ D_1 = -\frac{\tilde{K}_D(V_C)}{\sqrt{2\pi r_1}}\sin(\theta_1/2), \tag{8.141} $$

$$ D_2 = \frac{\tilde{K}_D(V_C)}{\sqrt{2\pi r_1}}\cos(\theta_1/2), \tag{8.142} $$

$$ B_1 = -\frac{\tilde{K}_B(V_C)}{\sqrt{2\pi r_1}}\sin(\theta_1/2), \tag{8.143} $$

$$ B_2 = \frac{\tilde{K}_B(V_C)}{\sqrt{2\pi r_1}}\cos(\theta_1/2), \tag{8.144} $$

where

$$ \tilde{K}_{III}(V_C) = \frac{\bar{c}_{44}^0}{c_{44}^0}[s-(k_{em}^0)^2]P_0\sqrt{\pi a}\Phi_1(1), \tag{8.145} $$

$$ \tilde{K}_D(V_C) = \frac{e_{15}^0}{c_{44}^0}P_0\sqrt{\pi a}\Phi_1(1), \tag{8.146} $$

$$\tilde{K}_B(V_C) = \frac{h_{15}^0}{c_{44}^0} P_0 \sqrt{\pi a} \Phi_1(1), \tag{8.147}$$

$$(k_{em}^0)^2 = \frac{e_{15}^0 c_1^0 + h_{15}^0 c_2^0}{\bar{c}_{44}^0} = \sqrt{1 - \left(\frac{c_{bg}^0}{c_T^0}\right)^2}. \tag{8.148}$$

As the magnetic field is shut off, the electrically permeable case for the Yoffe-type moving crack in a FGPM strip studied by Li and Weng (2002b) is recovered. The dependence of the dynamic stress intensity factor normalized by the quasi-static value on the crack velocity normalized by the Bleustein–Gulyaev wave speed is shown in Fig. 8.2.

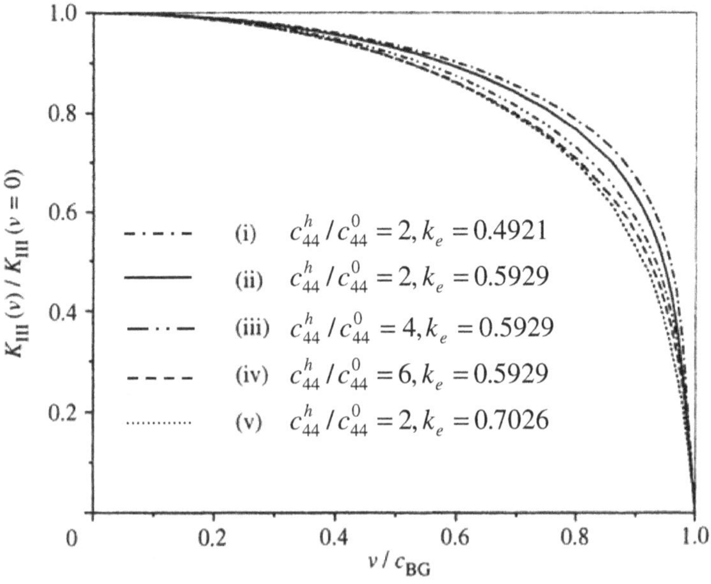

Fig. 8.2 Effect of crack velocity on stress intensity factor ($k = 1$, $a/h = 0.5$). (From Li and Weng, 2002b, with permission from the Royal Society).

The trend is akin to that for the propagation of a semi-infinite permeable crack in a homogeneous electromagnetic material as discussed in Chapter 7. It can be seen that the dynamic stress intensity factor for a FGPM decreases monotonically with the increase of the crack velocity and tends to zero as the crack velocity approaches the Bleustein–Gulyaev wave speed. Similar to the increase of the electromechanical coupling factor, the increase of the material property gradient helps to reduce the dynamic stress intensity factor.

8.4.2 *Dynamic energy release rate*

The two alternative representations (6.25) and (6.27) for the dynamic energy release rate introduced in Chapter 6 are valid for homogeneous or nonhomogeneous, linear or nonlinear, magneto-electro-thermo-elastic media, including FGMs, containing a propagating three-dimensional crack of arbitrary shape, that is,

$$
\begin{aligned}
\tilde{J}_0 &= \hat{J} \\
&= \lim_{\tilde{\Gamma}_0 \to 0} \frac{1}{\dot{A}} \int_{\tilde{\Gamma}_0} [\mathbf{n} \cdot (\boldsymbol{\sigma} + {}_{em}\boldsymbol{\sigma} + \mathbf{v} \otimes \mathbf{G}) \cdot \mathbf{v} - \mathbf{n} \cdot \boldsymbol{S} + (\tilde{\rho}\hat{k} + \tilde{\rho}\hat{h} + {}_{em}u^f)\mathbf{n} \cdot \boldsymbol{V}_C] d\tilde{\Gamma}, \\
&= \frac{1}{\dot{A}} \int_{\tilde{\Gamma}} [\mathbf{n} \cdot (\boldsymbol{\sigma} + {}_{em}\boldsymbol{\sigma} + \mathbf{v} \otimes \mathbf{G}) \cdot \mathbf{v} - \mathbf{n} \cdot \boldsymbol{S} + (\tilde{\rho}\hat{k} + \tilde{\rho}\hat{h} + {}_{em}u^f)\mathbf{n} \cdot \boldsymbol{V}_C] d\tilde{\Gamma} \\
&\quad - \lim_{\tilde{\Gamma}_0 \to 0} \frac{1}{\dot{A}} \int_{\tilde{V}_{\tilde{\Gamma}} - \tilde{V}_{\tilde{\Gamma}_0}} \frac{\tilde{\partial}}{\tilde{\partial} t} (\tilde{\rho}\hat{k} + \tilde{\rho}\hat{h} + {}_{em}u^f) d\tilde{V} + \lim_{\tilde{\Gamma}_0 \to 0} \frac{1}{\dot{A}} \int_{\tilde{V}_{\tilde{\Gamma}} - \tilde{V}_{\tilde{\Gamma}_0}} \tilde{\rho}\hat{\mathbf{f}} \cdot \mathbf{v} d\tilde{V} \\
&\quad - \lim_{\tilde{\Gamma}_0 \to 0} \frac{1}{\dot{A}} \int_{\tilde{V}_{\tilde{\Gamma}} - \tilde{V}_{\tilde{\Gamma}_0}} \frac{\tilde{\rho}}{\rho_0} \hat{\boldsymbol{E}} \cdot \boldsymbol{J}_e d\tilde{V} - \lim_{\tilde{\Gamma}_0 \to 0} \frac{1}{\dot{A}} \int_{\tilde{V}_{\tilde{\Gamma}} - \tilde{V}_{\tilde{\Gamma}_0}} \tilde{\rho}\hat{s}\dot{T} d\tilde{V}.
\end{aligned}
\tag{8.149}
$$

The invariant $\hat{J}$-integral method is more useful than the crack-front generalized $\tilde{J}$-integral method for numerical analysis. If there exists a steady-state solution for the propagation of a planar crack along the $\overline{\mathbf{E}}_1$-direction in a FGM without electricity conduction, the dynamic energy release rate can be expressed by the special form of the invariant $\hat{J}$-integral:

$$
\begin{aligned}
\tilde{J}_0 &= \hat{J} \\
&= -\frac{1}{B}\int_{\tilde{\Gamma}} \mathbf{n}\cdot(\boldsymbol{\sigma} + {}_{em}\boldsymbol{\sigma})\cdot \mathbf{u}\tilde{\nabla} d\tilde{\Gamma}\cdot \overline{\mathbf{E}}_1 + \frac{1}{B}\int_{\tilde{\Gamma}} \mathbf{n}\cdot(\tilde{\rho}\hat{k} + \tilde{\rho}\hat{h} + {}_{em}u^f)\mathbf{I} d\tilde{\Gamma}\cdot \overline{\mathbf{E}}_1 \\
&+ \frac{1}{B}\int_{\tilde{\Gamma}} [(\mathbf{n}\times\mathbf{E})\times\mathbf{D}]\cdot \mathbf{u}\tilde{\nabla} d\tilde{\Gamma}\cdot \overline{\mathbf{E}}_1 + \frac{1}{B}\int_{\tilde{\Gamma}} [(\mathbf{n}\times\mathbf{H})\times\mathbf{B}]\cdot \mathbf{u}\tilde{\nabla} d\tilde{\Gamma}\cdot \overline{\mathbf{E}}_1 \\
&+ \frac{1}{B}\int_{\tilde{\Gamma}} \mathbf{n}\cdot \mathbf{v}\otimes(\mathbf{P}\times\mathbf{B})\cdot \mathbf{u}\tilde{\nabla} d\tilde{\Gamma}\cdot \overline{\mathbf{E}}_1 \\
&- \frac{1}{B}\lim_{\tilde{\Gamma}_0\to 0}\int_{\tilde{V}_{\tilde{\Gamma}}-\tilde{V}_{\tilde{\Gamma}_0}} \tilde{\rho}\hat{\mathbf{f}}\cdot \mathbf{u}\tilde{\nabla} d\tilde{V}\cdot \overline{\mathbf{E}}_1 + \frac{1}{B}\lim_{\tilde{\Gamma}_0\to 0}\int_{\tilde{V}_{\tilde{\Gamma}}-\tilde{V}_{\tilde{\Gamma}_0}} \tilde{\rho}\hat{s}\tilde{\nabla} T d\tilde{V}\cdot \overline{\mathbf{E}}_1 \\
&- \frac{1}{B}\lim_{\tilde{\Gamma}_0\to 0}\int_{\tilde{V}_{\tilde{\Gamma}}-\tilde{V}_{\tilde{\Gamma}_0}} \frac{\partial}{\partial \tilde{X}_1}(\tilde{\rho}\hat{k} + \tilde{\rho}\hat{h} + {}_{em}u^f)\Big|_{\text{expl}} d\tilde{V}.
\end{aligned}
\tag{8.150}
$$

The last term involving the explicit derivative of the total energy density with respect to $\tilde{X}_1$ reflects the shielding or amplification influence of the material property gradient on crack propagation in FGMs.

8.4.3 *Path-domain independent integral*

It is known that the J-integral is generally not path independent for nonhomogeneous materials. Significant efforts have been made to modify the classical J-integral method to account for material inhomogeneity. Here, we introduce the $\hat{J}_K$-integral vector, the $\tilde{J}_K$-integral vector, and the energy-momentum tensor $\tilde{\mathbf{b}}$ as

$$
\begin{aligned}
\hat{J}_K &= \tilde{J}_K - \frac{1}{B}\lim_{\tilde{\Gamma}_0\to 0}\int_{\tilde{V}_{\tilde{\Gamma}}-\tilde{V}_{\tilde{\Gamma}_0}} \tilde{\rho}\hat{f}_i \frac{\partial u_i}{\partial \tilde{X}_K} d\tilde{V} + \frac{1}{B}\lim_{\tilde{\Gamma}_0\to 0}\int_{\tilde{V}_{\tilde{\Gamma}}-\tilde{V}_{\tilde{\Gamma}_0}} \tilde{\rho}\hat{s}\frac{\partial T}{\partial \tilde{X}_K} d\tilde{V} \\
&- \frac{1}{B}\lim_{\tilde{\Gamma}_0\to 0}\int_{\tilde{V}_{\tilde{\Gamma}}-\tilde{V}_{\tilde{\Gamma}_0}} \frac{\partial}{\partial \tilde{X}_K}(\tilde{\rho}\hat{h} + \tilde{\rho}\hat{k} + {}_{em}u^f)\Big|_{\text{expl}} d\tilde{V},
\end{aligned}
\tag{8.151}
$$

$$
\tilde{J}_K = \frac{1}{B}\int_{\tilde{\Gamma}} n_j b_{jK} d\tilde{\Gamma}, \tag{8.152}
$$

$$\begin{aligned}\tilde{\mathbf{b}} = &-[{}_t\boldsymbol{\sigma}+(\mathbf{D}\cdot\mathbf{E})\mathbf{I}-\mathbf{D}\otimes\mathbf{E}+(\mathbf{B}\cdot\mathbf{H})\mathbf{I}-\mathbf{B}\otimes\mathbf{H}\\ &-\mathbf{v}\otimes(\mathbf{P}\times\mathbf{B})]\cdot\mathbf{u}\tilde{\nabla}+(\tilde{\rho}\hat{h}+\tilde{\rho}\hat{k}+{}_{em}u^f)\mathbf{I}.\end{aligned} \tag{8.153}$$

It can be seen that the first component of the $\hat{J}_K$-integral vector reproduces expression (8.150) for the invariant $\hat{J}$-integral, which is an extension of the configuration force (material force) notation (Eshelby, 1951, 1956, 1970; Maugin and Trimarco, 1992; Gurtin, 2000). Several variations of path-domain independent integrals proposed for nonhomogeneous materials or graded materials (e.g., Eischen, 1987a–b; Honein and Herrmann, 1997; Gu *et al.*, 1999; Anlas *et al.*, 2000; Jin and Sun, 2007) can be retrieved as the electromagnetic fields are shut off, e.g.,

$$\begin{aligned}J_k^* = &\frac{1}{B}\int_{\tilde{\Gamma}}[(\tilde{\rho}\hat{h}+\tilde{\rho}\hat{k})n_k - n_j\sigma_{ji}u_{i,k}]d\tilde{\Gamma}\\ &-\frac{1}{B}\lim_{\tilde{\Gamma}_0\to 0}\int_{\tilde{V}_{\tilde{\Gamma}}-\tilde{V}_{\tilde{\Gamma}_0}}\tilde{\rho}\hat{f}_i u_{i,k}d\tilde{V}+\frac{1}{B}\lim_{\tilde{\Gamma}_0\to 0}\int_{\tilde{V}_{\tilde{\Gamma}}-\tilde{V}_{\tilde{\Gamma}_0}}\tilde{\rho}\hat{s}T_{,k}d\tilde{V}\\ &-\frac{1}{B}\lim_{\tilde{\Gamma}_0\to 0}\int_{\tilde{V}_{\tilde{\Gamma}}-\tilde{V}_{\tilde{\Gamma}_0}}(\tilde{\rho}\hat{h}+\tilde{\rho}\hat{k})_{,k}\Big|_{\text{expl}}d\tilde{V}.\end{aligned} \tag{8.154}$$

In contrast to homogeneous materials, the difference between the global and local $\tilde{J}_K$-integral vector for FGMs is caused by the gradient of material properties along the crack line, in addition to mechanical body force and temperature change. The domain integral terms in the expression (8.151) vanish only if mechanical body force and temperature change are negligible and the graded properties vary one-dimensionally perpendicular to the crack plane. For this special case, the path-domain independent $\hat{J}$-integral becomes path independent, that is,

$$\hat{J} = \tilde{J} = \frac{1}{B}\int_{\tilde{\Gamma}}\mathbf{n}\cdot\tilde{\mathbf{b}}d\tilde{\Gamma}\cdot\overline{\mathbf{E}}_1\,. \tag{8.155}$$

Moreover, the dynamic energy release rate for crack propagation in a FGM can be evaluated by the crack closure integral in the same way as discussed in Section 6.3, that is,

$$\tilde{J}_0 = \lim_{\delta a \to 0} \frac{1}{2\delta a} \int_a^{a+\delta a} {}_t\sigma_{2j}(\tilde{X}_1 - a, 0)\Delta u_j(\tilde{X}_1 - a - \delta a, 0^{\pm})d\tilde{X}_1 , \quad (8.156)$$

where $\Delta u_j(\tilde{X}_1 - a - \delta a, 0^{\pm}) = u_j(\tilde{X}_1 - a - \delta a, 0^{+}) - u_j(\tilde{X}_1 - a - \delta a, 0^{-})$ is the crack opening displacement at a distance $\delta a + a - \tilde{X}_1$ behind the crack tip.

For the Yoffe-type mode-III moving crack in a FGM, the dynamic energy release rate is thus calculated as

$$\begin{aligned} \tilde{J}_0 &= \lim_{\delta a \to 0} \frac{1}{2\delta a} \int_0^{\delta a} {}_t\sigma_{23}(\tilde{X}_1 - a, 0)\Delta w(\tilde{X}_1 - a - \delta a, 0^{\pm})d\tilde{X}_1 \\ &= \frac{1}{4}\tilde{K}_{III}(V_C)\tilde{K}_{III}^{COD}(V_C), \end{aligned} \quad (8.157)$$

where the mode-III dynamic crack opening displacement intensity factor at the right crack tip is given by

$$\begin{aligned} \tilde{K}_{III}^{COD}(V_C) &= \lim_{\tilde{X}_1 \to a^-} \sqrt{\frac{\pi}{2(a - \tilde{X}_1)}}\Delta w(\tilde{X}_1 - a, 0^{\pm}) \\ &= \frac{2}{\bar{c}_{44}^0\sqrt{1 - (V_C / c_T^0)^2}}\Big[\tilde{K}_{III}(V_C) \\ &\quad + \frac{e_{15}^0\mu_{11}^0 - h_{15}^0 g_{11}^0}{\kappa_{11}^0\mu_{11}^0 - (g_{11}^0)^2}\tilde{K}_D(V_C) + \frac{h_{15}^0\kappa_{11}^0 - e_{15}^0 g_{11}^0}{\kappa_{11}^0\mu_{11}^0 - (g_{11}^0)^2}\tilde{K}_B(V_C)\Big]. \end{aligned} \quad (8.158)$$

Therefore, the prediction that the dynamic energy release rate is an odd function of the dynamic electric displacement intensity factor and dynamic magnetic induction intensity factor is valid for FGMs, similar to homogeneous materials.

8.5 Remarks

Significant progress has been made in understanding the quasi-static and dynamic fracture behaviors of FGMs under combined magnetic, electric, thermal, and mechanical loadings, with generalization of classical fracture mechanics concepts such as intensity factors, energy release rate and *J*-integral to FGMs. Nevertheless, fracture mechanics of FGMs is still far from emerging as a mature engineering science discipline. Areas that require substantial efforts include numerical simulation, experimental characterization, mixed-mode fracture, creep-fatigue crack growth, environmentally assisted cracking, and higher-order theory. Multiscale modeling involving magneto-electro-thermo-mechanical coupling and dissipative effects may find increased usage in simulating fracture processes in FGMs. The development of efficient methods for analyzing flawed structure components made of FGMs is greatly needed. Correlation of theoretical prediction with experimental measurement under combined loadings is vital for successful applications of FGMs in various demanding areas such as aerospace, armor, and biomedical engineering.

Chapter 9

Magneto-Thermo-Viscoelastic Deformation and Fracture

9.1 Introduction

With increasing interests in the engineering applications of magnetosensitive polymers and polymer composites capable of large deformations, studies on nonlinear magneto-thermo-viscoelastic deformation and fracture are necessary for evaluating the reliability and durability of intelligent devices made of these advanced materials. It is well known that fracture in metals is influenced by plastic dissipation in the plastic zone, whereas fracture in polymers is accompanied by viscous dissipation in the bulk material. The need to incorporate the effect of viscous bulk dissipation on crack initiation and growth is the main motivation for the development of magneto-thermo-viscoelastic fracture mechanics.

As discussed in Chapter 3, nonequilibrium thermodynamics provides an effective way of studying irreversible processes involving energy dissipation. There are essentially two types of approaches to the derivation of constitutive, fracture, and strength models for nonlinear viscoelastic solids in the published literature (Schapery, 2000): functional thermodynamics and state-variable thermodynamics. In functional thermodynamics, the free energy is expressed as a functional of the histories of strain (stress), temperature, etc., whereas, in state-variable thermodynamics, the free energy is expressed as a function of current strain (stress), temperature, and other variables including so-called internal state variables. Recently, Chen (2009d) developed a nonlinear magneto-thermo-viscoelastic constitutive and fracture theory, which incorporates the augmented Helmholtz free energy as a functional of the histories of deformation, temperature, and magnetic induction in the reference configuration. The nonequilibrium thermodynamic approach affords a uniform treatment of complex time-dependent constitutive and fracture behaviors in the presence of multifield coupling and hysteresis effects.

In this chapter, we attempt to provide an insight into this rather new and developing area on nonlinear magneto-thermo-viscoelastic deformation and fracture. The presentation here is restricted to the quasi-magnetostatic approximation for a simple formulation. In Section 9.2, the local balance equations under combined magnetic, thermal and mechanical loadings are summarized. In Section 9.3, the free energy functional and entropy production inequality are introduced for memory-dependent magnetosensitive materials. In Section 9.4, nonlinear magneto-thermo-viscoelastic constitutive relations are formulated from the energy balance equation and the entropy production inequality. In Section 9.5, the generalized $\tilde{J}$-integral is constructed for use as a physically sound criterion for nonlinear magneto-thermo-viscoelastic fracture. In Section 9.6, applications to generalized plane crack problems are discussed and the mode-III fracture of a magnetostrictive solid in a bias magnetic field studied by Sabir and Maugin (1996) is revisited as a special case.

9.2 Local Balance Equations for Magnetic, Thermal, and Mechanical Field Quantities

A description of the balance laws in the continuum mechanics of electromagnetic solids can be found in Chapter 2. In the papers by Dorfmann and Ogden (2003, 2004), rather elegant and simple formulations of the governing equations and the constitutive relations were provided for the static situation of elastomer-like materials capable of large magnetoelastic deformations, based on a modified free energy function with the referential magnetic induction vector as the independent magnetic variable. Here, the governing equations and the constitutive relations are extended to nonlinear magneto-thermo-viscoelastic media under the quasi-magnetostatic approximation following the work by Chen (2009d).

The local balance equations under combined magnetic, thermal, and mechanical loadings are summarized below:

$$\nabla \cdot \mathbf{B} = 0, \tag{9.1}$$

$$\nabla \times \mathbf{H} = 0, \tag{9.2}$$

$$\frac{d\rho}{dt} = -\rho \nabla \cdot \mathbf{v}, \tag{9.3}$$

$$\rho \frac{d\mathbf{v}}{dt} = \nabla \cdot {}_t\boldsymbol{\sigma} + \rho \hat{\mathbf{f}}, \tag{9.4}$$

$${}_t\sigma_{ij} = {}_t\sigma_{ji}, \tag{9.5}$$

$$\rho \frac{d}{dt}\left(\hat{k} + \hat{e} + \frac{{}_m u^f}{\rho}\right) = -\nabla \cdot \mathbf{j}_q + \nabla \cdot ({}_t\boldsymbol{\sigma} \cdot \mathbf{v} - S) + \rho \hat{\mathbf{f}} \cdot \mathbf{v}, \tag{9.6}$$

where the total stress tensor ${}_t\boldsymbol{\sigma} = \boldsymbol{\sigma} + {}_m\boldsymbol{\sigma}$ is the sum of the Cauchy stress tensor $\boldsymbol{\sigma}$ and the magnetic stress tensor ${}_m\boldsymbol{\sigma} = -\mathbf{B} \otimes \mathbf{M} + \mathbf{B} \otimes \mathbf{B}/\mu_0 + (\mathbf{M} \cdot \mathbf{B} - {}_m u^f)\mathbf{I}$, ${}_m u^f = \mathbf{B} \cdot \mathbf{B}/2\mu_0$ is the energy density of the free magnetic field, and $S = (\mathbf{v} \times \mathbf{B}) \times \mathbf{H}$ is the Poynting vector in the co-moving frame.

Like thermoviscoelastic boundary-initial value problems, discussed in Chapter 3, these balance equations should be supplemented by constitutive relations together with appropriate boundary and initial conditions for proper mathematical formulation of magneto-thermo-viscoelastic boundary-initial value problems.

The boundary conditions are given by

$$\mathbf{n} \cdot [[\mathbf{B}]] = 0 \text{ across } S \ (t \geq 0), \tag{9.7}$$

$$\mathbf{n} \times [[\mathbf{H}]] = 0 \text{ across } S \ (t \geq 0), \tag{9.8}$$

$$\mathbf{n} \cdot {}_t\boldsymbol{\sigma} = \mathbf{t}_B(\mathbf{x}, t) \text{ on } S_\sigma \ (t \geq 0), \tag{9.9}$$

$$\mathbf{u} = \mathbf{u}_B(\mathbf{x},t) \text{ on } S_u \ (t \geq 0), \tag{9.10}$$

$$\mathbf{n} \cdot \mathbf{j}_q = q_B(\mathbf{x},t) \text{ on } S_q \ (t \geq 0), \tag{9.11}$$

$$T = T_B(\mathbf{x},t) \text{ on } S_T \ (t \geq 0), \tag{9.12}$$

where $S_{[\,]}$ refers to a certain part of the boundary: displacement is prescribed on S_u, traction on S_σ (the complement of S_u), temperature on S_θ, and heat flux on S_q (the complement of S_θ). Therefore, we have $S_u \cup S_\sigma = S$ and $S_T \cup S_q = S$. Other mixed boundary conditions may also be employed.

The initial conditions are taken as

$$\mathbf{u} = \mathbf{u}_0 \ (t < 0), \tag{9.13}$$

$$\dot{\mathbf{u}} = \mathbf{v}_0 \ (t = 0), \tag{9.14}$$

$$T = T_0 \ (t < 0). \tag{9.15}$$

$$\mathbf{B} = \mathbf{B}_0 \ (t < 0). \tag{9.16}$$

9.3 Free Energy and Entropy Production Inequality for Memory-Dependent Magnetosensitive Materials

By introduction of the augmented Helmholtz free energy, including the contribution of the energy of the free magnetic field, that is,

$$\tilde{h} = \hat{e} - T\hat{s} + \frac{{}_m u^f}{\rho}, \tag{9.17}$$

the local energy balance equation (9.6) becomes

$$\rho \frac{d}{dt}\left(\hat{k} + \tilde{h} + T\hat{s}\right) = -\nabla \cdot \mathbf{j}_q + \nabla \cdot ({}_t\boldsymbol{\sigma} \cdot \mathbf{v} - S) + \rho \hat{\mathbf{f}} \cdot \mathbf{v}\,. \tag{9.18}$$

In the reference configuration, V_R, the local energy balance equation can be rewritten as

$$\frac{d\hat{s}}{dt} = -\frac{1}{\rho_0}\nabla_R \cdot \frac{\mathbf{J}_q}{T} + \frac{1}{\rho_0}\mathbf{J}_q \cdot \nabla_R \frac{1}{T} + \frac{1}{2\rho_0 T}\, {}_t\mathbf{\Sigma} : \dot{\mathbf{C}}$$
$$+ \frac{1}{\rho_0 T}\hat{\mathbf{H}} \cdot \dot{\hat{\mathbf{B}}} - \frac{1}{T}\hat{s}\dot{T} - \frac{1}{T}\frac{d\tilde{h}}{dt} \tag{9.19}$$

and the entropy production inequality is expressed as

$$\frac{d_i\hat{s}}{dt} = \frac{d\hat{s}}{dt} + \frac{1}{\rho_0}\nabla_R \cdot \mathbf{J}_s \geq 0\,. \tag{9.20}$$

For memory-dependent magnetosensitive materials, the augmented Helmholtz free energy, including the contribution of the energy of the free magnetic field, is assumed to be a functional of the histories of deformation, temperature, temperature gradient, and magnetic induction in the reference configuration V_R, with respect to which the deformation gradient $\mathbf{F}$ is measured, that is,

$$\tilde{h} = \tilde{h}(\mathbf{C}(t-\tau), T(t-\tau), \nabla_R T(t-\tau), \hat{\mathbf{B}}(t-\tau); \mathbf{X})\,. \tag{9.21}$$

9.4 Coupled Magneto-Thermo-Viscoelastic Constitutive Relations

Since the entropy production inequality (9.20) is always valid, state equations should fulfill the following conditions:

$$\frac{\partial\tilde{h}}{\partial T_{,K}} = 0\,, \tag{9.22}$$

$${}_t\Sigma_{KL} = 2\rho_0\frac{\partial\tilde{h}}{\partial C_{KL}}\,, \tag{9.23}$$

$$\hat{s} = -\frac{\partial\tilde{h}}{\partial T}\,, \tag{9.24}$$

$$\hat{H}_K = \rho_0\frac{\partial\tilde{h}}{\partial\hat{B}_K}\,, \tag{9.25}$$

$$\hat{\Lambda} = -\frac{\partial \tilde{h}}{\partial t}, \tag{9.26}$$

$$\mathbf{J}_s = \frac{1}{T}\mathbf{J}_q, \tag{9.27}$$

$$\frac{d_i \hat{s}}{dt} = \frac{1}{\rho_0}\mathbf{J}_q \cdot \nabla_R \frac{1}{T} + \frac{\hat{\Lambda}}{T} \geq 0, \tag{9.28}$$

where $\hat{\Lambda}$ is the viscous dissipation rate, which is time-dependent.

From Eq. (9.22), the augmented Helmholtz free energy does not depend on the temperature gradient. Energy can be converted from one form to another due to mechanical, thermal, and magnetic coupling, accompanied by intrinsic dissipation associated with mechanical, thermal, and magnetic hysteresis. Since the inequality (9.28) must always be satisfied, kinetic laws for specific irreversible processes may be determined accordingly. Next, a special type of material behavior pertinent to finite magneto-thermo-viscoelasticity is illustrated as an extension of the coupled theory of thermoviscoelasticity at finite deformation discussed in Chapter 3.

The viscous dissipation rate satisfies the inequality

$$\hat{\Lambda} \geq 0. \tag{9.29}$$

It is proposed that the thermodynamic flux for heat conduction depends linearly on the corresponding thermodynamic force, that is,

$$\mathbf{J}_q = \hat{\mathbf{L}}^{qq} \cdot \nabla_R \frac{1}{T}, \tag{9.30}$$

where $\hat{\mathbf{L}}^{qq^T} = \hat{\mathbf{L}}^{qq}$ is positive definite.

Substituting Eqs. (9.24) and (9.30) into Eq. (9.19) yields the following heat transfer equation based on the augmented Helmholtz free energy functional:

$$-\frac{d}{dt}\left(\frac{\partial \tilde{h}}{\partial T}\right) = -\frac{1}{\rho_0 T}\nabla_R \cdot \left(\hat{\mathbf{L}}^{qq} \cdot \nabla_R \frac{1}{T}\right) + \frac{1}{T}\hat{\Lambda}. \tag{9.31}$$

With the use of the Lagrange strain measure $\mathbf{E}=(\mathbf{C}-\mathbf{I})/2$, the temperature deviation $\theta = T - T_0$, and the referential magnetic induction deviation $\hat{\mathbf{b}} = \hat{\mathbf{B}} - \hat{\mathbf{B}}_0$, expansion of the augmented Helmholtz free energy functional for materials with fading memory on an intrinsic time scale up to the second order yields

$$
\begin{aligned}
\rho_0 \tilde{h} &= \rho_0 \tilde{h}_0 + \int_{-\infty}^{\psi} L_{IJ}(\mathbf{X}, \psi - \psi') \frac{\partial \mathrm{E}_{IJ}(\mathbf{X}, \psi')}{\partial \psi'} d\psi' \\
&- \int_{-\infty}^{\psi} M(\mathbf{X}, \psi - \psi') \frac{\partial \theta(\mathbf{X}, \psi')}{\partial \psi'} d\psi' + \int_{-\infty}^{\psi} N_I^b(\mathbf{X}, \psi - \psi') \frac{\partial \hat{b}_I(\mathbf{X}, \psi')}{\partial \psi'} d\psi' \\
&+ \frac{1}{2} \int_{-\infty}^{\psi} \int_{-\infty}^{\psi} G_{IJKL}(\mathbf{X}, \psi - \psi', \psi - \psi'') \frac{\partial \mathrm{E}_{IJ}(\mathbf{X}, \psi')}{\partial \psi'} \frac{\partial \mathrm{E}_{KL}(\mathbf{X}, \psi'')}{\partial \psi''} d\psi' d\psi'' \\
&- \int_{-\infty}^{\psi} \int_{-\infty}^{\psi} \beta_{IJ}(\mathbf{X}, \psi - \psi', \psi - \psi'') \frac{\partial \mathrm{E}_{IJ}(\mathbf{X}, \psi')}{\partial \psi'} \frac{\partial \theta(\mathbf{X}, \psi'')}{\partial \psi''} d\psi' d\psi'' \\
&- \int_{-\infty}^{\psi} \int_{-\infty}^{\psi} f_{KIJ}^b(\mathbf{X}, \psi - \psi', \psi - \psi'') \frac{\partial \mathrm{E}_{IJ}(\mathbf{X}, \psi')}{\partial \psi'} \frac{\partial \hat{b}_K(\mathbf{X}, \psi'')}{\partial \psi''} d\psi' d\psi'' \\
&- \frac{1}{2T_0} \int_{-\infty}^{\psi} \int_{-\infty}^{\psi} C_H(\mathbf{X}, \psi - \psi', \psi - \psi'') \frac{\partial \theta(\mathbf{X}, \psi')}{\partial \psi'} \frac{\partial \theta(\mathbf{X}, \psi'')}{\partial \psi''} d\psi' d\psi'' \\
&- \int_{-\infty}^{\psi} \int_{-\infty}^{\psi} \gamma_I^b(\mathbf{X}, \psi - \psi', \psi - \psi'') \frac{\partial \hat{b}_I(\mathbf{X}, \psi')}{\partial \psi'} \frac{\partial \theta(\mathbf{X}, \psi'')}{\partial \psi''} d\psi' d\psi'' \\
&+ \frac{1}{2} \int_{-\infty}^{\psi} \int_{-\infty}^{\psi} \bar{\chi}_{IJ}^b(\mathbf{X}, \psi - \psi', \psi - \psi'') \frac{\partial \hat{b}_I(\mathbf{X}, \psi')}{\partial \psi'} \frac{\partial \hat{b}_J(\mathbf{X}, \psi'')}{\partial \psi''} d\psi' d\psi'',
\end{aligned}
\tag{9.32}
$$

where $\tilde{h}_0$ is the value of the augmented Helmholtz free energy in a reference state (i.e., $\mathbf{E}=0$, $T=T_0$, $\hat{\mathbf{B}}=\hat{\mathbf{B}}_0$), $\psi(t)=\int_0^t a(t')dt'$ is the intrinsic time, $a(t')$ is a shift function due to the effects of temperature, aging, etc., $G_{IJKL}(\mathbf{X},\psi-\psi',\psi-\psi'')=G_{KLIJ}(\mathbf{X},\psi-\psi'',\psi-\psi')$, $C_H(\mathbf{X},\psi-\psi',\psi-\psi'')=C_H(\mathbf{X},\psi-\psi'',\psi-\psi')$, and $\bar{\chi}_{IJ}^b(\mathbf{X},\psi-\psi',\psi-\psi'')=\bar{\chi}_{JI}^b(\mathbf{X},\psi-\psi'',\psi-\psi')$.

From Eqs. (9.23)–(9.25), the constitutive equations in finite magneto-thermo-viscoelasticity are obtained using the augmented Helmholtz free energy functional expansion (9.32) as

$$
\begin{aligned}
{}_t\Sigma_{IJ} &= L_{IJ}^0 + \int_{-\infty}^{\psi} G_{IJKL}(\mathbf{X},0,\psi-\psi'')\frac{\partial \mathrm{E}_{KL}(\mathbf{X},\psi'')}{\partial \psi''}d\psi'' \\
&- \int_{-\infty}^{\psi} \beta_{IJ}(\mathbf{X},0,\psi-\psi'')\frac{\partial \theta(\mathbf{X},\psi'')}{\partial \psi''}d\psi'' \\
&- \int_{-\infty}^{\psi} f_{KIJ}^{b}(\mathbf{X},0,\psi-\psi'')\frac{\partial \hat{b}_K(\mathbf{X},\psi'')}{\partial \psi''}d\psi'',
\end{aligned}
\tag{9.33}
$$

$$
\begin{aligned}
\rho_0\hat{s} &= M^0 + \int_{-\infty}^{\psi} \beta_{IJ}(\mathbf{X},\psi-\psi',0)\frac{\partial \mathrm{E}_{IJ}(\mathbf{X},\psi')}{\partial \psi'}d\psi' \\
&+ \frac{1}{T_0}\int_{-\infty}^{\psi} C_H(\mathbf{X},\psi-\psi',0)\frac{\partial \theta(\mathbf{X},\psi')}{\partial \psi'}d\psi' \\
&+ \int_{-\infty}^{\psi} \gamma_I^b(\mathbf{X},\psi-\psi',0)\frac{\partial \hat{b}_I(\mathbf{X},\psi')}{\partial \psi'}d\psi',
\end{aligned}
\tag{9.34}
$$

$$
\begin{aligned}
\hat{H}_I &= N_I^{b0} - \int_{-\infty}^{\psi} f_{IJK}^{b}(\mathbf{X},\psi-\psi',0)\frac{\partial \mathrm{E}_{JK}(\mathbf{X},\psi')}{\partial \psi'}d\psi' \\
&- \int_{-\infty}^{\psi} \gamma_I^b(\mathbf{X},0,\psi-\psi'')\frac{\partial \theta(\mathbf{X},\psi'')}{\partial \psi''}d\psi'' \\
&+ \int_{-\infty}^{\psi} \overline{\chi}_{IJ}^b(\mathbf{X},0,\psi-\psi'')\frac{\partial \hat{b}_J(\mathbf{X},\psi'')}{\partial \psi''}d\psi'',
\end{aligned}
\tag{9.35}
$$

where $G_{IJKL}(\mathbf{X},0,\psi-\psi'')$, $C_H(\mathbf{X},\psi-\psi',0)$, $\overline{\chi}_{IJ}^b(\mathbf{X},0,\psi-\psi'')$, $\beta_{IJ}(\mathbf{X},0,\psi-\psi'')$, $\beta_{IJ}(\mathbf{X},\psi-\psi',0)$, $f_{KIJ}^b(\mathbf{X},0,\psi-\psi'')$, $f_{IJK}^b(\mathbf{X},\psi-\psi',0)$, $\gamma_I^b(\mathbf{X},\psi-\psi',0)$, and $\gamma_I^b(\mathbf{X},0,\psi-\psi'')$ are appropriate memory functions.

The first terms L_{IJ}^0, M^0 and N_I^{b0} on the right-hand sides of Eqs. (9.33)–(9.35) stand for the values of ${}_t\Sigma_{IJ}$, $\rho_0\hat{s}$, and $\hat{H}_I$ in the reference state, the second terms for mechanical contribution, the third terms for thermal contribution, and the fourth terms for magnetic contribution. The dependence of the long-term property functions on aging time, temperature, etc. may be determined from short-term experiments with an accelerated test methodology. Physical aging refers to structural relaxation of the glassy state toward the metastable state, accompanied by changes in almost all physical properties (Hodge, 1995). The experimental observations by Maignan *et al.* (1998) and Dolinek and Jaglii (2002) show that the aging phenomenon exists in the samples for magnetization relaxation measurements and the magnetization data may be fitted with a stretched (fractional) exponential function. The concept of the intrinsic time (also called effective time or reduced time) is used to describe the equivalence of aging time, time, and temperature for polymeric materials (Struik, 1978; Ferry, 1980).

From Eq. (9.26), the viscous dissipation rate in finite magneto-thermo-viscoelasticity is obtained using the augmented Helmholtz free energy functional expansion (9.32) as

$$
\begin{aligned}
\rho_0\tilde{\Lambda} = &-\int_{-\infty}^{\psi}\frac{d\psi}{dt}\frac{\partial L_{IJ}(\mathbf{X},\psi-\psi')}{\partial\psi}\frac{\partial \mathrm{E}_{IJ}(\mathbf{X},\psi')}{\partial\psi'}d\psi' \\
&+\int_{-\infty}^{\psi}\frac{d\psi}{dt}\frac{\partial M(\mathbf{X},\psi-\psi')}{\partial\psi}\frac{\partial\theta(\mathbf{X},\psi')}{\partial\psi'}d\psi' \\
&-\int_{-\infty}^{\psi}\frac{d\psi}{dt}\frac{\partial N_I^b(\mathbf{X},\psi-\psi')}{\partial\psi}\frac{\partial\hat{b}_I(\mathbf{X},\psi')}{\partial\psi'}d\psi' \\
&-\frac{1}{2}\int_{-\infty}^{\psi}\int_{-\infty}^{\psi}\frac{d\psi}{dt}\frac{\partial G_{IJKL}(\mathbf{X},\psi-\psi',\psi-\psi'')}{\partial\psi}\frac{\partial \mathrm{E}_{IJ}(\mathbf{X},\psi')}{\partial\psi'}\frac{\partial \mathrm{E}_{KL}(\mathbf{X},\psi'')}{\partial\psi''}d\psi' d\psi'' \\
&+\int_{-\infty}^{\psi}\int_{-\infty}^{\psi}\frac{d\psi}{dt}\frac{\partial \beta_{IJ}(\mathbf{X},\psi-\psi',\psi-\psi'')}{\partial\psi}\frac{\partial \mathrm{E}_{IJ}(\mathbf{X},\psi')}{\partial\psi'}\frac{\partial\theta(\mathbf{X},\psi'')}{\partial\psi''}d\psi' d\psi'' \\
&+\int_{-\infty}^{\psi}\int_{-\infty}^{\psi}\frac{d\psi}{dt}\frac{\partial f_{KIJ}^b(\mathbf{X},\psi-\psi',\psi-\psi'')}{\partial\psi}\frac{\partial \mathrm{E}_{IJ}(\mathbf{X},\psi')}{\partial\psi'}\frac{\partial\hat{b}_K(\mathbf{X},\psi'')}{\partial\psi''}d\psi' d\psi'' \\
&+\frac{1}{2T_0}\int_{-\infty}^{\psi}\int_{-\infty}^{\psi}\frac{d\psi}{dt}\frac{\partial C_H(\mathbf{X},\psi-\psi',\psi-\psi'')}{\partial\psi}\frac{\partial\theta(\mathbf{X},\psi')}{\partial\psi'}\frac{\partial\theta(\mathbf{X},\psi'')}{\partial\psi''}d\psi' d\psi'' \\
&+\int_{-\infty}^{\psi}\int_{-\infty}^{\psi}\frac{d\psi}{dt}\frac{\partial \gamma_I^b(\mathbf{X},\psi-\psi',\psi-\psi'')}{\partial\psi}\frac{\partial\hat{b}_I(\mathbf{X},\psi')}{\partial\psi'}\frac{\partial\theta(\mathbf{X},\psi'')}{\partial\psi''}d\psi' d\psi'' \\
&-\frac{1}{2}\int_{-\infty}^{\psi}\int_{-\infty}^{\psi}\frac{d\psi}{dt}\frac{\partial \overline{\chi}_{IJ}^b(\mathbf{X},\psi-\psi',\psi-\psi'')}{\partial\psi}\frac{\partial\hat{b}_I(\mathbf{X},\psi')}{\partial\psi'}\frac{\partial\hat{b}_J(\mathbf{X},\psi'')}{\partial\psi''}d\psi' d\psi''.
\end{aligned}
\tag{9.36}
$$

The coupled heat transfer equation is obtained from Eq. (9.31) as

$$\begin{aligned}&\frac{d}{dt}[\int_{-\infty}^{\psi}\beta_{IJ}(\mathbf{X},\psi-\psi',0)\frac{\partial \mathrm{E}_{IJ}(\mathbf{X},\psi')}{\partial \psi'}d\psi'\\&+\frac{1}{T_0}\int_{-\infty}^{\psi}C_H(\mathbf{X},\psi-\psi',0)\frac{\partial \theta(\mathbf{X},\psi')}{\partial \psi'}d\psi'\\&+\int_{-\infty}^{\psi}\gamma_I^b(\mathbf{X},\psi-\psi',0)\frac{\partial \hat{b}_I(\mathbf{X},\psi')}{\partial \psi'}d\psi']\\&=\frac{1}{T_0^3}\nabla_R\cdot\left(\hat{\mathbf{L}}^{qq}\cdot\nabla_R\theta\right)+\frac{1}{T_0}\rho_0\hat{\Lambda},\end{aligned} \tag{9.37}$$

where the integral involving the strain history gives rise to a coupling between thermal and mechanical effects, and the integral involving the magnetic induction history gives rise to a coupling between thermal and magnetic effects.

9.5 Generalized $\tilde{J}$-Integral in Nonlinear Magneto-Thermo-Viscoelastic Fracture

Consider a three-dimensional cracked body (see Fig. 6.1) with the surface $\tilde{\Gamma}$ translating with the crack front moving at a speed $\mathbf{V}_C$. Using Eqs. (9.18), (9.20), (9.27), and (9.28) in nonlinear magneto-thermo-viscoelasticity without the requirement of a constitutive nature, except the existence of the free energy functional, the global form of energy balance leads to the following expression for the energy flux integral:

$$\begin{aligned}F(\tilde{\Gamma})&=\int_{\tilde{\Gamma}}[\mathbf{n}\cdot{}_t\boldsymbol{\sigma}\cdot\mathbf{v}-\mathbf{n}\cdot S+(\tilde{\rho}\tilde{h}+\tilde{\rho}\hat{k})\mathbf{n}\cdot\mathbf{V}_C]d\tilde{\Gamma}\\&=\int_{\partial\tilde{B}}(\mathbf{n}\cdot{}_t\boldsymbol{\sigma}\cdot\mathbf{v}-\mathbf{n}\cdot S)d\tilde{S}-\int_{\tilde{B}-\tilde{V}_{\tilde{\Gamma}}}\frac{\tilde{\partial}}{\tilde{\partial}t}(\tilde{\rho}\tilde{h}+\tilde{\rho}\hat{k})d\tilde{V}\\&+\int_{\tilde{B}-\tilde{V}_{\tilde{\Gamma}}}\tilde{\rho}\hat{\mathbf{f}}\cdot\mathbf{v}d\tilde{V}-\int_{\tilde{B}-\tilde{V}_{\tilde{\Gamma}}}\tilde{\rho}\tilde{s}\dot{T}d\tilde{V}-\int_{\tilde{B}-\tilde{V}_{\tilde{\Gamma}}}\tilde{\rho}\hat{\Lambda}d\tilde{V},\end{aligned} \tag{9.38}$$

The generalized $\tilde{J}$-integral is related to the energy flux integral by

$$\tilde{J}_{\tilde{\Gamma}}=\frac{F(\tilde{\Gamma})}{\dot{A}}, \tag{9.39}$$

where $\dot{A}$ is the crack area growth rate.

The rate of energy flow out of the body and into the crack front per unit crack advance provides the driving force for crack propagation in the presence of magneto-thermo-mechanical coupling and hysteresis effects, that is,

$$\begin{aligned}\tilde{J}_0 &= \lim_{\tilde{\Gamma}_0 \to 0} \left\{ \frac{F(\tilde{\Gamma}_0)}{\dot{A}} \right\} \\ &= \lim_{\tilde{\Gamma}_0 \to 0} \left\{ \frac{1}{\dot{A}} \int_{\tilde{\Gamma}_0} [\mathbf{n} \cdot {}_t\boldsymbol{\sigma} \cdot \mathbf{v} - \mathbf{n} \cdot \boldsymbol{S} + (\tilde{\rho}\tilde{h} + \tilde{\rho}\hat{k})\mathbf{n} \cdot \mathbf{V}_C] d\tilde{\Gamma} \right\}.\end{aligned} \tag{9.40}$$

It can be seen that the above expression for the crack driving force has a universal form for conservative or dissipative systems at small or large deformations under isothermal or nonisothermal conditions.

The relation between the global and local generalized $\tilde{J}$-integrals is obtained from Eq. (9.38) as

$$\begin{aligned}\tilde{J}_g = \tilde{J}_l &+ \frac{1}{\dot{A}} \int_{\tilde{V}_{\tilde{\Gamma}_g} - \tilde{V}_{\tilde{\Gamma}_l}} \frac{\tilde{\partial}}{\tilde{\partial} t} (\tilde{\rho}\tilde{h} + \tilde{\rho}\hat{k}) d\tilde{V} - \frac{1}{\dot{A}} \int_{\tilde{V}_{\tilde{\Gamma}_g} - \tilde{V}_{\tilde{\Gamma}_l}} \tilde{\rho}\hat{\mathbf{f}} \cdot \mathbf{v} d\tilde{V} \\ &+ \frac{1}{\dot{A}} \int_{\tilde{V}_{\tilde{\Gamma}_g} - \tilde{V}_{\tilde{\Gamma}_l}} \tilde{\rho}\tilde{s}\dot{T} d\tilde{V} + \frac{1}{\dot{A}} \int_{\tilde{V}_{\tilde{\Gamma}_g} - \tilde{V}_{\tilde{\Gamma}_l}} \tilde{\rho}\hat{\Lambda} d\tilde{V},\end{aligned} \tag{9.41}$$

where $\tilde{V}_g$ and $\tilde{V}_l$ are the volumes bounded by the surfaces $\tilde{\Gamma}_g$ and $\tilde{\Gamma}_l$, including crack faces.

Consequently, the difference between the global and local generalized $\tilde{J}$-integrals is caused by unsteady state, mechanical body force, temperature change, and viscous dissipation rate. Thus, the generalized $\tilde{J}$-integral loses path independence, even for steady-state crack growth, due to the occurrence of viscous bulk dissipation.

For the accuracy of numerical evaluation by means of finite element analysis, an equivalent path-domain integral expression is given by

$$
\begin{aligned}
\tilde{J}_0 = \hat{J} = &\frac{1}{\hat{A}}\int_{\tilde{\Gamma}}[\mathbf{n}\cdot{}_t\boldsymbol{\sigma}\cdot\mathbf{v} - \mathbf{n}\cdot\boldsymbol{S} + (\tilde{\rho}\tilde{h} + \tilde{\rho}\hat{k})\mathbf{n}\cdot\boldsymbol{V}_C]d\tilde{\Gamma} \\
&- \lim_{\tilde{\Gamma}_0\to 0}\frac{1}{\hat{A}}\int_{\tilde{V}_{\tilde{\Gamma}}-\tilde{V}_{\tilde{\Gamma}_0}}\frac{\tilde{\partial}}{\tilde{\partial} t}(\tilde{\rho}\tilde{h} + \tilde{\rho}\hat{k})d\tilde{V} + \lim_{\tilde{\Gamma}_0\to 0}\frac{1}{\hat{A}}\int_{\tilde{V}_{\tilde{\Gamma}}-\tilde{V}_{\tilde{\Gamma}_0}}\tilde{\rho}\hat{\mathbf{f}}\cdot\mathbf{v}d\tilde{V} \\
&- \lim_{\tilde{\Gamma}_0\to 0}\frac{1}{\hat{A}}\int_{\tilde{V}_{\tilde{\Gamma}}-\tilde{V}_{\tilde{\Gamma}_0}}\tilde{\rho}\hat{s}\dot{T}d\tilde{V} - \frac{1}{\hat{A}}\int_{\tilde{V}_{\tilde{\Gamma}}-\tilde{V}_{\tilde{\Gamma}_0}}\tilde{\rho}\hat{\Lambda}d\tilde{V}.
\end{aligned}
\tag{9.42}
$$

Because of the addition of the domain integral terms reflecting the influence of unsteady state, mechanical body force, temperature change, and viscous dissipation rate, the $\hat{J}$-integral is invariant, that is, path-domain independent.

For a flat, straight, through-crack, if a field quantity is invariant in a reference frame traveling with the crack tip at a uniform speed $\mathbf{V}_C = V_C\overline{\mathbf{E}}_1$, the field quantity depends on t only through the combination $\tilde{\mathbf{X}} = \mathbf{X} - \mathbf{V}_C t$. Under the condition that there exists a steady-state solution for crack propagation in a magneto-thermo-viscoelastic homogeneous medium or FGM, the above expression for the path-domain independent $\hat{J}$-integral becomes

$$
\begin{aligned}
\hat{J} = \tilde{J}_{\tilde{\Gamma}} &- \frac{1}{B}\lim_{\tilde{\Gamma}_0\to 0}\int_{\tilde{V}_{\tilde{\Gamma}}-\tilde{V}_{\tilde{\Gamma}_0}}\left.\frac{\partial(\tilde{\rho}\tilde{h} + \tilde{\rho}\hat{k})}{\partial\tilde{X}_1}\right|_{\text{expl}} d\tilde{V} \\
&- \frac{1}{B}\lim_{\tilde{\Gamma}_0\to 0}\int_{\tilde{V}_{\tilde{\Gamma}}-\tilde{V}_{\tilde{\Gamma}_0}}\tilde{\rho}\hat{f}_i\frac{\partial u_i}{\partial\tilde{X}_1}d\tilde{V} + \frac{1}{B}\lim_{\tilde{\Gamma}_0\to 0}\int_{\tilde{V}_{\tilde{\Gamma}}-\tilde{V}_{\tilde{\Gamma}_0}}\tilde{\rho}\hat{s}\frac{\partial T}{\partial\tilde{X}_1}d\tilde{V} \\
&- \frac{1}{BV_C}\int_{\tilde{V}_{\tilde{\Gamma}}-\tilde{V}_{\tilde{\Gamma}_0}}\tilde{\rho}\hat{\Lambda}d\tilde{V},
\end{aligned}
\tag{9.43}
$$

$$
\tilde{J}_{\tilde{\Gamma}} = \frac{1}{B}\int_{\tilde{\Gamma}}\mathbf{n}\cdot\tilde{\mathbf{b}}d\tilde{\Gamma}\cdot\overline{\mathbf{E}}_1 = \tilde{\mathbf{J}}_{\tilde{\Gamma}}\cdot\overline{\mathbf{E}}_1, \tag{9.44}
$$

$$
\tilde{\mathbf{b}} = -[{}_t\boldsymbol{\sigma} + (\mathbf{B}\cdot\mathbf{H})\mathbf{I} - \mathbf{B}\otimes\mathbf{H}]\cdot\mathbf{u}\tilde{\nabla} + (\tilde{\rho}\tilde{h} + \tilde{\rho}\hat{k})\mathbf{I}, \tag{9.45}
$$

where B is the thickness along the crack front.

The domain integral term involving the explicit derivative of the total energy density with respect to $\tilde{X}_1$ reflects the influence of material inhomogeneity on crack propagation. The generalized $\tilde{J}$-integral can be taken as the projection of the generalized $\tilde{J}_K$-integral vector along the crack advance direction. For the special case of steady-state crack propagation under isothermal conditions in the absence of mechanical body force, viscous bulk dissipation, and material property variation along the crack line, the generalized $\tilde{J}$-integral becomes path independent. With $(\mathbf{n}\times\mathbf{H})\times\mathbf{B}=(\mathbf{n}\cdot\mathbf{B})\mathbf{H}-(\mathbf{H}\cdot\mathbf{B})\mathbf{n}$, the expression for the generalized $\tilde{J}$-integral can be rewritten as

$$\tilde{J}_{\tilde{\Gamma}}=-\frac{1}{B}\int_{\tilde{\Gamma}}\mathbf{n}\cdot{}_{t}\boldsymbol{\sigma}\cdot\mathbf{u}\tilde{\nabla}\cdot\overline{\mathbf{E}}_1 d\tilde{\Gamma}+\frac{1}{B}\int_{\tilde{\Gamma}}[(\mathbf{n}\times\mathbf{H})\times\mathbf{B}]\cdot\mathbf{u}\tilde{\nabla}\cdot\overline{\mathbf{E}}_1 d\tilde{\Gamma}$$

$$+\frac{1}{B}\int_{\tilde{\Gamma}}(\tilde{\rho}\tilde{h}+\tilde{\rho}\hat{k})\mathbf{n}\cdot\overline{\mathbf{E}}_1 d\tilde{\Gamma}. \tag{9.46}$$

It is noted that the generalized $\tilde{J}$-integral and the energy-momentum tensor $\tilde{\mathbf{b}}$ constructed with the use of the augmented Helmholtz free energy, including the contribution of the energy of the free magnetic field, are different from those obtained with the use of the magnetic enthalpy, including or excluding the contribution of the energy of the free magnetic field (Sabir and Maugin, 1996).

9.6 Generalized Plane Crack Problem and Revisit of Mode-III Fracture of a Magnetostrictive Solid in a Bias Magnetic Field

For a generalized plane crack problem in a magnetosensitive solid, we choose the contour as shown in Fig. 9.1. A reference frame is affixed to the crack tip advancing at instantaneous speed V_C. As discussed in previous chapters, this is a convenient choice because $n_1=0$ along the segments parallel to the $\tilde{X}_1$-axis. The contour is shrunk onto the crack tip by first letting $\delta_2\to 0$ and then $\delta_1\to 0$. There is no contribution to $\tilde{J}_0$ from the segments parallel to the $\tilde{X}_2$-axis and the segments along the crack faces.

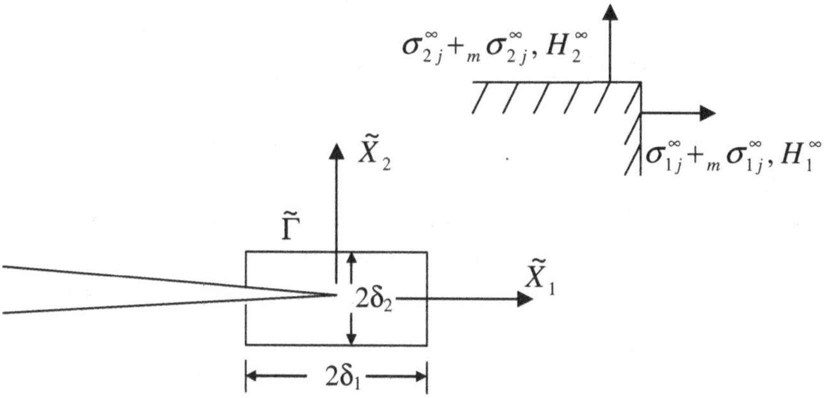

Fig. 9.1 A generalized plane crack problem in a magnetosensitive material. (From Chen, 2009d, with permission from Elsevier.)

Consequently, $\tilde{J}_0$ can be computed by evaluating only the first and second terms on the right-hand side of Eq. (9.46) along the segments parallel to the $\tilde{X}_1$-axis, that is,

$$\begin{aligned}
\tilde{J}_0 = &-2 \lim_{\delta_1 \to 0} \int_{-\delta_1}^{\delta_1} {}_t\sigma_{2j}(\tilde{X}_1, 0^+, t) \frac{\partial u_j(\tilde{X}_1, 0^+, t)}{\partial \tilde{X}_1} d\tilde{X}_1 \\
&+ 2 \lim_{\delta_1 \to 0} \int_{-\delta_1}^{\delta_1} B_2(\tilde{X}_1, 0^+, t) H_1(\tilde{X}_1, 0^+, t) \frac{\partial u_1(\tilde{X}_1, 0^+, t)}{\partial \tilde{X}_1} d\tilde{X}_1 \\
&- 2 \lim_{\delta_1 \to 0} \int_{-\delta_1}^{\delta_1} B_1(\tilde{X}_1, 0^+, t) H_1(\tilde{X}_1, 0^+, t) \frac{\partial u_2(\tilde{X}_1, 0^+, t)}{\partial \tilde{X}_1} d\tilde{X}_1 \qquad (9.47) \\
&- 2 \lim_{\delta_1 \to 0} \int_{-\delta_1}^{\delta_1} B_3(\tilde{X}_1, 0^+, t) H_3(\tilde{X}_1, 0^+, t) \frac{\partial u_2(\tilde{X}_1, 0^+, t)}{\partial \tilde{X}_1} d\tilde{X}_1 \\
&+ 2 \lim_{\delta_1 \to 0} \int_{-\delta_1}^{\delta_1} B_2(\tilde{X}_1, 0^+, t) H_3(\tilde{X}_1, 0^+, t) \frac{\partial u_3(\tilde{X}_1, 0^+, t)}{\partial \tilde{X}_1} d\tilde{X}_1 .
\end{aligned}$$

Since fracture mechanics analysis incorporating nonlinear magneto-thermo-viscoelastic material response is rather complex, numerical methods are necessary for solving this class of problems. There is also a pressing need for comprehensive sets of material data, as systematic experimental work under combined magnetic, thermal, and mechanical loadings is not yet available. To date, most applications to magnetosensitive materials still use conventional fracture mechanics methodology without time dependence.

For mode-III fracture of an isotropic magnetostrictive solid placed in a bias static magnetic field along the crack front, studied by Sabir and Maugin (1996), that is, $\mathbf{H}^0 = H^0\overline{\mathbf{E}}_3$, the crack-tip generalized $\tilde{J}$-integral is calculated from Eq. (9.47) as

$$\tilde{J}_0 = \frac{1}{2}\mu(K^{\mathrm{S}})^2 + \frac{1}{2}bH^0K^{\mathrm{H}}K^{\mathrm{S}}, \tag{9.48}$$

where H^0 is the intensity of the bias static magnetic field, μ is the shear modulus, b is a magnetostriction constant, K^{H} is the magnetic field intensity factor, and K^{S} is the strain intensity factor.

Hence, the crack-tip generalized $\tilde{J}$-integral is an odd function of the magnetic field intensity factor, indicating that the magnetic field either promotes or impedes crack propagation, depending on its direction. A fracture criterion with use of the generalized $\tilde{J}$-integral as the characterizing parameter overcomes the difficulties encountered by other treatments and helps understand the fracture behaviors of both conservative and dissipative material systems subjected to combined magnetic, thermal, and mechanical loadings.

Chapter 10

Electro-Thermo-Viscoelastic Deformation and Fracture

10.1 Introduction

With the growing demand of electroactive polymeric materials for various engineering needs, such as robotic arms and adaptive control systems (see for example, Bar-Cohen, 2002; Dorfman and Ogden, 2005–2006; Vu and Steinmann, 2007), considerable attention has been drawn to the time-dependent response of these smart material systems with novel electronic structures and molecular architecture. Like the magnetosensitive polymers and polymer composites studied in Chapter 9, proper determination of the constitutive relations and fracture criteria is also essential for design analysis and durability assessment of electroactive polymer actuators and sensors under aggressive operation conditions. In addition to piezoelectric, pyroelectric, dielectric, and electrostrictive properties, the hysteresis effect should be considered in analyzing the deformation and fracture behavior of electronic electroactive polymers (e.g., ferroelectric polymers, electroviscoelastic elastomers) and ionic electroactive polymers (e.g., conductive polymers, ionic polymer-metal composites, responsive gels), among others. Crack initiation and growth has a pronounced effect on how such electromechanical devices behave over time. For crack propagation in the presence of electro-thermo-mechanical coupling and hysteresis effects, we seek a physically meaningful quantity whose critical value can be used in a fracture criterion. Due to its great importance for practical applications, the subject of nonlinear electro-thermo-viscoelastic deformation and fracture is addressed separately here.

This chapter commences with the local balance equations and associated boundary and initial conditions for electrosensitive materials subjected to combined electric, thermal, and mechanical loadings. It is then followed by the introduction of the free energy and entropy production inequality for memory-dependent electrosensitive materials in consideration of the augmented Helmholtz free energy, including the contribution of the energy of the free electric field, as a functional of the histories of deformation, temperature, and electric displacement in the reference configuration. This gives rise to nonlinear electro-thermo-viscoelastic constitutive relations, including, as a special case, finite electro-thermo-viscoelasticity for materials with fading memory on an intrinsic time scale. Next, the generalized $\tilde{J}$ -integral is formulated as a physically sound criterion for nonlinear electro-thermo-viscoelastic fracture. Then, the analogy between the nonlinear magneto- and electro-thermo-viscoelastic constitutive and fracture theories is summarized. Finally, reduction to Dorfmann–Ogden nonlinear magneto- and electro-elasticity is discussed.

10.2 Local Balance Equations for Electric, Thermal, and Mechanical Field Quantities

The local balance equations for nonlinear electro-thermo-viscoelastic media under the quasi-electrostatic approximation are summarized below:

$$\nabla \cdot \mathbf{D} = q_f , \tag{10.1}$$

$$\nabla \times \mathbf{E} = 0, \tag{10.2}$$

$$-\frac{\partial q_f}{\partial t} = \nabla \cdot \mathbf{j}_e , \tag{10.3}$$

$$\frac{d\rho}{dt} = -\rho \nabla \cdot \mathbf{v} , \tag{10.4}$$

$$\rho \frac{d\mathbf{v}}{dt} = \nabla \cdot {}_t\boldsymbol{\sigma} + \rho \hat{\mathbf{f}}\,, \tag{10.5}$$

$${}_t\sigma_{ij} = {}_t\sigma_{ji}\,, \tag{10.6}$$

$$\rho \frac{d}{dt}\left(\hat{k} + \hat{e} + \frac{{}_e u^f}{\rho} \right) = -\nabla \cdot \mathbf{j}_q + \nabla \cdot ({}_t\boldsymbol{\sigma} \cdot \mathbf{v} - \boldsymbol{S}) + \rho \hat{\mathbf{f}} \cdot \mathbf{v}\,, \tag{10.7}$$

where q_f is the free body charge density, $\mathbf{j}_e$ is the total electric current, the total stress tensor ${}_t\boldsymbol{\sigma} = \boldsymbol{\sigma} + {}_e\boldsymbol{\sigma}$ is a sum of the Cauchy stress tensor $\boldsymbol{\sigma}$ and the electric stress tensor ${}_e\boldsymbol{\sigma} = \mathbf{P} \otimes \mathbf{E} + \varepsilon_0 \mathbf{E} \otimes \mathbf{E} - {}_e u^f \mathbf{I}$, ${}_e u^f = \varepsilon_0 \mathbf{E} \cdot \mathbf{E}/2$ is the energy density of the free electric field, and $\boldsymbol{S} = (\mathbf{v} \times \mathbf{D}) \times \mathbf{E}$ is the Poynting vector in the co-moving frame.

It can be seen that the electric displacement is not divergence-free in the presence of free body electric charges, in contrast to the magnetic induction. The mathematical boundary-initial value problems for nonlinear electro-thermo-viscoelastic media subjected to combined electric, thermal, and mechanical loadings can be formulated with Gauss's law (10.1), Faraday's law (10.2), the electric charge balance equation (10.3), the mass balance equation (10.4), the linear momentum balance equation (10.5), the angular momentum balance equation (10.6), and the energy balance equation (10.7), together with constitutive relations as well as appropriate boundary and initial conditions.

The boundary conditions are given by

$$\mathbf{n} \cdot [[\mathbf{D}]] = \varpi_f \quad (t \geq 0), \tag{10.8}$$

$$\mathbf{n} \times [[\mathbf{E}]] = 0 \quad (t \geq 0), \tag{10.9}$$

$$\mathbf{n} \cdot [[\mathbf{j}_e - q_f \mathbf{v}]] = 0 \quad (t \geq 0), \tag{10.10}$$

$$\mathbf{n} \cdot {}_t\boldsymbol{\sigma} = \mathbf{t}_B(\mathbf{x}, t) \text{ on } S_\sigma \quad (t \geq 0), \tag{10.11}$$

$$\mathbf{u} = \mathbf{u}_B \text{ on } S_u \quad (t \geq 0), \tag{10.12}$$

$$\mathbf{n}\cdot\mathbf{j}_q = q_B(\mathbf{x},t) \text{ on } S_q \ (t\geq 0), \tag{10.13}$$

$$T = T_B(\mathbf{x},t) \text{ on } S_T \ (t\geq 0), \tag{10.14}$$

where ϖ_f is the free surface charge density, $S_u \cup S_\sigma = S$, and $S_T \cup S_q = S$. Other mixed boundary conditions may also be used.

The initial conditions are taken as

$$\mathbf{u} = \mathbf{u}_0 \ (t<0), \tag{10.15}$$

$$\dot{\mathbf{u}} = \mathbf{v}_0 \ (t=0), \tag{10.16}$$

$$T = T_0 \ (t<0), \tag{10.17}$$

$$\mathbf{D} = \mathbf{D}_0 \ (t<0). \tag{10.18}$$

10.3 Free Energy and Entropy Production Inequality for Memory-Dependent Electrosensitive Materials

By introducing the augmented Helmholtz free energy, including the contribution of the energy of the free electric field, that is,

$$\tilde{h} = \hat{e} - T\hat{s} + \frac{{}_e u^f}{\rho}, \tag{10.19}$$

the local energy balance equation (10.7) becomes

$$\rho\frac{d}{dt}\left(\hat{k} + \tilde{h} + T\hat{s}\right) = -\nabla\cdot\mathbf{j}_q + \nabla\cdot({}_t\boldsymbol{\sigma}\cdot\mathbf{v} - \mathbf{S}) + \rho\hat{\mathbf{f}}\cdot\mathbf{v}. \tag{10.20}$$

In the reference configuration, V_R, the local energy balance equation can be rewritten as

$$\begin{aligned}\frac{d\hat{s}}{dt} &= -\frac{1}{\rho_0}\nabla_R\cdot\frac{\mathbf{J}_q}{T} + \frac{1}{\rho_0}\mathbf{J}_q\cdot\nabla_R\frac{1}{T} + \frac{1}{2\rho_0 T}{}_t\boldsymbol{\Sigma}:\dot{\mathbf{C}} \\ &+ \frac{1}{\rho_0 T}\hat{\mathbf{E}}\cdot\dot{\hat{\mathbf{D}}} + \frac{1}{\rho_0 T}\hat{\mathbf{E}}\cdot\boldsymbol{J}_e - \frac{1}{T}\hat{s}\dot{T} - \frac{1}{T}\frac{d\tilde{h}}{dt}.\end{aligned} \tag{10.21}$$

In the same reference configuration, V_R, the entropy production inequality is expressed as

$$\frac{d_i \hat{s}}{dt} = \frac{d\hat{s}}{dt} + \frac{1}{\rho_0} \nabla_R \cdot \mathbf{J}_s \geq 0. \tag{10.22}$$

For memory-dependent electrosensitive materials, the augmented Helmholtz free energy, including the contribution of the energy of the free electric field, is assumed to be a functional of the histories of deformation, temperature, temperature gradient, and electric displacement in the reference configuration, V_R, with respect to which the deformation gradient $\mathbf{F}$ is measured, that is,

$$\tilde{h} = \tilde{h}(\mathbf{C}(t-\tau), T(t-\tau), \nabla_R T(t-\tau), \hat{\mathbf{D}}(t-\tau); \mathbf{X}). \tag{10.23}$$

10.4 Coupled Electro-Thermo-Viscoelastic Constitutive Relations

Since the entropy production inequality (10.22) is always valid, the state equations should fulfill the following conditions:

$$\frac{\partial \tilde{h}}{\partial T_{,K}} = 0, \tag{10.24}$$

$${}_t\Sigma_{KL} = 2\rho_0 \frac{\partial \tilde{h}}{\partial C_{KL}}, \tag{10.25}$$

$$\hat{s} = -\frac{\partial \tilde{h}}{\partial T}, \tag{10.26}$$

$$\hat{E}_K = \rho_0 \frac{\partial \tilde{h}}{\partial \hat{D}_K}, \tag{10.27}$$

$$\hat{\Lambda} = -\frac{\partial \tilde{h}}{\partial t}, \tag{10.28}$$

$$\mathbf{J}_s = \frac{1}{T}\mathbf{J}_q, \tag{10.29}$$

$$\frac{d_i\hat{s}}{dt} = \frac{1}{\rho_0}\mathbf{J}_q \cdot \nabla_R \frac{1}{T} + \frac{1}{\rho_0 T}\hat{\mathbf{E}} \cdot \boldsymbol{J}_e + \frac{\hat{\Lambda}}{T} \geq 0, \tag{10.30}$$

where $\hat{\Lambda}$ is viscous dissipation rate, which is time-dependent.

From Eq. (10.24), the augmented Helmholtz free energy does not depend on the temperature gradient. Energy can be converted from one form to another due to mechanical, thermal, and electric coupling, accompanied by intrinsic dissipation associated with mechanical, thermal, and electric hysteresis. Since inequality (10.30) must always be satisfied, the kinetic laws for specific irreversible processes may be determined accordingly. Next, finite electro-thermo-viscoelasticity is illustrated, as was finite magneto-thermo-viscoelasticity in Chapter 9.

The viscous dissipation rate satisfies the inequality

$$\hat{\Lambda} \geq 0. \tag{10.31}$$

In the reference configuration, V_R, it is proposed that the thermodynamic fluxes for heat conduction and electricity conduction depend linearly on the corresponding thermodynamic forces with the Onsager reciprocity relations, that is,

$$\mathbf{J}_q = \hat{\mathbf{L}}^{qq} \cdot \nabla_R \frac{1}{T} + \frac{1}{T}\hat{\mathbf{L}}^{qe} \cdot \hat{\mathbf{E}}, \tag{10.32}$$

$$\boldsymbol{J}_e = \hat{\mathbf{L}}^{eq} \cdot \nabla_R \frac{1}{T} + \frac{1}{T}\hat{\mathbf{L}}^{ee} \cdot \hat{\mathbf{E}}, \tag{10.33}$$

where the coefficient matrix

$$\begin{bmatrix} \hat{\mathbf{L}}^{qq} & \hat{\mathbf{L}}^{qe} \\ \hat{\mathbf{L}}^{eq} & \hat{\mathbf{L}}^{ee} \end{bmatrix}^T = \begin{bmatrix} \hat{\mathbf{L}}^{qq} & \hat{\mathbf{L}}^{qe} \\ \hat{\mathbf{L}}^{eq} & \hat{\mathbf{L}}^{ee} \end{bmatrix}, \tag{10.34}$$

is positive definite.

Substituting Eqs. (10.26), (10.32), and (10.33) into Eq. (10.21) yields the following heat transfer equation based on the augmented Helmholtz free energy functional:

$$-\frac{d}{dt}\left(\frac{\partial\tilde{h}}{\partial T}\right)=-\frac{1}{\rho_0 T}\nabla_R\cdot\left(\hat{\mathbf{L}}^{qq}\cdot\nabla_R\frac{1}{T}+\frac{1}{T}\hat{\mathbf{L}}^{qe}\cdot\hat{\mathbf{E}}\right)$$
$$+\frac{1}{\rho_0 T}\hat{\mathbf{E}}\cdot(\hat{\mathbf{L}}^{eq}\cdot\nabla_R\frac{1}{T}+\frac{1}{T}\hat{\mathbf{L}}^{ee}\cdot\hat{\mathbf{E}})+\frac{1}{T}\hat{\Lambda}. \tag{10.35}$$

Substituting Eq. (10.33) into Eq. (10.3) gives the coupled electric charge balance equation:

$$\frac{dQ_f}{dt}+\nabla_R\cdot(\hat{\mathbf{L}}^{eq}\cdot\nabla_R\frac{1}{T}+\frac{1}{T}\hat{\mathbf{L}}^{ee}\cdot\hat{\mathbf{E}})=0, \tag{10.36}$$

where $Q_f = jq_f$.

Using the Lagrange strain measure $\mathbf{E}=(\mathbf{C}-\mathbf{I})/2$, the temperature deviation $\theta=T-T_0$, and the referential electric displacement deviation $\hat{\mathbf{d}}=\hat{\mathbf{D}}-\hat{\mathbf{D}}_0$, expansion of the augmented Helmholtz free energy functional for materials with fading memory on an intrinsic time scale up to the second order yields

$$\rho_0\tilde{h}=\rho_0\tilde{h}_0+\int_{-\infty}^{\psi}L_{IJ}(\mathbf{X},\psi-\psi')\frac{\partial\mathrm{E}_{IJ}(\mathbf{X},\psi')}{\partial\psi'}d\psi'$$
$$-\int_{-\infty}^{\psi}M(\mathbf{X},\psi-\psi')\frac{\partial\theta(\mathbf{X},\psi')}{\partial\psi'}d\psi'+\int_{-\infty}^{\psi}N_I^d(\mathbf{X},\psi-\psi')\frac{\partial\hat{d}_I(\mathbf{X},\psi')}{\partial\psi'}d\psi'$$
$$+\frac{1}{2}\int_{-\infty}^{\psi}\int_{-\infty}^{\psi}G_{IJKL}(\mathbf{X},\psi-\psi',\psi-\psi'')\frac{\partial\mathrm{E}_{IJ}(\mathbf{X},\psi')}{\partial\psi'}\frac{\partial\mathrm{E}_{KL}(\mathbf{X},\psi'')}{\partial\psi''}d\psi'd\psi''$$
$$-\int_{-\infty}^{\psi}\int_{-\infty}^{\psi}\beta_{IJ}(\mathbf{X},\psi-\psi',\psi-\psi'')\frac{\partial\mathrm{E}_{IJ}(\mathbf{X},\psi')}{\partial\psi'}\frac{\partial\theta(\mathbf{X},\psi'')}{\partial\psi''}d\psi'd\psi''$$
$$-\int_{-\infty}^{\psi}\int_{-\infty}^{\psi}f_{KIJ}^d(\mathbf{X},\psi-\psi',\psi-\psi'')\frac{\partial\mathrm{E}_{IJ}(\mathbf{X},\psi')}{\partial\psi'}\frac{\partial\hat{d}_K(\mathbf{X},\psi'')}{\partial\psi''}d\psi'd\psi''$$

$$-\frac{1}{2T_0}\int_{-\infty}^{\psi}\int_{-\infty}^{\psi} C_H(\mathbf{X},\psi-\psi^{'},\psi-\psi^{''})\frac{\partial\theta(\mathbf{X},\psi^{'})}{\partial\psi^{'}}\frac{\partial\theta(\mathbf{X},\psi^{''})}{\partial\psi^{''}}d\psi^{'}d\psi^{''}$$

$$-\int_{-\infty}^{\psi}\int_{-\infty}^{\psi}\gamma_I^d(\mathbf{X},\psi-\psi^{'},\psi-\psi^{''})\frac{\partial\hat{d}_I(\mathbf{X},\psi^{'})}{\partial\psi^{'}}\frac{\partial\theta(\mathbf{X},\psi^{''})}{\partial\psi^{''}}d\psi^{'}d\psi^{''}$$

$$+\frac{1}{2}\int_{-\infty}^{\psi}\int_{-\infty}^{\psi}\overline{\chi}_{IJ}^d(\mathbf{X},\psi-\psi^{'},\psi-\psi^{''})\frac{\partial\hat{d}_I(\mathbf{X},\psi^{'})}{\partial\psi^{'}}\frac{\partial\hat{d}_J(\mathbf{X},\psi^{''})}{\partial\psi^{''}}d\psi^{'}d\psi^{''},$$

(10.37)

where $\tilde{h}_0$ is the value of the augmented Helmholtz free energy in a reference state (i.e., $\mathbf{E}=0, T=T_0, \hat{\mathbf{D}}=\hat{\mathbf{D}}_0$), $\psi(t)=\int_0^t a(t^{'})dt^{'}$ is the intrinsic time, $a(t^{'})$ is a shift function due to the effects of temperature, aging, etc., $G_{IJKL}(\mathbf{X},\psi-\psi^{'},\psi-\psi^{''})=G_{KLIJ}(\mathbf{X},\psi-\psi^{''},\psi-\psi^{'})$, $C_H(\mathbf{X},\psi-\psi^{'},\psi-\psi^{''})=C_H(\mathbf{X},\psi-\psi^{''},\psi-\psi^{'})$, and $\overline{\chi}_{IJ}^d(\mathbf{X},\psi-\psi^{'},\psi-\psi^{''})=\overline{\chi}_{JI}^d(\mathbf{X},\psi-\psi^{''},\psi-\psi^{'})$.

From Eqs. (10.25)–(10.27), the constitutive equations in finite electro-thermo-viscoelasticity are obtained using the augmented Helmholtz free energy functional expansion (10.37) as

$$\begin{aligned}{}_t\Sigma_{IJ} = L_{IJ}^0 &+ \int_{-\infty}^{\psi} G_{IJKL}(\mathbf{X},0,\psi-\psi^{''})\frac{\partial \mathrm{E}_{KL}(\mathbf{X},\psi^{''})}{\partial\psi^{''}}d\psi^{''}\\ &-\int_{-\infty}^{\psi}\beta_{IJ}(\mathbf{X},0,\psi-\psi^{''})\frac{\partial\theta(\mathbf{X},\psi^{''})}{\partial\psi^{''}}d\psi^{''}\\ &-\int_{-\infty}^{\psi} f_{KIJ}^d(\mathbf{X},0,\psi-\psi^{''})\frac{\partial\hat{d}_K(\mathbf{X},\psi^{''})}{\partial\psi^{''}}d\psi^{''},\end{aligned} \tag{10.38}$$

$$\begin{aligned}\rho_0\hat{s} = M^0 &+ \int_{-\infty}^{\psi}\beta_{IJ}(\mathbf{X},\psi-\psi^{'},0)\frac{\partial \mathrm{E}_{IJ}(\mathbf{X},\psi^{'})}{\partial\psi^{'}}d\psi^{'}\\ &+\frac{1}{T_0}\int_{-\infty}^{\psi} C_H(\mathbf{X},\psi-\psi^{'},0)\frac{\partial\theta(\mathbf{X},\psi^{'})}{\partial\psi^{'}}d\psi^{'}\\ &+\int_{-\infty}^{\psi}\gamma_I^d(\mathbf{X},\psi-\psi^{'},0)\frac{\partial\hat{d}_I(\mathbf{X},\psi^{'})}{\partial\psi^{'}}d\psi^{'},\end{aligned} \tag{10.39}$$

$$\hat{E}_I = N_I^{d0} - \int_{-\infty}^{\psi} f_{IJK}^d(\mathbf{X}, \psi - \psi', 0) \frac{\partial \mathrm{E}_{JK}(\mathbf{X}, \psi')}{\partial \psi'} d\psi'$$

$$- \int_{-\infty}^{\psi} \gamma_I^d(\mathbf{X}, 0, \psi - \psi'') \frac{\partial \theta(\mathbf{X}, \psi'')}{\partial \psi''} d\psi'' \tag{10.40}$$

$$+ \int_{-\infty}^{\psi} \overline{\chi}_{IJ}^d(\mathbf{X}, 0, \psi - \psi'') \frac{\partial \hat{d}_J(\mathbf{X}, \psi'')}{\partial \psi''} d\psi'',$$

where $G_{IJKL}(\mathbf{X},0,\psi-\psi'')$, $C_H(\mathbf{X},\psi-\psi',0)$, $\overline{\chi}_{IJ}^d(\mathbf{X},0,\psi-\psi'')$, $\beta_{IJ}(\mathbf{X},0,\psi-\psi'')$, $\beta_{IJ}(\mathbf{X},\psi-\psi',0)$, $f_{KIJ}^d(\mathbf{X},0,\psi-\psi'')$, $f_{IJK}^d(\mathbf{X},\psi-\psi',0)$, $\gamma_I^d(\mathbf{X},\psi-\psi',0)$, and $\gamma_I^d(\mathbf{X},0,\psi-\psi'')$ are appropriate memory functions.

The first terms, L_{IJ}^0, M^0, and N_I^{d0}, on the right-hand sides of Eqs. (10.38)–(10.40) stand for the values of ${}_t\Sigma_{IJ}$, $\rho_0\hat{s}$, and $\hat{E}_I$ in the reference state, the second terms for mechanical contribution, the third terms for thermal contribution, and the fourth terms for electric contribution. The dependence of the long-term property functions on aging time, temperature, etc. may be determined from short-term experiments with an accelerated test methodology. It has been reported that piezoelectric and dielectric properties follow the stretched exponential law (e.g., Zhang *et al.*, 1996, 1997; Koh *et al.*, 2006). The number of material properties required for coupled multifield analysis depends on the material type. For an electrosensitive material with transverse isotropy, there are 18 independent properties: dielectric permittivity (2), heat capacity (1), compliance (5), piezoelectric coefficients (3), pyroelectric coefficients (1), thermal expansion coefficients (2), thermal conductivity (2), and electrical conductivity (2).

From Eq. (10.28), the viscous dissipation rate in finite electro-thermo-viscoelasticity is obtained with the use of the augmented Helmholtz free energy functional expansion (10.37) as

$$
\begin{aligned}
\rho_0\tilde{\Lambda} = & -\int_{-\infty}^{\psi}\frac{d\psi}{dt}\frac{\partial L_{IJ}(\mathbf{X},\psi-\psi')}{\partial\psi}\frac{\partial \mathrm{E}_{IJ}(\mathbf{X},\psi')}{\partial\psi'}d\psi' \\
& +\int_{-\infty}^{\psi}\frac{d\psi}{dt}\frac{\partial M(\mathbf{X},\psi-\psi')}{\partial\psi}\frac{\partial\theta(\mathbf{X},\psi')}{\partial\psi'}d\psi' \\
& -\int_{-\infty}^{\psi}\frac{d\psi}{dt}\frac{\partial N_I^d(\mathbf{X},\psi-\psi')}{\partial\psi}\frac{\partial\hat{b}_I(\mathbf{X},\psi')}{\partial\psi'}d\psi' \\
& -\frac{1}{2}\int_{-\infty}^{\psi}\int_{-\infty}^{\psi}\frac{d\psi}{dt}\frac{\partial G_{IJKL}(\mathbf{X},\psi-\psi',\psi-\psi'')}{\partial\psi}\frac{\partial \mathrm{E}_{IJ}(\mathbf{X},\psi')}{\partial\psi'}\frac{\partial \mathrm{E}_{KL}(\mathbf{X},\psi'')}{\partial\psi''}d\psi' d\psi'' \\
& +\int_{-\infty}^{\psi}\int_{-\infty}^{\psi}\frac{d\psi}{dt}\frac{\partial \beta_{IJ}(\mathbf{X},\psi-\psi',\psi-\psi'')}{\partial\psi}\frac{\partial \mathrm{E}_{IJ}(\mathbf{X},\psi')}{\partial\psi'}\frac{\partial\theta(\mathbf{X},\psi'')}{\partial\psi''}d\psi' d\psi'' \\
& +\int_{-\infty}^{\psi}\int_{-\infty}^{\psi}\frac{d\psi}{dt}\frac{\partial f_{KIJ}^d(\mathbf{X},\psi-\psi',\psi-\psi'')}{\partial\psi}\frac{\partial \mathrm{E}_{IJ}(\mathbf{X},\psi')}{\partial\psi'}\frac{\partial\hat{d}_K(\mathbf{X},\psi'')}{\partial\psi''}d\psi' d\psi'' \\
& +\frac{1}{2T_0}\int_{-\infty}^{\psi}\int_{-\infty}^{\psi}\frac{d\psi}{dt}\frac{\partial C_H(\mathbf{X},\psi-\psi',\psi-\psi'')}{\partial\psi}\frac{\partial\theta(\mathbf{X},\psi')}{\partial\psi'}\frac{\partial\theta(\mathbf{X},\psi'')}{\partial\psi''}d\psi' d\psi'' \\
& +\int_{-\infty}^{\psi}\int_{-\infty}^{\psi}\frac{d\psi}{dt}\frac{\partial \gamma_I^d(\mathbf{X},\psi-\psi',\psi-\psi'')}{\partial\psi}\frac{\partial\hat{d}_I(\mathbf{X},\psi')}{\partial\psi'}\frac{\partial\theta(\mathbf{X},\psi'')}{\partial\psi''}d\psi' d\psi'' \\
& -\frac{1}{2}\int_{-\infty}^{\psi}\int_{-\infty}^{\psi}\frac{d\psi}{dt}\frac{\partial \bar{\chi}_{IJ}^d(\mathbf{X},\psi-\psi',\psi-\psi'')}{\partial\psi}\frac{\partial\hat{d}_I(\mathbf{X},\psi')}{\partial\psi'}\frac{\partial\hat{d}_J(\mathbf{X},\psi'')}{\partial\psi''}d\psi' d\psi''.
\end{aligned}
\tag{10.41}
$$

The coupled heat transfer equation is obtained from Eq. (10.35) as

$$
\begin{aligned}
& \frac{d}{dt}[\int_{-\infty}^{\psi}\beta_{IJ}(\mathbf{X},\psi-\psi',0)\frac{\partial \mathrm{E}_{IJ}(\mathbf{X},\psi')}{\partial\psi'}d\psi' \\
& +\frac{1}{T_0}\int_{-\infty}^{\psi}C_H(\mathbf{X},\psi-\psi',0)\frac{\partial\theta(\mathbf{X},\psi')}{\partial\psi'}d\psi' \\
& +\int_{-\infty}^{\psi}\gamma_I^d(\mathbf{X},\psi-\psi',0)\frac{\partial\hat{d}_I(\mathbf{X},\psi')}{\partial\psi'}d\psi'] \\
& =-\frac{1}{T_0}\nabla_R\cdot(-\frac{1}{T_0^2}\hat{\mathbf{L}}^{qq}\cdot\nabla_R\theta+\frac{1}{T_0}\hat{\mathbf{L}}^{qe}\cdot\hat{\mathbf{E}}) \\
& +\frac{1}{T_0}\hat{\mathbf{E}}\cdot(-\frac{1}{T_0^2}\hat{\mathbf{L}}^{eq}\cdot\nabla_R\theta+\frac{1}{T_0}\hat{\mathbf{L}}^{ee}\cdot\hat{\mathbf{E}})+\frac{1}{T_0}\rho_0\hat{\Lambda},
\end{aligned}
\tag{10.42}
$$

where the integral involving the strain history gives rise to a coupling between thermal and mechanical effects, and the integral involving the electric displacement history gives rise to a coupling between thermal and electric effects.

The coupled electric charge balance equation is obtained from Eq. (10.36) as

$$\frac{dQ_f}{dt} + \nabla_R \cdot \left(-\frac{1}{T_0^2}\hat{\mathbf{L}}^{eq} \cdot \nabla_R \theta + \frac{1}{T_0}\hat{\mathbf{L}}^{ee} \cdot \hat{\mathbf{E}}\right) = 0 . \tag{10.43}$$

10.5 Generalized $\tilde{J}$-Integral in Nonlinear Electro-Thermo-Viscoelastic Fracture

Consider a three-dimensional body $\tilde{B}$ that contains an extending crack with the surface $\tilde{\Gamma}$ translating with the crack front moving at a speed $\mathbf{V}_C$ (see Fig. 6.1). Using Eqs. (10.20), (10.22), (10.29), and (10.30) in nonlinear electro-thermo-viscoelasticity, without the requirement of a constitutive nature except the existence of the free energy functional, the global form of energy balance leads to the following expression for the energy flux integral:

$$\begin{aligned} F(\tilde{\Gamma}) &= \int_{\tilde{\Gamma}} [\mathbf{n} \cdot_t \boldsymbol{\sigma} \cdot \mathbf{v} - \mathbf{n} \cdot S + (\tilde{\rho}\tilde{h} + \tilde{\rho}\hat{k})\mathbf{n} \cdot \mathbf{V}_C] d\tilde{\Gamma} \\ &= \int_{\partial \tilde{B}} (\mathbf{n} \cdot_t \boldsymbol{\sigma} \cdot \mathbf{v} - \mathbf{n} \cdot S) d\tilde{S} - \int_{\tilde{B}-\tilde{V}_{\tilde{\Gamma}}} \frac{\tilde{\partial}}{\tilde{\partial} t} (\tilde{\rho}\tilde{h} + \tilde{\rho}\hat{k}) d\tilde{V} \\ &+ \int_{\tilde{B}-\tilde{V}_{\tilde{\Gamma}}} \tilde{\rho}\hat{\mathbf{f}} \cdot \mathbf{v} d\tilde{V} - \int_{\tilde{B}-\tilde{V}_{\tilde{\Gamma}}} \tilde{\rho}\hat{s}\dot{T} d\tilde{V} - \int_{\tilde{B}-\tilde{V}_{\tilde{\Gamma}}} \frac{\tilde{\rho}}{\rho_0} \hat{\mathbf{E}} \cdot \boldsymbol{J}_e d\tilde{V} - \int_{\tilde{B}-\tilde{V}_{\tilde{\Gamma}}} \tilde{\rho}\hat{\Lambda} d\tilde{V} . \end{aligned} \tag{10.44}$$

The generalized $\tilde{J}$-integral is related to the energy flux integral by

$$\tilde{J}_{\tilde{\Gamma}} = \frac{F(\tilde{\Gamma})}{\dot{A}}, \tag{10.45}$$

where $\dot{A}$ is the crack area growth rate.

The rate of energy flow out of the body and into the crack front per unit crack advance provides the driving force for crack propagation in the presence of electro-thermo-mechanical coupling and hysteresis effects, that is,

$$\tilde{J}_0 = \lim_{\tilde{\Gamma}_0 \to 0} \left\{ \frac{1}{\dot{A}} \int_{\tilde{\Gamma}_0} [\mathbf{n} \cdot_{,t} \boldsymbol{\sigma} \cdot \mathbf{v} - \mathbf{n} \cdot S + (\tilde{\rho}\tilde{h} + \tilde{\rho}\hat{k}) \mathbf{n} \cdot \mathbf{V}_C] d\tilde{\Gamma} \right\}. \quad (10.46)$$

It appears that expression (10.46) for the crack driving force in electrosensitive materials has the same universal form as expression (9.40) for the crack driving force in magnetosensitive materials, both of which can be taken as a generalization of the conventional J-integral method, the dynamic contour integral method, and the crack-tip model for viscoelastic crack initiation and growth discussed in Chapter 1.

The relationship between the global and local generalized $\tilde{J}$-integrals is obtained from Eq. (10.44) as

$$\begin{aligned} \tilde{J}_g = \tilde{J}_l &+ \frac{1}{\dot{A}} \int_{\tilde{V}_{\tilde{\Gamma}_g} - \tilde{V}_{\tilde{\Gamma}_l}} \frac{\tilde{\partial}}{\tilde{\partial} t} (\tilde{\rho}\tilde{h} + \tilde{\rho}\hat{k}) d\tilde{V} - \frac{1}{\dot{A}} \int_{\tilde{V}_{\tilde{\Gamma}_g} - \tilde{V}_{\tilde{\Gamma}_l}} \tilde{\rho}\hat{\mathbf{f}} \cdot \mathbf{v} d\tilde{V} \\ &+ \frac{1}{\dot{A}} \int_{\tilde{V}_{\tilde{\Gamma}_g} - \tilde{V}_{\tilde{\Gamma}_l}} \tilde{\rho}\hat{s}\dot{T} d\tilde{V} + \frac{1}{\dot{A}} \int_{\tilde{V}_g - \tilde{V}_l} \frac{\tilde{\rho}}{\rho_0} \hat{\mathbf{E}} \cdot \boldsymbol{J}_e d\tilde{V} + \frac{1}{\dot{A}} \int_{\tilde{V}_{\tilde{\Gamma}_g} - \tilde{V}_{\tilde{\Gamma}_l}} \tilde{\rho}\hat{\Lambda} d\tilde{V}, \end{aligned} \quad (10.47)$$

where $\tilde{V}_g$ and $\tilde{V}_l$ are the volumes bounded by the surfaces $\tilde{\Gamma}_g$ and $\tilde{\Gamma}_l$, including the crack faces.

Thus, the difference between the global and local generalized $\tilde{J}$-integrals is caused by unsteady state, mechanical body force, temperature change, electricity conduction, and viscous dissipation rate. The Joule heating in conductive or semiconductive polymeric materials contributes to the loss of path independence of the generalized $\tilde{J}$-integral.

For the accuracy of numerical evaluation by means of finite element analysis, an equivalent invariant integral expression is given by

$$\tilde{J}_0 = \hat{J} = \frac{1}{\dot{A}}\int_{\tilde{\Gamma}}[\mathbf{n}\cdot{}_t\boldsymbol{\sigma}\cdot\mathbf{v} - \mathbf{n}\cdot\boldsymbol{S} + (\tilde{\rho}\tilde{h} + \tilde{\rho}\hat{k})\mathbf{n}\cdot\boldsymbol{V}_C]d\tilde{\Gamma}$$
$$-\lim_{\tilde{\Gamma}_0\to 0}\frac{1}{\dot{A}}\int_{\tilde{V}_{\tilde{\Gamma}}-\tilde{V}_{\tilde{\Gamma}_0}}\frac{\tilde{\partial}}{\tilde{\partial}t}(\tilde{\rho}\tilde{h} + \tilde{\rho}\hat{k})d\tilde{V} + \lim_{\tilde{\Gamma}_0\to 0}\frac{1}{\dot{A}}\int_{\tilde{V}_{\tilde{\Gamma}}-\tilde{V}_{\tilde{\Gamma}_0}}\tilde{\rho}\hat{\mathbf{f}}\cdot\mathbf{v}d\tilde{V}$$
$$-\lim_{\tilde{\Gamma}_0\to 0}\frac{1}{\dot{A}}\int_{\tilde{V}_{\tilde{\Gamma}}-\tilde{V}_{\tilde{\Gamma}_0}}\tilde{\rho}\tilde{s}\dot{T}d\tilde{V} - \frac{1}{\dot{A}}\int_{\tilde{B}-\tilde{V}_{\tilde{\Gamma}}}\frac{\tilde{\rho}}{\rho_0}\hat{\mathbf{E}}\cdot\boldsymbol{J}_e d\tilde{V} - \frac{1}{\dot{A}}\int_{\tilde{V}_{\tilde{\Gamma}}-\tilde{V}_{\tilde{\Gamma}_0}}\tilde{\rho}\hat{\Lambda}d\tilde{V}. \tag{10.48}$$

For a flat, straight, through-crack, if a field quantity is invariant in a reference frame traveling with the crack tip at a uniform speed $\mathbf{V}_C = V_C\overline{\mathbf{E}}_1$, the field quantity depends on t only through the combination $\tilde{\mathbf{X}} = \mathbf{X} - \mathbf{V}_C t$. Under the condition that there exists a steady-state solution for crack propagation in an electro-thermo-viscoelastic homogeneous medium or FGM, the above expression for the path-domain independent $\hat{J}$-integral becomes

$$\hat{J} = \tilde{J}_{\tilde{\Gamma}} - \frac{1}{B}\lim_{\tilde{\Gamma}_0\to 0}\int_{\tilde{V}_{\tilde{\Gamma}}-\tilde{V}_{\tilde{\Gamma}_0}}\left.\frac{\partial(\tilde{\rho}\tilde{h} + \tilde{\rho}\hat{k})}{\partial\tilde{X}_1}\right|_{\text{expl}} d\tilde{V}$$
$$-\frac{1}{B}\lim_{\tilde{\Gamma}_0\to 0}\int_{\tilde{V}_{\tilde{\Gamma}}-\tilde{V}_{\tilde{\Gamma}_0}}\tilde{\rho}\hat{f}_i\frac{\partial u_i}{\partial\tilde{X}_1}d\tilde{V} + \frac{1}{B}\lim_{\tilde{\Gamma}_0\to 0}\int_{\tilde{V}_{\tilde{\Gamma}}-\tilde{V}_{\tilde{\Gamma}_0}}\tilde{\rho}\tilde{s}\frac{\partial T}{\partial\tilde{X}_1}d\tilde{V} \tag{10.49}$$
$$-\frac{1}{BV_C}\int_{\tilde{B}-\tilde{V}_{\tilde{\Gamma}}}\frac{\tilde{\rho}}{\rho_0}\hat{\mathbf{E}}\cdot\boldsymbol{J}_e d\tilde{V} - \frac{1}{BV_C}\int_{\tilde{V}_{\tilde{\Gamma}}-\tilde{V}_{\tilde{\Gamma}_0}}\tilde{\rho}\hat{\Lambda}d\tilde{V},$$

$$\tilde{J}_{\tilde{\Gamma}} = \frac{1}{B}\int_{\tilde{\Gamma}}\mathbf{n}\cdot\tilde{\mathbf{b}}d\tilde{\Gamma}\cdot\overline{\mathbf{E}}_1 = \tilde{\mathbf{J}}_{\tilde{\Gamma}}\cdot\overline{\mathbf{E}}_1, \tag{10.50}$$

$$\tilde{\mathbf{b}} = -[{}_t\boldsymbol{\sigma} + (\mathbf{D}\cdot\mathbf{E})\mathbf{I} - \mathbf{D}\otimes\mathbf{E}]\cdot\mathbf{u}\tilde{\nabla} + (\tilde{\rho}\tilde{h} + \tilde{\rho}\hat{k})\mathbf{I}. \tag{10.51}$$

The domain integral term involving the explicit derivative of the total energy density with respect to $\tilde{X}_1$ reflects the influence of material inhomogeneity on crack propagation. The generalized $\tilde{J}$-integral can be taken as the projection of the generalized $\tilde{J}_K$-integral vector along the crack advance direction. It is noted the generalized $\tilde{J}$-integral is path dependent due to the occurrence of viscous dissipation in the bulk material, even for steady-state crack propagation under isothermal conditions in the absence of material property variation, mechanical body force, and electricity conduction. The $\hat{J}$-integral becomes invariant because of the addition of the domain integral terms to account for the effects of property variation, mechanical body force, temperature change, electricity conduction, and viscous dissipation rate in the bulk material.

10.6 Analogy between Nonlinear Magneto- and Electro-Thermo-Viscoelastic Constitutive and Fracture Theories

The nonequilibrium thermodynamic approach enables derivation of nonlinear magneto- and electro-thermo-viscoelastic constitutive and fracture theories in a unified way. The analogy is summarized in Table 10.1. The referential electric displacement can be taken as the thermodynamic dual of the referential electric field under the quasi-electrostatic approximation, whereas the referential magnetic induction can be taken as the thermodynamic dual of the referential magnetic field under the quasi-magnetostatic approximation. The thermodynamic driving force for crack propagation in electro- and magneto-sensitive materials can be expressed as the crack-front generalized $\tilde{J}$-integral, which has a universal form for conservative or dissipative systems at small or large deformations under isothermal or nonisothermal conditions. A fracture criterion based on the generalized $\tilde{J}$-integral thus formulated, without the requirement of a constitutive nature except the existence of the free energy functional, is a generalization of the conventional J-integral method, the dynamic contour integral method, the configuration force (material force) method, and the crack-tip model for viscoelastic crack initiation and growth discussed in Chapter 1.

Both magneto-thermo-viscoelastic fracture and electro-thermo-viscoelastic fracture are time-dependent, involving viscous bulk dissipation which contributes to the difference between the global and local generalized $\tilde{J}$-integrals. The fully dynamic framework for magneto-electro-thermo-elastic fracture presented in Chapter 6 may also be generalized with the inclusion of viscous dissipation in the bulk material.

Table 10.1 Analogy between nonlinear magneto- and electro-thermo-viscoelastic constitutive and fracture theories

	Nonlinear magneto-thermo-viscoelastic constitutive and fracture theory	**Nonlinear electro-thermo-viscoelastic constitutive and fracture theory**
Augmented Helmholtz free energy	$\tilde{h} = \hat{e} - T\hat{s} +_m u^f / \rho$ $= \tilde{h}(\mathbf{C}(t-\tau), T(t-\tau), \hat{\mathbf{B}}(t-\tau); \mathbf{X})$	$\tilde{h} = \hat{e} - T\hat{s} +_e u^f / \rho$ $= \tilde{h}(\mathbf{C}(t-\tau), T(t-\tau), \hat{\mathbf{D}}(t-\tau); \mathbf{X})$
Constitutive equations	${}_t\Sigma_{KL} = 2\rho_0 \frac{\partial \tilde{h}}{\partial C_{KL}}$ $\hat{s} = -\frac{\partial \tilde{h}}{\partial T}$ $\hat{H}_K = \rho_0 \frac{\partial \tilde{h}}{\partial \hat{B}_K}$ $\hat{\Lambda} = -\frac{\partial \tilde{h}}{\partial t}$	${}_t\Sigma_{KL} = 2\rho_0 \frac{\partial \tilde{h}}{\partial C_{KL}}$ $\hat{s} = -\frac{\partial \tilde{h}}{\partial T}$ $\hat{E}_K = \rho_0 \frac{\partial \tilde{h}}{\partial \hat{D}_K}$ $\hat{\Lambda} = -\frac{\partial \tilde{h}}{\partial t}$
Crack driving force	$\tilde{J}_0 = \lim_{\tilde{\Gamma}_0 \to 0} \{\frac{1}{\dot{A}} \int_{\tilde{\Gamma}_0} [\mathbf{n} \cdot_t \boldsymbol{\sigma} \cdot \mathbf{v} - \mathbf{n} \cdot S$ $+ (\tilde{\rho}\tilde{h} + \tilde{\rho}\hat{k})\mathbf{n} \cdot \mathbf{V}_C] d\tilde{\Gamma}\}$	$\tilde{J}_0 = \lim_{\tilde{\Gamma}_0 \to 0} \{\frac{1}{\dot{A}} \int_{\tilde{\Gamma}_0} [\mathbf{n} \cdot_t \boldsymbol{\sigma} \cdot \mathbf{v} - \mathbf{n} \cdot S$ $+ (\tilde{\rho}\tilde{h} + \tilde{\rho}\hat{k})\mathbf{n} \cdot \mathbf{V}_C] d\tilde{\Gamma}\}$
Poynting vector	$\boldsymbol{S} = (\mathbf{v} \times \mathbf{B}) \times \mathbf{H}$	$\boldsymbol{S} = (\mathbf{v} \times \mathbf{D}) \times \mathbf{E}$
Energy-momentum tensor	$\tilde{\mathbf{b}} = -_t\boldsymbol{\sigma} \cdot \mathbf{u}\tilde{\nabla} + \mathbf{B} \otimes \mathbf{H} \cdot \mathbf{u}\tilde{\nabla}$ $- (\mathbf{B} \cdot \mathbf{H})\mathbf{u}\tilde{\nabla} + (\tilde{\rho}\tilde{h} + \tilde{\rho}\hat{k})\mathbf{I}$	$\tilde{\mathbf{b}} = -_t\boldsymbol{\sigma} \cdot \mathbf{u}\tilde{\nabla} + \mathbf{D} \otimes \mathbf{E} \cdot \mathbf{u}\tilde{\nabla}$ $- (\mathbf{D} \cdot \mathbf{E})\mathbf{u}\tilde{\nabla} + (\tilde{\rho}\tilde{h} + \tilde{\rho}\hat{k})\mathbf{I}$

10.7 Reduction to Dorfmann–Ogden Nonlinear Magneto- and Electro-elasticity

The coupled theories of nonlinear magneto- and electro-thermo-viscoelasticity formulated based on nonequilibrium thermodynamics can be reduced to the refined theories of nonlinear magneto- and electro-elastic deformations developed by Dorfmann and Ogden (2004, 2006) as the augmented Helmholtz free energy is taken as a function of basic variables such as deformation, magnetic induction, or electric displacement in the reference configuration. Within this theoretical framework, the boundary-initial value problems can be formulated in a simple and elegant way for proper evaluation of the performance of deformable electro- and magneto-sensitive materials. Alternative formulations which consider the augmented Gibbs free energy to be a function or functional of stress, magnetic field, or electric field in the reference configuration can also be established (Dorfmann and Ogden, 2005–2006; Chen, 2010). New examples of nonlinear constitutive equations for practical applications may be implemented as user subroutines in commercial finite element analysis software packages such as ABAQUS or ANSYS. Nevertheless, there is still a shortage of systematic experimental work to obtain comprehensive sets of much-needed data. It is hoped that the general formulation presented here may provide fundamental guidelines for future experimental and computational work.

Chapter 11

Nonlinear Field Theory of Fracture Mechanics for Paramagnetic and Ferromagnetic Materials

11.1 Introduction

From the viewpoint of global energy balance, Griffith (1921) proposed a fracture theory of brittle materials based on the theorem of minimum potential energy by introducing a specific surface energy on the crack faces, marking an epoch of fracture mechanics as described in Chapter 1. The energy-based approach is fundamental to thermodynamics and continuum mechanics, and is not unique to crack problems. Thermodynamics/thermomechanics has been widely used to study thermoelasticity, electrodynamics, viscoelasticity, inelasticity, plasticity, damage, and fracture (e.g., Schapery, 1964, 1969, 1997, 1999, 2000; Coleman and Gurtin, 1967; Crochet and Naghdi, 1969; Rice, 1971, 1978; Cost, 1973; Eringen, 1980; Christensen, 1982; Truesdell, 1984; Maugin, 1988, 1992; Gurney and Hunt, 1967; Gurney, 1994; Lemaitre, 1996; Fung and Tong, 2001; Makowski and Stumpf, 2001; Truesdell and Noll, 2004; Dorfmann and Ogden, 2003–2006; Chen, 2007, 2009a–e, 2010; Horstemeyer and Bammann, 2010). Magneto- and electro-thermo-mechanical coupling and dissipative effects accompanying crack propagation bring about new challenges in extending conventional fracture mechanics approaches.

In this chapter, a nonlinear field theory of fracture mechanics, which includes magneto-thermo-mechanical coupling and dissipative effects, is formulated from the global energy balance equation and the non-negative global dissipation requirement, following the work of Chen (2009e) as a generalization of the Griffith global energy balance approach. In Section 11.2, the global energy balance equation and the non-negative global dissipation requirement are given for crack propagation under combined magnetic, thermal, and mechanical loadings in the quasi-magnetostatic approximation. In Section 11.3, the Hamiltonian density and the thermodynamic requirement on constitutive laws are provided, based on two types of nonequilibrium thermodynamic approaches: generalized functional thermodynamics and generalized state-variable thermodynamics. In Section 11.4, the thermodynamically consistent time-dependent fracture criterion is expressed in terms of the generalized energy release rate as the thermodynamic driving force conjugate to the crack variable. In Section 11.5, the generalized energy release rate method is proposed for crack propagation in the presence of time-dependent or loading path/history-dependent dissipation in the bulk material. In Section 11.6, the generalized $\tilde{J}$ -integral method is proposed, with the crack-front generalized $\tilde{J}$ -integral equivalent to the generalized energy release rate and the global generalized $\tilde{J}$ -integral including additional contributions due to unsteady, thermal, and dissipative effects. In Section 11.7, the extended essential work of fracture method is proposed, with the specific essential work of fracture equivalent to the crack resistance and the nonessential work of fracture associated with kinetic energy change, temperature change, and time or loading path/history-dependent bulk dissipation.

11.2 Global Energy Balance Equation and Non-Negative Global Dissipation Requirement

Consider a cracked body V_t subjected to combined magnetic, thermal, and mechanical loadings under the quasi-magnetostatic approximation. Following the conservation law of energy, the global energy balance equation over the cracked body V_t is given by

$$\frac{dE^i}{dt}+\frac{dE^k}{dt}=\dot{W}+\dot{Q}\,, \tag{11.1}$$

where E^i is the internal energy, E^k is the kinetic energy, $\dot{W}$ is the power applied by external forces, and $\dot{Q}$ is the heat exchange rate.

The internal energy E^i and the kinetic energy E^k over the cracked body V_t are defined as

$$E^i \equiv \int_{V_t} \rho \hat{e} dV , \tag{11.2}$$

$$E^k \equiv \int_{V_t} \rho \hat{k} dV , \tag{11.3}$$

where $\hat{e}$ is the internal energy per unit mass and $\hat{k} = \mathbf{v} \cdot \mathbf{v}/2$ is the kinetic energy per unit mass.

The power applied by external forces is given by

$$\dot{W} = \int_{\partial V_t} \mathbf{t}_{(n)} \cdot \mathbf{v} dS + \int_{V_t} \rho \hat{\mathbf{f}} \cdot \mathbf{v} dV + \int_{V_t} {}_m w dV , \tag{11.4}$$

where ${}_m w = {}_m\mathbf{f} \cdot \mathbf{v} - \mathbf{M} \cdot \dot{\mathbf{B}}$ is the magnetic power density.

Using the Poynting vector $\boldsymbol{S} = (\mathbf{v} \times \mathbf{B}) \times \mathbf{H}$ in the co-moving frame, Eq. (11.4) leads to

$$\dot{W} = \int_{\partial V_t} \mathbf{n} \cdot (\boldsymbol{\sigma} + {}_m\boldsymbol{\sigma}) \cdot \mathbf{v} dS + \int_{V_t} \rho \hat{\mathbf{f}} \cdot \mathbf{v} dV - \int_{\partial V_t} \mathbf{n} \cdot S dS - \frac{d}{dt} \int_{V_t} {}_m u^f dV . \tag{11.5}$$

The heat flux $\mathbf{j}_q$ is introduced to describe the heat exchange rate through the boundary ∂V_t as

$$\dot{Q} = -\int_{\partial V_t} \mathbf{n} \cdot \mathbf{j}_q dS . \tag{11.6}$$

Hence, the global energy balance equation over the cracked body V_t becomes

$$\begin{aligned} \frac{d}{dt} \int_{V_t} \rho(\hat{k} + \hat{e}) dV = &-\int_{V_t} \nabla \cdot \mathbf{j}_q dV + \int_{V_t} (\nabla \cdot \boldsymbol{\sigma} + \rho \hat{\mathbf{f}} + {}_m\hat{\mathbf{f}}) \cdot \mathbf{v} dV \\ &+ \int_{V_t} \boldsymbol{\sigma} : \mathbf{v}\nabla dV - \int_{V_t} \mathbf{M} \cdot \dot{\mathbf{B}} dV . \end{aligned} \tag{11.7}$$

By introducing the Helmholtz free energy per unit mass,

$$\hat{h} \equiv \hat{e} - T\hat{s}\ , \tag{11.8}$$

Eq. (11.7) becomes

$$\begin{aligned}\int_{V_t} \rho T \frac{d\hat{s}}{dt} dV &= -\int_{V_t} \nabla \cdot \mathbf{j}_q dV + \int_{V_t} (\nabla \cdot \boldsymbol{\sigma} + \rho \hat{\mathbf{f}} +_m \hat{\mathbf{f}}) \cdot \mathbf{v} dV \\ &+ \int_{V_t} (\boldsymbol{\sigma} : \mathbf{v}\nabla - \mathbf{M} \cdot \dot{\mathbf{B}} - \rho \hat{s} \dot{T}) dV - \frac{d}{dt} \int_{V_t} \rho (\hat{k} + \hat{h}) dV.\end{aligned} \tag{11.9}$$

The entropy production is defined as

$$\frac{d_i \hat{s}}{dt} \equiv \frac{d\hat{s}}{dt} + \frac{1}{\rho} \nabla \cdot \mathbf{j}_s . \tag{11.10}$$

The non-negative global dissipation requirement is given by

$$\Phi \equiv \int_{V_t} \rho T \frac{d_i \hat{s}}{dt} dV \geq 0 . \tag{11.11}$$

Using Eqs. (11.9) and (11.10), the non-negative global dissipation requirement (11.11) becomes

$$\begin{aligned}\int_{V_t} \rho T \frac{d_i \hat{s}}{dt} dV &= -\int_{\partial V_t} \mathbf{n} \cdot (\mathbf{j}_q - T\mathbf{j}_s) dS - \int_{V_t} \mathbf{j}_s \cdot \nabla T dV \\ &+ \int_{V_t} \rho \dot{\mathbf{v}} \cdot \mathbf{v} dV + \frac{1}{2} \int_{V_t} j^{-1} {}_t\boldsymbol{\Sigma} : \dot{\mathbf{C}} dV + \int_{V_t} j^{-1} \hat{\mathbf{H}} \cdot \dot{\hat{\mathbf{B}}} dV \\ &- \int_{V_t} \rho \hat{s} \dot{T} dV - \frac{d}{dt} \int_{V_t} \rho \left(\hat{k} + \hat{h} + \frac{{}_m u^f}{\rho} \right) dV \geq 0,\end{aligned} \tag{11.12}$$

where ${}_t\boldsymbol{\Sigma} = j\mathbf{F}^{-1}\, {}_t\boldsymbol{\sigma} \mathbf{F}^{-T}$ is the second Piola–Kirchhoff total stress tensor, ${}_t\boldsymbol{\sigma} = (\boldsymbol{\sigma} +_m \boldsymbol{\sigma})$ is symmetric, and $\mathbf{C} = \mathbf{F}^T \mathbf{F}$ is the right Cauchy–Green deformation tensor, $\hat{\mathbf{H}} = \mathbf{H} \cdot \mathbf{F}$, $\hat{\mathbf{B}} = j\mathbf{F}^{-1} \cdot \mathbf{B}$.

11.3 Hamiltonian Density and Thermodynamically Admissible Conditions

11.3.1 *Generalized functional thermodynamics*

In generalized functional thermodynamics, the augmented Helmholtz free energy for memory-dependent magnetosensitive materials, $\tilde{h} = \hat{h} + {}_m u^f / \rho$, is assumed to be a functional of the histories of deformation, temperature, temperature gradient, and referential magnetic induction with the crack parameter, A, as a state variable:

$$\tilde{h} = \tilde{h}(\mathbf{C}(t-\tau), T(t-\tau), \nabla_R T(t-\tau), \hat{\mathbf{B}}(t-\tau); A, \mathbf{X}) . \tag{11.13}$$

The corresponding Hamiltonian density is given by

$$\begin{aligned} &\mathcal{H}(\mathbf{v}, \mathbf{C}(t-\tau), T(t-\tau), \nabla_R T(t-\tau), \hat{\mathbf{B}}(t-\tau); A, \mathbf{X}) = \\ &\rho\hat{k}(\mathbf{v}; A, \mathbf{X}) + \rho\tilde{h}(\mathbf{C}(t-\tau), T(t-\tau), \nabla_R T(t-\tau), \hat{\mathbf{B}}(t-\tau); A, \mathbf{X}). \end{aligned} \tag{11.14}$$

where $\rho\hat{k}(\mathbf{v}; A, \mathbf{X})$ is the kinetic energy density.

In order that the non-negative global dissipation requirement (11.12) is always valid, it is necessary and sufficient that state equations fulfill the following thermodynamically admissible conditions:

$$\frac{\partial \tilde{h}}{\partial T_{,K}} = 0, \tag{11.15}$$

$${}_t\Sigma_{KL} = 2\rho_0 \frac{\partial \tilde{h}}{\partial C_{KL}}, \tag{11.16}$$

$$\hat{H}_K = \rho_0 \frac{\partial \tilde{h}}{\partial \hat{B}_K}, \tag{11.17}$$

$$\hat{s} = -\frac{\partial \tilde{h}}{\partial T}, \tag{11.18}$$

$$\mathbf{j}_q = T\mathbf{j}_s, \tag{11.19}$$

$$\Phi = \int_{V_t} \rho T \frac{d_i \hat{s}}{dt} dV = \int_{V_t} T\mathbf{j}_q \cdot \nabla \frac{1}{T} dV + \int_{V_t} \rho \hat{\Lambda} dV + \tilde{G}\dot{A} \geq 0, \tag{11.20}$$

where the time-dependent dissipation rate, $\hat{\Lambda}$, and the thermodynamic force, $\tilde{G}$, conjugate to the crack variable, A, are given by

$$\hat{\Lambda} = -\frac{\partial \tilde{h}}{\partial t}, \tag{11.21}$$

$$\tilde{G} = -\frac{\partial}{\partial A} \int_{V_t} \mathcal{H} dV . \tag{11.22}$$

It can be seen that the total dissipation originates from heat conduction, time-dependent bulk dissipation and crack propagation. Since the non-negative global dissipation requirement (11.20) should always be satisfied, the kinetic laws for specific irreversible processes may be determined accordingly, that is,

$$\Phi_T = \int_{V_t} T\mathbf{j}_q \cdot \nabla \frac{1}{T} dV \geq 0, \tag{11.23}$$

$$\Phi_\Lambda = \int_{V_t} \rho \hat{\Lambda} dV \geq 0, \tag{11.24}$$

$$\Phi_A = \tilde{G}\dot{A} \geq 0. \tag{11.25}$$

It is proposed that the thermodynamic flux for heat conduction depends linearly on the corresponding thermodynamic force, that is,

$$\mathbf{j}_q = \mathbf{L}^{qq} \cdot \nabla \frac{1}{T}, \tag{11.26}$$

where $\mathbf{L}^{qq}$ is positive definite.

Substituting Eqs. (11.10, 11.18, 11.19, 11.26) into Eq. (11.20) yields the coupled heat transfer equation for the cracked body as

$$\int_{V_t} \rho T \frac{d}{dt}\left(-\frac{\partial \tilde{h}}{\partial T}\right) dV = -\int_{V_t} \nabla \cdot \left(\mathbf{L}^{qq} \cdot \nabla \frac{1}{T}\right) dV + \int_{V_t} \rho \hat{\Lambda} dV + \tilde{G}\dot{A}. \quad (11.27)$$

The time-dependent dissipation rate in the bulk material satisfies the following inequality:

$$\hat{\Lambda} \geq 0. \quad (11.28)$$

11.3.2 *Generalized state-variable thermodynamics*

In generalized state-variable thermodynamics, the augmented Helmholtz free energy for magnetosensitive materials with dissipative reconfigurations, $\tilde{h} = \hat{h} + {}_m u^f / \rho$, is taken to be a function of current deformation, temperature, temperature gradient, referential magnetic induction, and a set of state variables (scalar, vectorial, or tensorial) at the micro/mesoscale, $\alpha^{(m)}$ $(m = 1,2,\cdots)$, and at the macroscale, $\mathrm{A}^{(n)}$ $(n = 1,2,\cdots)$:

$$\tilde{h} = \tilde{h}(\mathbf{C}, T, \nabla_R T, \hat{\mathbf{B}}; \alpha^{(m)}, \mathrm{A}^{(n)}, \mathbf{X}). \quad (11.29)$$

The corresponding Hamiltonian density is given by

$$\begin{aligned} \mathcal{H}(\mathbf{v}, \mathbf{C}, T, \nabla_R T, \hat{\mathbf{B}}; \alpha^{(m)}, \mathrm{A}^{(n)}, \mathbf{X}) &= \rho \hat{k}(\mathbf{v}; \alpha^{(m)}, \mathrm{A}^{(n)}, \mathbf{X}) \\ &+ \rho \tilde{h}(\mathbf{C}, T, \nabla_R T, \hat{\mathbf{B}}; \alpha^{(m)}, \mathrm{A}^{(n)}, \mathbf{X}). \end{aligned} \quad (11.30)$$

In order that the non-negative global dissipation requirement (11.12) is always valid, it is necessary and sufficient that the state equations fulfill the following thermodynamically admissible conditions:

$$\frac{\partial \tilde{h}}{\partial T_{,K}} = 0, \quad (11.31)$$

$${}_t\Sigma_{KL} = 2\rho_0 \frac{\partial \tilde{h}}{\partial C_{KL}}, \quad (11.32)$$

$$\hat{H}_K = \rho_0 \frac{\partial \tilde{h}}{\partial \hat{B}_K}, \tag{11.33}$$

$$\hat{s} = -\frac{\partial \tilde{h}}{\partial T}, \tag{11.34}$$

$$\mathbf{j}_q = T\mathbf{j}_s, \tag{11.35}$$

$$\begin{aligned} \Phi &= \int_{V_t} \rho T \frac{d_i \hat{s}}{dt} dV \\ &= \int_{V_t} T\mathbf{j}_q \cdot \nabla \frac{1}{T} dV + \sum_m \int_{V_t} j^{-1} \tilde{g}^{(m)} \dot{\alpha}^{(m)} dV + \sum_n \tilde{G}^{(n)} \dot{\mathrm{A}}^{(n)} \geq 0, \end{aligned} \tag{11.36}$$

where the thermodynamic force, $\tilde{g}^{(m)}$, for configuration changes at the micro/mesoscale and the thermodynamic force, $\tilde{G}^{(n)}$, for configuration changes at the macroscale, are given by

$$\tilde{g}^{(m)} = -\rho_0 \frac{\partial \tilde{h}}{\partial \alpha^{(m)}}, \tag{11.37}$$

$$\tilde{G}^{(n)} = -\frac{\partial}{\partial \mathrm{A}^{(n)}} \int_{V_t} \mathcal{H} dV . \tag{11.38}$$

The total dissipation originates from heat conduction and intrinsic dissipative reconfigurations at different scales. For example, the ferromagnetic domain-wall motion corresponds to the change of the associated state variable at the microscale, resulting in the intrinsic bulk dissipation rate. Damage evolution corresponds to the change of the associated state variable at the mesoscale, resulting in the generalized energy density release rate. Crack propagation corresponds to the change of the associated state variable at the macroscale, resulting in the generalized energy release rate. Since the non-negative global dissipation requirement (11.36) should always be satisfied, the kinetic laws for specific irreversible processes may be determined accordingly, that is,

$$\Phi_T = \int_{V_t} T\mathbf{j}_q \cdot \nabla \frac{1}{T} dV \geq 0, \tag{11.39}$$

$$\Phi_\alpha = \sum_m \int_{V_t} j^{-1} \tilde{g}^{(m)} \dot{\alpha}^{(m)} dV \geq 0, \tag{11.40}$$

$$\Phi_{\mathrm{A}} = \sum_n \tilde{G}^{(n)} \dot{\mathrm{A}}^{(n)} \geq 0. \tag{11.41}$$

It is proposed that the thermodynamic flux for heat conduction depends linearly on the corresponding thermodynamic force, that is,

$$\mathbf{j}_q = \mathbf{L}^{qq} \cdot \nabla \frac{1}{T}, \tag{11.42}$$

where $\mathbf{L}^{qq}$ is positive definite.

Substituting Eqs. (11.10, 11.34, 11.35, 11.42) into (11.36) yields the coupled heat transfer equation for the cracked body:

$$\begin{aligned} \int_{V_t} \rho T \frac{d}{dt}\left(-\frac{\partial \tilde{h}}{\partial T}\right) dV &= -\int_{V_t} \nabla \cdot \left(\mathbf{L}^{qq} \cdot \nabla \frac{1}{T}\right) dV \\ &+ \sum_m \int_{V_t} j^{-1} \tilde{g}^{(m)} \dot{\alpha}^{(m)} dV + \sum_n \tilde{G}^{(n)} \dot{\mathrm{A}}^{(n)}. \end{aligned} \tag{11.43}$$

By analogy with the laws for plasticity and damage (Lemaitre, 1996), the evolution laws for dissipative reconfigurations in the bulk material may be derived from a dissipation potential, $\mathcal{D}$, with the normality rule as

$$\dot{\alpha}^{(m)} = \dot{\lambda} \frac{\partial \mathcal{D}}{\partial \tilde{g}^{(m)}}, \tag{11.44}$$

where the multiplier $\dot{\lambda} > 0$ can be determined by the loading condition for rate-independent or rate-dependent cases, respectively.

11.4 Thermodynamically Consistent Time-Dependent Fracture Criterion

In terms of the generalized energy release rate, $\tilde{G}$, defined in Eq. (11.22) or (11.38) as the thermodynamic driving force conjugate to the crack variable, A, crack propagation under combined magnetic, thermal, and mechanical loadings is determined by

$$\tilde{G} = \tilde{R}\,, \tag{11.45}$$

where $\tilde{R}$ is the crack resistance.

The time dependence of the thermodynamically consistent fracture criterion is reflected by the dependence of the explicitly defined free energy functional on the histories of its arguments for magnetosensitive materials with memory, whereas the loading path/history dependence of the thermodynamically consistent fracture criterion is reflected by the dependence of the explicitly defined free energy function on the associated state variables for magnetosensitive materials with dissipative reconfigurations.

For fatigue crack growth under cyclic loading, the crack growth rate may be governed by

$$\frac{da}{dN} = f(\Delta\tilde{G}, r)\,, \tag{11.46}$$

where da/dN is the crack growth per cycle, $\Delta\tilde{G} = (\tilde{G}_{\max} - \tilde{G}_{\min})$, and $r = \tilde{G}_{\min}/\tilde{G}_{\max}$.

11.5 Generalized Energy Release Rate versus Bulk Dissipation Rate

From the non-negative global dissipation requirement (11.20) or (11.36), it can be seen that the global dissipation has three sources: heat conduction, time or loading path/history-dependent material response, and crack propagation. Heat conduction leads to the thermal dissipation term. Structural relaxation or reconfiguration in the bulk material contributes to the time or loading path/history-dependent bulk dissipation term. The energy released during crack growth results in the surface dissipation term.

Substituting Eqs. (11.2, 11.6, 11.8, 11.10, 11.19, 11.20) or Eqs. (11.2, 11.6, 11.8, 11.10, 11.35, 11.36) into the global energy balance equation (11.1) yields

$$\dot{W} = \frac{d}{dt}\int_{V_t} \rho\hat{h}dV + \frac{d}{dt}\int_{V_t} \rho\hat{k}dV + \int_{V_t} \rho\hat{s}\dot{T}dV + \int_{V_t} \rho\hat{\Lambda}dV + \tilde{G}\dot{A}, \quad (11.47)$$

$$\dot{W} = \frac{d}{dt}\int_{V_t} \rho\hat{h}dV + \frac{d}{dt}\int_{V_t} \rho\hat{k}dV + \int_{V_t} \rho\hat{s}\dot{T}dV + \sum_m \int_{V_t} j^{-1}\tilde{g}^{(m)}\dot{\alpha}^{(m)}dV + \tilde{G}\dot{A}. \quad (11.48)$$

For stable crack growth, substituting Eq. (11.45) into Eqs. (11.47) or (11.48) yields

$$\dot{W} = \frac{dH}{dt} + \frac{dE^k}{dt} + \int_{V_t} \rho\hat{s}\dot{T}dV + \int_{V_t} \rho\hat{\Lambda}dV + \tilde{R}\dot{A}, \quad (11.49)$$

$$\dot{W} = \frac{dH}{dt} + \frac{dE^k}{dt} + \int_{V_t} \rho\hat{s}\dot{T}dV + \sum_m \int_{V_t} j^{-1}\tilde{g}^{(m)}\dot{\alpha}^{(m)}dV + \tilde{R}\dot{A}, \quad (11.50)$$

where $H \equiv \int_{V_t} \rho\hat{h}dV$ is the Helmholtz free energy over the cracked body.

With inclusion of magneto-thermo-mechanical coupling and dissipative effects, the generalized energy release rate method is applicable to crack propagation in a broad class of magnetosensitive materials with time dependence or loading path/history dependence. Equation (11.45) with the definition of the generalized energy release rate given by Eq. (11.22) or (11.38) can be taken as a generalization of the strain energy release rate criterion. Equation (11.46) in terms of the generalized energy release rate difference during a loading cycle can be easily adopted to describe the fatigue crack growth rate in the presence of remanent magnetization and remanent deformation under cyclic magnetic, thermal, and mechanical loadings. Equation (11.49) or (11.50) is a generalization of the rate-dependent criterion for viscoelastic or viscoplastic crack growth.

11.6 Local Generalized $\tilde{J}$-Integral versus Global Generalized $\tilde{J}$-Integral

Consider a three-dimensional cracked body $\tilde{V}_{\tilde{\Gamma}}$ bounded by a surface $\tilde{\Gamma}$ in a reference frame ($\tilde{\mathbf{X}} = \mathbf{X} - \mathbf{V}_C t$) traveling with the crack front at speed $\mathbf{V}_C$. With the use of Eqs. (11.2, 11.5, 11.8, 11.10, 11.19, 11.20), the global energy balance equation (11.1) for crack propagation in magnetosensitive materials with memory can be rewritten as

$$\begin{aligned} F(\tilde{\Gamma}) &\equiv \int_{\tilde{\Gamma}} [\mathbf{n}\cdot_t \boldsymbol{\sigma}\cdot\mathbf{v} - \mathbf{n}\cdot S + (\tilde{\rho}\tilde{h} + \tilde{\rho}\hat{k})\mathbf{n}\cdot\mathbf{V}_C] d\tilde{\Gamma} \\ &= \int_{\tilde{V}_{\tilde{\Gamma}}} \frac{\tilde{\partial}}{\tilde{\partial} t}(\tilde{\rho}\tilde{h} + \tilde{\rho}\hat{k}) d\tilde{V} - \int_{\tilde{V}_{\tilde{\Gamma}}} \tilde{\rho}\hat{\mathbf{f}}\cdot\mathbf{v} d\tilde{V} + \int_{\tilde{V}_{\tilde{\Gamma}}} \tilde{\rho}\hat{s}\dot{T} d\tilde{V} \\ &+ \int_{\tilde{V}_{\tilde{\Gamma}}} \tilde{\rho}\hat{\Lambda} d\tilde{V} + \tilde{G}\dot{A}. \end{aligned} \tag{11.51}$$

By contrast, using Eqs. (11.2, 11.5, 11.8, 11.10, 11.35, 11.36), the global energy balance equation (11.1) for crack propagation in magnetosensitive materials with dissipative reconfigurations is rewritten as

$$\begin{aligned} F(\tilde{\Gamma}) &\equiv \int_{\tilde{\Gamma}} [\mathbf{n}\cdot_t \boldsymbol{\sigma}\cdot\mathbf{v} - \mathbf{n}\cdot S + (\tilde{\rho}\tilde{h} + \tilde{\rho}\hat{k})\mathbf{n}\cdot\mathbf{V}_C] d\tilde{\Gamma} \\ &= \int_{\tilde{V}_{\tilde{\Gamma}}} \frac{\tilde{\partial}}{\tilde{\partial} t}(\tilde{\rho}\tilde{h} + \tilde{\rho}\hat{k}) d\tilde{V} - \int_{\tilde{V}_{\tilde{\Gamma}}} \tilde{\rho}\hat{\mathbf{f}}\cdot\mathbf{v} d\tilde{V} + \int_{\tilde{V}_{\tilde{\Gamma}}} \tilde{\rho}\hat{s}\dot{T} d\tilde{V} \\ &+ \sum_m \int_{\tilde{V}_{\tilde{\Gamma}}} j^{-1}\tilde{g}^{(m)}\dot{\alpha}^{(m)} d\tilde{V} + \tilde{G}\dot{A}. \end{aligned} \tag{11.52}$$

The generalized $\tilde{J}$-integral is defined as

$$\tilde{J}_{\tilde{\Gamma}} \equiv \frac{F(\tilde{\Gamma})}{\dot{A}} = \frac{1}{\dot{A}} \int_{\tilde{\Gamma}} [\mathbf{n}\cdot_t \boldsymbol{\sigma}\cdot\mathbf{v} - \mathbf{n}\cdot S + (\tilde{\rho}\tilde{h} + \tilde{\rho}\hat{k})\mathbf{n}\cdot\mathbf{V}_C] d\tilde{\Gamma}. \tag{11.53}$$

From expression (11.53), the crack-front generalized $\tilde{J}$-integral is the energy flux towards the crack front per unit crack advance, which is equivalent to the generalized energy release rate, $\tilde{G}$, serving as the crack driving force, that is,

$$\tilde{J}_0 = \lim_{\tilde{\Gamma}\to 0}\left\{\frac{F(\tilde{\Gamma})}{\dot{A}}\right\} = \tilde{G}. \tag{11.54}$$

The relationship between the global and local generalized $\tilde{J}$-integrals is obtained as

$$\begin{aligned}\tilde{J}_g = \tilde{J}_l &+ \frac{1}{\dot{A}}\int_{\tilde{V}_{\tilde{\Gamma}_g}-\tilde{V}_{\tilde{\Gamma}_l}} \frac{\tilde{\partial}}{\tilde{\partial} t}(\tilde{\rho}\tilde{h} + \tilde{\rho}\hat{k})d\tilde{V} - \frac{1}{\dot{A}}\int_{\tilde{V}_{\tilde{\Gamma}_g}-\tilde{V}_{\tilde{\Gamma}_l}} \tilde{\rho}\hat{\mathbf{f}}\cdot\mathbf{v}d\tilde{V} \\ &+ \frac{1}{\dot{A}}\int_{\tilde{V}_{\tilde{\Gamma}_g}-\tilde{V}_{\tilde{\Gamma}_l}} \tilde{\rho}\hat{s}\dot{T}d\tilde{V} + \frac{1}{\dot{A}}\int_{\tilde{V}_{\tilde{\Gamma}_g}-\tilde{V}_{\tilde{\Gamma}_l}} \tilde{\rho}\hat{\Lambda}d\tilde{V},\end{aligned} \tag{11.55}$$

$$\begin{aligned}\tilde{J}_g = \tilde{J}_l &+ \frac{1}{\dot{A}}\int_{\tilde{V}_{\tilde{\Gamma}_g}-\tilde{V}_{\tilde{\Gamma}_l}} \frac{\tilde{\partial}}{\tilde{\partial} t}(\tilde{\rho}\tilde{h} + \tilde{\rho}\hat{k})d\tilde{V} - \frac{1}{\dot{A}}\int_{\tilde{V}_{\tilde{\Gamma}_g}-\tilde{V}_{\tilde{\Gamma}_l}} \tilde{\rho}\hat{\mathbf{f}}\cdot\mathbf{v}d\tilde{V} \\ &+ \frac{1}{\dot{A}}\int_{\tilde{V}_{\tilde{\Gamma}_g}-\tilde{V}_{\tilde{\Gamma}_l}} \tilde{\rho}\hat{s}\dot{T}d\tilde{V} + \frac{1}{\dot{A}}\sum_m\int_{\tilde{V}_{\tilde{\Gamma}_g}-\tilde{V}_{\tilde{\Gamma}_l}} j^{-1}\tilde{g}^{(m)}\dot{\alpha}^{(m)}d\tilde{V},\end{aligned} \tag{11.56}$$

where $\tilde{V}_{\tilde{\Gamma}_g}$ and $\tilde{V}_{\tilde{\Gamma}_l}$ are the volumes bounded by the closed surfaces $\tilde{\Gamma}_g$ and $\tilde{\Gamma}_l$, including the crack faces.

Thus, the difference between the global generalized $\tilde{J}$-integral and the local generalized $\tilde{J}$-integral is caused by unsteady state, mechanical body force, temperature change, and time or loading path/history-dependent bulk dissipation rate. With the addition of the domain integral terms to the generalized $\tilde{J}$-integral, an invariant integral representation of the generalized energy release rate serving as the crack driving force is obtained as

$$\begin{aligned}\hat{J} \equiv \tilde{J}_{\tilde{\Gamma}} &- \frac{1}{\dot{A}}\int_{\tilde{V}_{\tilde{\Gamma}}} \frac{\tilde{\partial}}{\tilde{\partial} t}(\tilde{\rho}\tilde{h} + \tilde{\rho}\hat{k})d\tilde{V} + \frac{1}{\dot{A}}\int_{\tilde{V}_{\tilde{\Gamma}}} \tilde{\rho}\hat{\mathbf{f}}\cdot\mathbf{v}d\tilde{V} \\ &- \frac{1}{\dot{A}}\int_{\tilde{V}_{\tilde{\Gamma}}} \tilde{\rho}\hat{s}\dot{T}d\tilde{V} - \frac{1}{\dot{A}}\int_{\tilde{V}_{\tilde{\Gamma}}} \tilde{\rho}\hat{\Lambda}d\tilde{V} \\ &= \tilde{G},\end{aligned} \tag{11.57}$$

$$\begin{aligned}\hat{J} &\equiv \tilde{J}_{\tilde{\Gamma}} - \frac{1}{\dot{A}} \int_{\tilde{V}_{\tilde{\Gamma}}} \frac{\tilde{\partial}}{\tilde{\partial} t} (\tilde{\rho}\tilde{h} + \tilde{\rho}\hat{k}) d\tilde{V} + \frac{1}{\dot{A}} \int_{\tilde{V}_{\tilde{\Gamma}}} \tilde{\rho}\hat{\mathbf{f}} \cdot \mathbf{v} d\tilde{V} \\ &- \frac{1}{\dot{A}} \int_{\tilde{V}_{\tilde{\Gamma}}} \tilde{\rho}\hat{s}\dot{T} d\tilde{V} - \frac{1}{\dot{A}} \sum_m \int_{\tilde{V}_{\tilde{\Gamma}}} j^{-1} \tilde{g}^{(m)} \dot{\alpha}^{(m)} dV \\ &= \tilde{G}.\end{aligned} \tag{11.58}$$

By introducing the generalized energy-momentum tensor

$$\tilde{\mathbf{b}} \equiv -{}_{t}\boldsymbol{\sigma} \cdot \mathbf{u}\tilde{\nabla} + \mathbf{B} \otimes \mathbf{H} \cdot \mathbf{u}\tilde{\nabla} - (\mathbf{B} \cdot \mathbf{H}) \mathbf{u}\tilde{\nabla} + (\tilde{\rho}\tilde{h} + \tilde{\rho}\hat{k}) \mathbf{I}, \tag{11.59}$$

the generalized $\tilde{J}$-integral for steady-state propagation of a planar crack in the $\overline{\mathbf{E}}_1$-direction can be expressed as

$$\tilde{J}_{\tilde{\Gamma}} = \frac{1}{B} \int_{\tilde{\Gamma}} \mathbf{n} \cdot \tilde{\mathbf{b}} d\tilde{\Gamma} \cdot \overline{\mathbf{E}}_1 = \tilde{\mathbf{J}}_{\tilde{\Gamma}} \cdot \overline{\mathbf{E}}_1, \tag{11.60}$$

where $\tilde{\mathbf{J}}_{\tilde{\Gamma}}$ is the generalized $\tilde{J}_K$-integral vector.

It is noted that the generalized $\tilde{J}$-integral for magnetosensitive materials with memory or dissipative reconfigurations has the same form, implying that it is universally independent of the material's constitutive nature.

11.7 Essential Work of Fracture versus Nonessential Work of Fracture

Integrating Eq. (11.49) or (11.50) over the time domain gives the following expression for the total work

$$\Delta W = \Delta H + \Delta E^k + \int_{t_0}^{t} \int_{V_t} \rho\hat{s}\dot{T} dV dt + \int_{t_0}^{t} \int_{V_t} \rho\hat{\Lambda} dV dt + \int_{t_0}^{t} \tilde{R}\dot{A} dt, \tag{11.61}$$

$$\Delta W = \Delta H + \Delta E^k + \int_{t_0}^{t} \int_{V_t} \rho\hat{s}\dot{T} dV dt + \int_{t_0}^{t} \sum_m \int_{V_t} j^{-1} \tilde{g}^{(m)} \dot{\alpha}^{(m)} dV dt + \int_{t_0}^{t} \tilde{R}\dot{A} dt. \tag{11.62}$$

From Eq. (11.61) or (11.62), the total work, W_f, from the start of loading until final fracture can be partitioned into the essential work of fracture, W_e, and the nonessential work of fracture, W_{ne}, as

$$W_f = W_e + W_{ne}, \tag{11.63}$$

$$W_e \equiv \int_{t_0}^{t_f} w_e \dot{A} dt = \int_{t_0}^{t_f} \tilde{R} \dot{A} dt, \tag{11.64}$$

$$W_{ne} = \Delta H + \Delta E^k + \int_{t_0}^{t_f} \int_{V_t} \rho \hat{s} \dot{T} dV dt + \int_{t_0}^{t_f} \int_{V_t} \rho \hat{A} dV dt, \tag{11.65}$$

or

$$W_{ne} = \Delta H + \Delta E^k + \int_{t_0}^{t_f} \int_{V_t} \rho \hat{s} \dot{T} dV dt + \int_{t_0}^{t} \sum_m \int_{V_t} j^{-1} \tilde{g}^{(m)} \dot{\alpha}^{(m)} dV dt, \tag{11.66}$$

where w_e is the specific essential work of fracture.

The essential work of fracture is a material property due to its equivalence to the crack resistance, and the nonessential work of fracture is geometry dependent due to its association with kinetic energy change, temperature change, and time or loading path/history-dependent bulk dissipation. Hence, this formulation provides a fundamental basis for extending the simple yet elegant EWF method, as described in Chapter 1, to fracture characterization of magnetosensitive materials involving dynamic, thermal, hysteresis, and other dissipative effects. The critical generalized energy release rate, $\tilde{G}_c$, the critical crack-front generalized $\tilde{J}$-integral, $\tilde{J}_c$, and the specific essential work of fracture, w_e, are equivalent as a measure of fracture toughness.

Chapter 12

Nonlinear Field Theory of Fracture Mechanics for Piezoelectric and Ferroelectric Materials

12.1 Introduction

The preceding chapter presents a general and straightforward formulation of a nonlinear field theory of fracture mechanics for paramagnetic and ferromagnetic materials based on the fundamental principles of thermodynamics. By analogy with magneto-thermo-mechanical coupling and dissipative effects, electro-thermo-mechanical coupling and dissipative effects also bring about new challenges in generalizing the Griffith global energy balance approach and the conventional J-integral method to fracture characterization of electrosensitive materials for a wide variety of applications. For example, it has been realized that domain switching plays an important role in the apparent fracture toughness variation for ferroelectrics, but existing work is predominantly limited to small-scale switching conditions, as reviewed in Chapter 4. It becomes necessary to calculate separately the energy release rate and the rate of bulk dissipation for the fracture of switchable ferroelectrics and electroactive polymers when the effects of bulk dissipation exhibit dependence on geometry and cannot be lumped into a material parameter like the plane-strain fracture toughness.

This chapter focuses on the parallel development of a nonlinear field theory of fracture mechanics for piezoelectric and ferroelectric materials, accounting for the total dissipation associated with heat conduction, electricity conduction, time or loading path/history-dependent bulk dissipation, and crack propagation. In Section 12.2, the nonlinear field equations for a cracked body in the presence of electro-thermo-mechanical coupling and dissipative effects are summarized. In Section 12.3, a thermodynamically consistent time-dependent fracture criterion under combined electric, thermal, and mechanical loadings is obtained from the global energy balance equation and the non-negative global dissipation requirement. In Section 12.4, on the basis of the developed theory, the generalized energy release rate method, the generalized $\tilde{J}$-integral method, and the extended essential work of fracture method are proposed for fracture characterization of piezoelectric and ferroelectric materials, and the interrelation of these methods and their correlations with conventional methods are discussed.

12.2 Nonlinear Field Equations

Nonlinear field equations consist of the balance equations irrespective of material constitution and configuration as well as the constitutive laws characterizing the material nature and configuration change.

12.2.1 *Balance equations*

The Maxwell equations and mass, linear momentum, and angular momentum balance equations under the quasi-electrostatic approximation are summarized in Table 12.1, in comparison with those under the quasi-magnetostatic approximation. The global energy balance equation and the non-negative global dissipation requirement for a cracked body V_t under combined electric, thermal, and mechanical loadings are given in Table 12.2, in comparison with their counterparts under combined magnetic, thermal, and mechanical loadings as described in Chapter 11.

Table 12.1 Balance equations in quasi-electrostatic or quasi-magnetostatic approximation

	Combined magnetic, thermal, and mechanical loadings	**Combined electric, thermal, and mechanical loadings**
Maxwell equations	$\nabla \cdot \mathbf{B} = 0$ $\nabla \times \mathbf{H} = 0$	$\nabla \cdot \mathbf{D} = q_f$ $\nabla \times \mathbf{E} = 0$ $-\dfrac{\partial q_f}{\partial t} = \nabla \cdot \mathbf{j}_e$
Mass balance	$\dfrac{d}{dt}\int_{V_t} \rho dV = 0$ $\dfrac{d\rho}{dt} = -\rho \nabla \cdot \mathbf{v}$	$\dfrac{d}{dt}\int_{V_t} \rho dV = 0$ $\dfrac{d\rho}{dt} = -\rho \nabla \cdot \mathbf{v}$
Linear momentum balance	$\dfrac{d}{dt}\int_{V_t} \rho \mathbf{v} dV = \int_{\partial V_t} \mathbf{n} \cdot \boldsymbol{\sigma} dS$ $+ \int_{V_t} (\rho \hat{\mathbf{f}} + {}_m\mathbf{f}) dV$ $\rho \dfrac{d\mathbf{v}}{dt} = \nabla \cdot (\boldsymbol{\sigma} + {}_m\boldsymbol{\sigma}) + \rho \hat{\mathbf{f}}$	$\dfrac{d}{dt}\int_{V_t} \rho \mathbf{v} dV = \int_{\partial V_t} \mathbf{n} \cdot \boldsymbol{\sigma} dS$ $+ \int_{V_t} (\rho \hat{\mathbf{f}} + {}_e\mathbf{f}) dV$ $\rho \dfrac{d\mathbf{v}}{dt} = \nabla \cdot (\boldsymbol{\sigma} + {}_e\boldsymbol{\sigma}) + \rho \hat{\mathbf{f}}$
Angular momentum balance	$\dfrac{d}{dt}\int_{V_t} \mathbf{r} \times \rho \mathbf{v} dV = \int_{\partial V_t} [\mathbf{r} \times (\mathbf{n} \cdot \boldsymbol{\sigma})] dS$ $+ \int_{V_t} (\mathbf{r} \times \rho \hat{\mathbf{f}} + \mathbf{r} \times_m \hat{\mathbf{f}} + {}_m\mathbf{c}) dV$ $\varepsilon_{kij}(\sigma_{ij} + {}_m\sigma_{ij}) = 0$	$\dfrac{d}{dt}\int_{V_t} \mathbf{r} \times \rho \mathbf{v} dV = \int_{\partial V_t} [\mathbf{r} \times (\mathbf{n} \cdot \boldsymbol{\sigma})] dS$ $+ \int_{V_t} (\mathbf{r} \times \rho \hat{\mathbf{f}} + \mathbf{r} \times_e \hat{\mathbf{f}} + {}_e\mathbf{c}) dV$ $\varepsilon_{kij}(\sigma_{ij} + {}_e\sigma_{ij}) = 0$

Table 12.2 Global energy balance equation and non-negative global dissipation requirement

	Combined magnetic, thermal and mechanical loadings	**Combined electric, thermal and mechanical loadings**
Global energy balance equation	$\frac{d}{dt}\int_{V_t}\rho(\hat{k}+\hat{e})dV=-\int_{V_t}\nabla\cdot\mathbf{j}_q dV$ $+\int_{V_t}(\nabla\cdot\boldsymbol{\sigma}+\rho\hat{\mathbf{f}}+{}_m\hat{\mathbf{f}})\cdot\mathbf{v}dV$ $+\int_{V_t}\boldsymbol{\sigma}:\mathbf{v}\nabla dV-\int_{V_t}\mathbf{M}\cdot\dot{\mathbf{B}}dV$	$\frac{d}{dt}\int_{V_t}\rho(\hat{k}+\hat{e})dV=-\int_{V_t}\nabla\cdot\mathbf{j}_q dV$ $+\int_{V_t}(\nabla\cdot\boldsymbol{\sigma}+\rho\hat{\mathbf{f}}+{}_e\hat{\mathbf{f}})\cdot\mathbf{v}dV$ $+\int_{V_t}\boldsymbol{\sigma}:\mathbf{v}\nabla dV+\int_{V_t}\rho\mathbf{E}\cdot\dot{\boldsymbol{\pi}}dV$ $+\int_{V_t}\mathbf{E}\cdot(\mathbf{j}_e-q_f\mathbf{v})dV$
Non-negative global dissipation requirement	$\int_{V_t}\rho T\frac{d_i\hat{s}}{dt}dV$ $=-\int_{\partial V_t}\mathbf{n}\cdot(\mathbf{j}_q-T\mathbf{j}_s)dS$ $-\int_{V_t}\mathbf{j}_s\cdot\nabla TdV+\int_{V_t}\rho\dot{\mathbf{v}}\cdot\mathbf{v}dV$ $+\frac{1}{2}\int_{V_t}j^{-1}{}_t\boldsymbol{\Sigma}:\dot{\mathbf{C}}dV$ $+\int_{V_t}j^{-1}\hat{\mathbf{H}}\cdot\dot{\hat{\mathbf{B}}}dV-\int_{V_t}\rho\hat{s}\dot{T}dV$ $-\frac{d}{dt}\int_{V_t}\rho\left(\hat{k}+\hat{h}+\frac{{}_mu^f}{\rho}\right)dV\geq 0$	$\int_{V_t}\rho T\frac{d_i\hat{s}}{dt}dV$ $=-\int_{\partial V_t}\mathbf{n}\cdot(\mathbf{j}_q-T\mathbf{j}_s)dS$ $-\int_{V_t}\mathbf{j}_s\cdot\nabla TdV+\int_{V_t}\rho\dot{\mathbf{v}}\cdot\mathbf{v}dV$ $+\int_{V_t}\mathbf{E}\cdot(\mathbf{j}_e-q_f\mathbf{v})dV$ $+\frac{1}{2}\int_{V_t}j^{-1}{}_t\boldsymbol{\Sigma}:\dot{\mathbf{C}}dV$ $+\int_{V_t}j^{-1}\hat{\mathbf{E}}\cdot\dot{\hat{\mathbf{D}}}dV-\int_{V_t}\rho\hat{s}\dot{T}dV$ $-\frac{d}{dt}\int_{V_t}\rho\left(\hat{k}+\hat{h}+\frac{{}_eu^f}{\rho}\right)dV\geq 0$

12.2.2 *Constitutive laws*

In generalized functional thermodynamics, the augmented Helmholtz free energy for memory-dependent electrosensitive materials, $\tilde{h}=\hat{h}+{}_eu^f/\rho$, is assumed to be a functional of the histories of deformation, temperature, temperature gradient, and referential electric displacement, with the crack parameter, A, as a state variable:

$$\tilde{h}=\tilde{h}(\mathbf{C}(t-\tau),T(t-\tau),\nabla_R T(t-\tau),\hat{\mathbf{D}}(t-\tau);A,\mathbf{X}). \qquad (12.1)$$

The corresponding Hamiltonian density is given by

$$\begin{aligned}&\mathcal{H}(\mathbf{v},\mathbf{C}(t-\tau),T(t-\tau),\nabla_R T(t-\tau),\hat{\mathbf{D}}(t-\tau);A,\mathbf{X})=\\&\rho\hat{k}(\mathbf{v};A,\mathbf{X})+\rho\tilde{h}(\mathbf{C}(t-\tau),T(t-\tau),\nabla_R T(t-\tau),\hat{\mathbf{D}}(t-\tau);A,\mathbf{X}).\end{aligned} \tag{12.2}$$

In generalized state-variable thermodynamics, the augmented Helmholtz free energy for electrosensitive materials with dissipative reconfigurations, $\tilde{h}=\hat{h}+{}_e u^f/\rho$, is assumed to be a function of current deformation, temperature, temperature gradient, referential electric displacement, and a set of state variables (scalar, vectorial, or tensorial) at the micro/mesoscale, $\alpha^{(m)}$ $(m=1,2,\cdots)$, and at the macroscale, $\mathrm{A}^{(n)}$ $(n=1,2,\cdots)$:

$$\tilde{h}=\tilde{h}(\mathbf{C},T,\nabla_R T,\hat{\mathbf{D}};\alpha^{(m)},\mathrm{A}^{(n)},\mathbf{X}). \tag{12.3}$$

The corresponding Hamiltonian density is given by

$$\begin{aligned}&\mathcal{H}(\mathbf{v},\mathbf{C},T,\nabla_R T,\hat{\mathbf{D}};\alpha^{(m)},\mathrm{A}^{(n)},\mathbf{X})=\rho\hat{k}(\mathbf{v};\alpha^{(m)},\mathrm{A}^{(n)},\mathbf{X})\\&+\rho\tilde{h}(\mathbf{C},T,\nabla_R T,\hat{\mathbf{D}};\alpha^{(m)},\mathrm{A}^{(n)},\mathbf{X}).\end{aligned} \tag{12.4}$$

The Hamiltonain density and the thermodynamically admissible conditions for electrosensitive materials with memory or dissipative reconfigurations are given in Table 12.3 for comparison.

12.3 Thermodynamically Consistent Time-Dependent Fracture Criterion

By analogy with Section 11.4, in terms of the generalized energy release rate, $\tilde{G}$, as the thermodynamic driving force conjugate to the crack variable, A, crack propagation under combined electric, thermal, and mechanical loadings is determined by

$$\tilde{G}=\tilde{R}, \tag{12.5}$$

where $\tilde{R}$ is the crack resistance.

Table 12.3 Hamiltonian density and thermodynamically admissible conditions for electrosensitive materials with memory or dissipative reconfigurations

	Electrosensitive materials with memory	**Electrosensitive materials with dissipative reconfigurations**
Hamiltonian density	$\mathcal{H}(\mathbf{v}, \mathbf{C}(t-\tau), T(t-\tau), \nabla_R T(t-\tau), \hat{\mathbf{D}}(t-\tau); A, \mathbf{X}) = \rho\hat{k}(\mathbf{v}; A, \mathbf{X}) + \rho\tilde{h}(\mathbf{C}(t-\tau), T(t-\tau), \nabla_R T(t-\tau), \hat{\mathbf{D}}(t-\tau); A, \mathbf{X})$	$\mathcal{H}(\mathbf{v}, \mathbf{C}, T, \nabla_R T, \hat{\mathbf{D}}; \alpha^{(m)}, \mathrm{A}^{(n)}, \mathbf{X}) = \rho\hat{k}(\mathbf{v}; \alpha^{(m)}, \mathrm{A}^{(n)}, \mathbf{X}) + \rho\tilde{h}(\mathbf{C}, T, \nabla_R T, \hat{\mathbf{D}}; \alpha^{(m)}, \mathrm{A}^{(n)}, \mathbf{X})$
Constitutive laws	$_t\Sigma_{KL} = 2\rho_0 \frac{\partial \tilde{h}}{\partial C_{KL}}$ $\hat{s} = -\frac{\partial \tilde{h}}{\partial T}$ $\hat{E}_K = \rho_0 \frac{\partial \tilde{h}}{\partial \hat{D}_K}$ $\hat{\Lambda} = -\frac{\partial \tilde{h}}{\partial t}$ $\tilde{G} = -\frac{\partial}{\partial A}\int_{V_t} \mathcal{H} dV$	$_t\Sigma_{KL} = 2\rho_0 \frac{\partial \tilde{h}}{\partial C_{KL}}$ $\hat{s} = -\frac{\partial \tilde{h}}{\partial T}$ $\hat{E}_K = \rho_0 \frac{\partial \tilde{h}}{\partial \hat{D}_K}$ $\tilde{g}^{(m)} = -\rho_0 \frac{\partial \tilde{h}}{\partial \alpha^{(m)}}$ $\tilde{G}^{(n)} = -\frac{\partial}{\partial \mathrm{A}^{(n)}}\int_{V_t} \mathcal{H} dV$
Transport laws	$\mathbf{j}_q = \mathbf{L}^{qq} \cdot \nabla \frac{1}{T} + \frac{1}{T}\mathbf{L}^{qe} \cdot \mathbf{E}$ $\mathbf{j}_e - q_f \mathbf{v} = \mathbf{L}^{eq} \cdot \nabla \frac{1}{T} + \frac{1}{T}\mathbf{L}^{ee} \cdot \mathbf{E}$	$\mathbf{j}_q = \mathbf{L}^{qq} \cdot \nabla \frac{1}{T} + \frac{1}{T}\mathbf{L}^{qe} \cdot \mathbf{E}$ $\mathbf{j}_e - q_f \mathbf{v} = \mathbf{L}^{eq} \cdot \nabla \frac{1}{T} + \frac{1}{T}\mathbf{L}^{ee} \cdot \mathbf{E}$
Bulk dissipation rate	$\hat{\Lambda} \geq 0$	$\sum_m \tilde{g}^{(m)} \dot{\alpha}^{(m)} \geq 0$
Crack propagation	$\tilde{G}\dot{A} \geq 0$	$\sum_n \tilde{G}^{(n)} \dot{\mathrm{A}}^{(n)} \geq 0$

The time dependence of the thermodynamically consistent fracture criterion is reflected by the dependence of the explicitly defined free energy functional on the histories of its arguments for electrosensitive materials with memory, whereas the loading path/history dependence of the thermodynamically consistent fracture criterion is reflected by the dependence of the explicitly defined free energy function on the associated state variables for electrosensitive materials with dissipative reconfigurations.

For fatigue crack growth under cyclic loading, the crack growth rate may be governed by

$$\frac{da}{dN} = f(\Delta\tilde{G}, r), \qquad (12.6)$$

where da/dN is the crack growth per cycle, $\Delta\tilde{G} = (\tilde{G}_{\max} - \tilde{G}_{\min})$, and $r = \tilde{G}_{\min}/\tilde{G}_{\max}$.

12.4 Correlation with Conventional Fracture Mechanics Approaches

Like magnetosensitive materials with time or loading path/history dependence as described in Section 11.5, the global energy balance equation for electrosensitive materials with time or loading path/history dependence may be rewritten as

$$\begin{aligned}\dot{W} &= \frac{d}{dt}\int_{V_t} \rho\hat{h}dV + \frac{d}{dt}\int_{V_t} \rho\hat{k}dV + \int_{V_t} \rho\hat{s}\dot{T}dV \\ &+ \int_{V_t} \mathbf{E}\cdot(\mathbf{j}_e - q_f\mathbf{v})dV + \int_{V_t} \rho\hat{\Lambda}dV + \tilde{G}\dot{A},\end{aligned} \qquad (12.7)$$

$$\begin{aligned}\dot{W} &= \frac{d}{dt}\int_{V_t} \rho\hat{h}dV + \frac{d}{dt}\int_{V_t} \rho\hat{k}dV + \int_{V_t} \rho\hat{s}\dot{T}dV \\ &+ \int_{V_t} \mathbf{E}\cdot(\mathbf{j}_e - q_f\mathbf{v})dV + \sum_m \int_{V_t} j^{-1}\tilde{g}^{(m)}\dot{\alpha}^{(m)}dV + \tilde{G}\dot{A}.\end{aligned} \qquad (12.8)$$

For stable crack growth, substituting Eq. (12.5) into Eq. (12.7) or (12.8) yields

$$\dot{W} = \frac{dH}{dt} + \frac{dE^k}{dt} + \int_{V_t} \rho\hat{s}\dot{T}dV + \int_{V_t} \mathbf{E}\cdot(\mathbf{j}_e - q_f\mathbf{v})dV \\ + \int_{V_t} \rho\hat{\Lambda}dV + \tilde{R}\dot{A}, \tag{12.9}$$

$$\dot{W} = \frac{dH}{dt} + \frac{dE^k}{dt} + \int_{V_t} \rho\hat{s}\dot{T}dV + \int_{V_t} \mathbf{E}\cdot(\mathbf{j}_e - q_f\mathbf{v})dV \\ + \sum_m \int_{V_t} j^{-1}\tilde{g}^{(m)}\dot{\alpha}^{(m)}dV + \tilde{R}\dot{A}. \tag{12.10}$$

The generalized energy release rate serves as the thermodynamic driving force for quasi-static and dynamic crack propagation in homogeneous or nonhomogeneous, conservative or dissipative materials, which is analogous to the generalized energy density release rate for damage evolution. Hence, the thermodynamically consistent formulation based on the global energy balance equation and the non-negative global dissipation requirement unifies the way to handle crack propagation and damage evolution. Equation (12.5) is a generalization of the strain energy release rate criterion. Equation (12.6) is an extension of the fatigue crack growth criterion under cyclic mechanical loading. Equation (12.9) or (12.10) is a generalization of the rate-dependent criterion for viscoelastic or viscoplastic crack growth.

Consider a three-dimensional cracked body $\tilde{V}_{\tilde{\Gamma}}$ bounded by a surface $\tilde{\Gamma}$ in a reference frame ($\tilde{\mathbf{X}} = \mathbf{X} - \mathbf{V}_C t$) traveling with the crack front at speed $\mathbf{V}_C$. The global energy balance equation (12.7) or (12.8) for crack propagation in electrosensitive materials with memory or dissipative reconfigurations can be rewritten as

$$F(\tilde{\Gamma}) \equiv \int_{\tilde{\Gamma}} [\mathbf{n}\cdot_t \boldsymbol{\sigma}\cdot\mathbf{v} - \mathbf{n}\cdot S + (\tilde{\rho}\tilde{h} + \tilde{\rho}\hat{k})\mathbf{n}\cdot\mathbf{V}_C]d\tilde{\Gamma} \\ = \int_{\tilde{V}_{\tilde{\Gamma}}} \frac{\tilde{\partial}}{\tilde{\partial}t}(\tilde{\rho}\tilde{h} + \tilde{\rho}\hat{k})d\tilde{V} - \int_{\tilde{V}_{\tilde{\Gamma}}} \tilde{\rho}\hat{\mathbf{f}}\cdot\mathbf{v}d\tilde{V} + \int_{\tilde{V}_{\tilde{\Gamma}}} \tilde{\rho}\hat{s}\dot{T}d\tilde{V} \\ + \int_{\tilde{V}_{\tilde{\Gamma}}} \frac{\tilde{\rho}}{\rho}\mathbf{E}\cdot(\mathbf{j}_e - q_f\mathbf{v})d\tilde{V} + \int_{\tilde{V}_{\tilde{\Gamma}}} \tilde{\rho}\hat{\Lambda}d\tilde{V} + \tilde{G}\dot{A}, \tag{12.11}$$

$$
\begin{aligned}
F(\tilde{\Gamma}) &\equiv \int_{\tilde{\Gamma}} [\mathbf{n}\cdot_t \boldsymbol{\sigma}\cdot\mathbf{v} - \mathbf{n}\cdot \boldsymbol{S} + (\tilde{\rho}\tilde{h} + \tilde{\rho}\hat{k})\mathbf{n}\cdot\mathbf{V}_C] d\tilde{\Gamma} \\
&= \int_{\tilde{V}_{\tilde{\Gamma}}} \frac{\tilde{\partial}}{\tilde{\partial} t}(\tilde{\rho}\tilde{h} + \tilde{\rho}\hat{k}) d\tilde{V} - \int_{\tilde{V}_{\tilde{\Gamma}}} \tilde{\rho}\hat{\mathbf{f}}\cdot\mathbf{v} d\tilde{V} + \int_{\tilde{V}_{\tilde{\Gamma}}} \tilde{\rho}\hat{s}\dot{T} d\tilde{V} \\
&\quad + \int_{\tilde{V}_{\tilde{\Gamma}}} \frac{\tilde{\rho}}{\rho}\mathbf{E}\cdot(\mathbf{j}_e - q_f\mathbf{v}) d\tilde{V} + \sum_m \int_{\tilde{V}_{\tilde{\Gamma}}} j^{-1}\tilde{g}^{(m)}\dot{\alpha}^{(m)} d\tilde{V} + \tilde{G}\dot{A}.
\end{aligned} \tag{12.12}
$$

The generalized $\tilde{J}$-integral is related to the energy flux in the same way as defined in Eq. (11.53), that is,

$$
\tilde{J}_{\tilde{\Gamma}} \equiv \frac{F(\tilde{\Gamma})}{\dot{A}} = \frac{1}{\dot{A}}\int_{\tilde{\Gamma}} [\mathbf{n}\cdot_t \boldsymbol{\sigma}\cdot\mathbf{v} - \mathbf{n}\cdot\boldsymbol{S} + (\tilde{\rho}\tilde{h} + \tilde{\rho}\hat{k})\mathbf{n}\cdot\mathbf{V}_C] d\tilde{\Gamma}. \tag{12.13}
$$

It is noted that the generalized $\tilde{J}$-integral for electro- or magneto-sensitive materials with memory or dissipative reconfigurations has an identical form. This formulation further extends the generalized $\tilde{J}$-integral concept developed in Sections 9.5 and 10.5 for nonlinear magneto- and electro-thermo-viscoelastic fracture.

As the surface $\tilde{\Gamma} \to 0$, the crack-front generalized $\tilde{J}$-integral is related to the generalized energy release rate $\tilde{G}$ by

$$
\tilde{J}_0 = \lim_{\tilde{\Gamma}\to 0}\left\{\frac{F(\tilde{\Gamma})}{\dot{A}}\right\} = \tilde{G}. \tag{12.14}
$$

The relationship between the global and local generalized $\tilde{J}$-integrals is obtained as

$$
\begin{aligned}
\tilde{J}_g &= \tilde{J}_l + \frac{1}{\dot{A}}\int_{\tilde{V}_{\tilde{\Gamma}_g} - \tilde{V}_{\tilde{\Gamma}_l}} \frac{\tilde{\partial}}{\tilde{\partial} t}(\tilde{\rho}\tilde{h} + \tilde{\rho}\hat{k}) d\tilde{V} - \frac{1}{\dot{A}}\int_{\tilde{V}_{\tilde{\Gamma}_g} - \tilde{V}_{\tilde{\Gamma}_l}} \tilde{\rho}\hat{\mathbf{f}}\cdot\mathbf{v} d\tilde{V} \\
&\quad + \frac{1}{\dot{A}}\int_{\tilde{V}_{\tilde{\Gamma}_g} - \tilde{V}_{\tilde{\Gamma}_l}} \tilde{\rho}\hat{s}\dot{T} d\tilde{V} + \int_{\tilde{V}_{\tilde{\Gamma}_g} - \tilde{V}_{\tilde{\Gamma}_l}} \frac{\tilde{\rho}}{\rho}\mathbf{E}\cdot(\mathbf{j}_e - q_f\mathbf{v}) d\tilde{V} \\
&\quad + \frac{1}{\dot{A}}\int_{\tilde{V}_{\tilde{\Gamma}_g} - \tilde{V}_{\tilde{\Gamma}_l}} \tilde{\rho}\hat{\Lambda} d\tilde{V},
\end{aligned} \tag{12.15}
$$

$$\tilde{J}_g = \tilde{J}_l + \frac{1}{\dot{A}} \int_{\tilde{V}_{\tilde{\Gamma}_g} - \tilde{V}_{\tilde{\Gamma}_l}} \frac{\tilde{\partial}}{\tilde{\partial} t} (\tilde{\rho} \tilde{h} + \tilde{\rho} \hat{k}) d\tilde{V} - \frac{1}{\dot{A}} \int_{\tilde{V}_{\tilde{\Gamma}_g} - \tilde{V}_{\tilde{\Gamma}_l}} \tilde{\rho} \hat{\mathbf{f}} \cdot \mathbf{v} d\tilde{V}$$
$$+ \frac{1}{\dot{A}} \int_{\tilde{V}_{\tilde{\Gamma}_g} - \tilde{V}_{\tilde{\Gamma}_l}} \tilde{\rho} \hat{s} \dot{T} d\tilde{V} + \int_{\tilde{V}_{\tilde{\Gamma}_g} - \tilde{V}_{\tilde{\Gamma}_l}} \frac{\tilde{\rho}}{\rho} \mathbf{E} \cdot (\mathbf{j}_e - q_f \mathbf{v}) d\tilde{V} \tag{12.16}$$
$$+ \frac{1}{\dot{A}} \sum_m \int_{\tilde{V}_{\tilde{\Gamma}_g} - \tilde{V}_{\tilde{\Gamma}_l}} j^{-1} \tilde{g}^{(m)} \dot{\alpha}^{(m)} d\tilde{V},$$

where $\tilde{V}_{\tilde{\Gamma}_g}$ and $\tilde{V}_{\tilde{\Gamma}_l}$ are the volumes bounded by the closed surfaces $\tilde{\Gamma}_g$ and $\tilde{\Gamma}_l$, including the crack faces.

Thus, the difference between the global generalized $\tilde{J}$-integral and local generalized $\tilde{J}$-integral is caused by unsteady state, mechanical body force, temperature change, electricity conduction, and time or loading path/history-dependent bulk dissipation rate. As a generalization of the conventional J-integral method, the dynamic contour integral method, and Schapery's crack-tip model for viscoelastic facture, as described in Chapter 1, the generalized $\tilde{J}$-integral method is applicable to arbitrary transient crack problems in the presence of electro-thermo-mechanical coupling and dissipative effects.

By introducing the generalized energy-momentum tensor

$$\tilde{\mathbf{b}} \equiv -{}_t\boldsymbol{\sigma} \cdot \mathbf{u}\tilde{\nabla} + \mathbf{D} \otimes \mathbf{E} \cdot \mathbf{u}\tilde{\nabla} - (\mathbf{D} \cdot \mathbf{E})\mathbf{u}\tilde{\nabla} + (\tilde{\rho}\tilde{h} + \tilde{\rho}\hat{k})\mathbf{I}, \tag{12.17}$$

the generalized $\tilde{J}$-integral for steady-state propagation of a planar crack in the $\overline{\mathbf{E}}_1$-direction can be expressed as

$$\tilde{J}_{\tilde{\Gamma}} = \frac{1}{B} \int_{\tilde{\Gamma}} \mathbf{n} \cdot \tilde{\mathbf{b}} d\tilde{\Gamma} \cdot \overline{\mathbf{E}}_1 = \tilde{\mathbf{J}}_{\tilde{\Gamma}} \cdot \overline{\mathbf{E}}_1, \tag{12.18}$$

where $\tilde{\mathbf{J}}_{\tilde{\Gamma}}$ is the generalized $\tilde{J}_K$-integral vector.

Hence, the special form of the generalized $\tilde{J}$-integral for steady-state crack propagation is related to the generalized energy-momentum tensor in the same way as the configuration force (material force) method.

With the addition of the domain integral terms to the generalized $\tilde{J}$-integral, an invariant integral representation of the generalized energy release rate serving as the crack driving force is obtained as

$$
\begin{aligned}
\hat{J} &\equiv \tilde{J}_{\tilde{\Gamma}} - \frac{1}{\dot{A}} \int_{\tilde{V}_{\tilde{\Gamma}}} \frac{\tilde{\partial}}{\tilde{\partial} t} (\tilde{\rho}\tilde{h} + \tilde{\rho}\hat{k}) d\tilde{V} + \frac{1}{\dot{A}} \int_{\tilde{V}_{\tilde{\Gamma}}} \tilde{\rho}\hat{\mathbf{f}} \cdot \mathbf{v} d\tilde{V} - \frac{1}{\dot{A}} \int_{\tilde{V}_{\tilde{\Gamma}}} \tilde{\rho}\hat{s}\dot{T} d\tilde{V} \\
&\quad - \int_{\tilde{V}_{\tilde{\Gamma}}} \frac{\tilde{\rho}}{\rho} \mathbf{E} \cdot (\mathbf{j}_e - q_f \mathbf{v}) d\tilde{V} - \frac{1}{\dot{A}} \int_{\tilde{V}_{\tilde{\Gamma}}} \tilde{\rho}\hat{\Lambda} d\tilde{V} \\
&= \tilde{G},
\end{aligned} \tag{12.19}
$$

$$
\begin{aligned}
\hat{J} &\equiv \tilde{J}_{\tilde{\Gamma}} - \frac{1}{\dot{A}} \int_{\tilde{V}_{\tilde{\Gamma}}} \frac{\tilde{\partial}}{\tilde{\partial} t} (\tilde{\rho}\tilde{h} + \tilde{\rho}\hat{k}) d\tilde{V} + \frac{1}{\dot{A}} \int_{\tilde{V}_{\tilde{\Gamma}}} \tilde{\rho}\hat{\mathbf{f}} \cdot \mathbf{v} d\tilde{V} - \frac{1}{\dot{A}} \int_{\tilde{V}_{\tilde{\Gamma}}} \tilde{\rho}\hat{s}\dot{T} d\tilde{V} \\
&\quad - \int_{\tilde{V}_{\tilde{\Gamma}}} \frac{\tilde{\rho}}{\rho} \mathbf{E} \cdot (\mathbf{j}_e - q_f \mathbf{v}) d\tilde{V} - \frac{1}{\dot{A}} \sum_m \int_{\tilde{V}_{\tilde{\Gamma}}} j^{-1} \tilde{g}^{(m)} \dot{\alpha}^{(m)} d\tilde{V} \\
&= \tilde{G}.
\end{aligned} \tag{12.20}
$$

Hence, the invariant $\hat{J}$-integral is an extension of the path-domain independent integral method for nonhomogeneous or graded materials. Time or loading path/history-dependent bulk dissipation rate and electricity conduction bring about additional domain integral terms.

Integrating Eq. (12.9) or (12.10) over the time domain gives the following expression for the total work:

$$
\begin{aligned}
\Delta W &= \Delta H + \Delta E^k + \int_{t_0}^{t} \int_{V_t} \rho \hat{s} \dot{T} dV dt + \int_{t_0}^{t} \int_{V_t} \mathbf{E} \cdot (\mathbf{j}_e - q_f \mathbf{v}) dV dt \\
&\quad + \int_{t_0}^{t} \int_{V_t} \rho \hat{\Lambda} dV dt + \int_{t_0}^{t} \tilde{R} \dot{A} dt,
\end{aligned} \tag{12.21}
$$

$$
\begin{aligned}
\Delta W &= \Delta H + \Delta E^k + \int_{t_0}^{t} \int_{V_t} \rho \hat{s} \dot{T} dV dt + + \int_{t_0}^{t} \int_{V_t} \mathbf{E} \cdot (\mathbf{j}_e - q_f \mathbf{v}) dV dt \\
&\quad + \int_{t_0}^{t} \sum_m \int_{V_t} j^{-1} \tilde{g}^{(m)} \dot{\alpha}^{(m)} dV dt + \int_{t_0}^{t} \tilde{R} \dot{A} dt.
\end{aligned} \tag{12.22}
$$

From Eq. (12.21) or (12.22), the total work, W_f, from the start of loading until final fracture can be partitioned into the essential work of fracture, W_e, and the nonessential work of fracture, W_{ne}, as

$$
W_f = W_e + W_{ne}, \tag{12.23}
$$

$$
W_e \equiv \int_{t_0}^{t_f} w_e \dot{A} dt = \int_{t_0}^{t_f} \tilde{R} \dot{A} dt, \tag{12.24}
$$

$$W_{ne} = \Delta H + \Delta E^k + \int_{t_0}^{t_f} \int_{V_t} \rho \hat{s} \dot{T} dV dt + \int_{t_0}^{t_f} \int_{V_t} \mathbf{E} \cdot (\mathbf{j}_e - q_f \mathbf{v}) dV dt + \int_{t_0}^{t_f} \int_{V_t} \rho \hat{\Lambda} dV dt, \tag{12.25}$$

$$W_{ne} = \Delta H + \Delta E^k + \int_{t_0}^{t_f} \int_{V_t} \rho \hat{s} \dot{T} dV dt + \int_{t_0}^{t_f} \int_{V_t} \mathbf{E} \cdot (\mathbf{j}_e - q_f \mathbf{v}) dV dt + \int_{t_0}^{t} \sum_m \int_{V_t} j^{-1} \tilde{g}^{(m)} \dot{\alpha}^{(m)} dV dt, \tag{12.26}$$

where w_e is the specific essential work of fracture.

The essential work of fracture is a material property due to its equivalence to the crack resistance, and the nonessential work of fracture is geometry dependent due to its association with kinetic energy change, temperature change, electricity conduction, and time or loading path/history-dependent bulk dissipation. The separation of the total work, W_f, from the start of loading until final fracture into the essential work of fracture, W_e, and the nonessential work of fracture, W_{ne}, allows for the extension of the simple yet elegant EWF method to quasi-static and impact fracture characterization of electroactive polymers, switchable ferroelectrics, and piezoelectric semiconductors. The critical generalized energy release rate, $\tilde{G}_c$, the critical crack-front generalized $\tilde{J}$-integral, $\tilde{J}_c$, and the specific essential work of fracture, w_e, are equivalent as a measure of fracture toughness.

Chapter 13

Applications to Fracture Characterization

13.1 Introduction

In previous chapters we have dealt with the current status of conventional fracture mechanics and the new formulation of a nonlinear field theory of fracture mechanics for electromagnetic materials. Although standardized procedures for fracture toughness measurements of metallic and plastic materials have been published by a variety of standards organizations, such as the American Society for Testing and Materials (ASTM), the British Standards Institution (BSI), and the European Structural Integrity Society (ESIS), recent advances in multifunctional smart materials have created new frontiers due to the occurrence of magneto-electro-thermo-mechanical coupling and dissipative effects accompanying crack propagation. In this chapter, the generalization of fracture characterization techniques to electromagnetic materials is examined, with explanations of concepts which are central to the development of these techniques and discussions of areas in which future work is needed.

13.2 Energy Release Rate Method and its Generalization

The Griffith–Irwin–Orowan theory, as reviewed in Sections 1.1 to 1.3, lays a fundamental basis for evaluating the amount of energy required to extend a crack per unit area, with the energy release rate given by

$$G = -\left(\frac{\partial U}{\partial A}\right)_{\Delta}, \tag{13.1}$$

where U is the strain energy stored in the system and Δ is the load-point displacement.

From Irwin's crack closure analysis (Irwin, 1957), if a crack extends by a small amount δa, the energy released in the process is equal to the work required to close the crack to its original length, that is,

$$G = \lim_{\delta a \to 0} \frac{1}{2\delta a} \int_0^{\delta a} \sigma_{2i}(x_1, 0) \Delta u_i (x_1 - \delta a) dx_1. \tag{13.2}$$

The virtual crack closure or crack extension technique has been implemented in finite element analysis to calculate the energy release rate (e.g., Parks, 1974, 1977; Hellen, 1975; Rybicki and Kanninen, 1977; deLorenzi, 1982, 1985; Jih and Sun, 1990; Krueger, 2004).

In the regime of LEFM, the energy release rate is related to mode-I, II and III stress intensity factors by

$$\begin{aligned} G &= \lim_{\delta a \to 0} \frac{1}{2\delta a} \int_0^{\delta a} \sigma_{2i}(x_1, 0) \Delta u_i (x_1 - \delta a) dx_1 \\ &= \frac{1}{E'}(K_I^2 + K_{II}^2) + \frac{1+\nu}{E} K_{III}^2, \end{aligned} \tag{13.3}$$

where $E' = E$ for plane stress and $E' = E/(1-\nu^2)$ for plane strain, E is Young's modulus, and ν is Poisson's ratio.

Equation (13.3) allows the evaluation of the energy release rate via the stress intensity factor method (ASTM Standard E399 and D5045, British Standard BS5447). The specimen size requirement to obtain a valid measurement of K_{IC} as the critical plane-strain value at crack initiation is given by

$$B, a, (W-a) > 2.5\left(\frac{K_{IC}}{\sigma_y}\right)^2, \tag{13.4}$$

where B is the specimen thickness, a is the crack length, W is the specimen width, and σ_y is the yield strength.

By contrast, Eq. (13.1) may result in a simple expression of G_{IC} that can be related to the elastic strain energy U stored in the system for mode-I fracture (Williams, 1987):

$$G_{IC} = \frac{U}{BW\phi}, \tag{13.5}$$

where B is the specimen thickness, W is the specimen width, and ϕ is a correction factor which is determined by the specimen compliance C:

$$\phi = \frac{C}{\dfrac{dC}{d(a/W)}}. \tag{13.6}$$

Numerical values of ϕ can be obtained for different specimen geometries. Equation (13.5) enables the direct evaluation of G_{IC} from the slope of the linear relationship between U and $BW\phi$ for a series of specimens with different initial crack lengths. The reader may refer to the book by Williams (1987) and the book chapter by Mai *et al.* (2000) for further information.

In Chapters 11 and 12, the generalized energy release rate method was proposed for quasi-static and dynamic fracture characterization of conservative and dissipative magneto- or electro-sensitive materials with the generalized energy release rate defined as

$$\tilde{G} = -\frac{\partial}{\partial A}\int_{V_t} \mathcal{H} dV, \tag{13.7}$$

where $\mathcal{H} = \rho\hat{k} + \rho\tilde{h}$ is the Hamiltonian density, $\rho\hat{k}$ is the kinetic energy density, and $\rho\tilde{h}$ is the augmented Helmholtz free energy density, including the contribution of the energy of the free electromagnetic fields.

In terms of the generalized energy release rate as the crack driving force, the thermodynamically consistent time-dependent fracture criterion is given by

$$\tilde{G} = \tilde{R}, \tag{13.8}$$

where $\tilde{R}$ is the crack resistance.

As an extension of Irwin's crack closure integral, the generalized energy release rate can be evaluated via the following crack closure integral:

$$\tilde{G} = \lim_{\delta a \to 0} \frac{1}{2\delta a} \int_0^{\delta a} {}_t\sigma_{2j}(\tilde{X}_1, 0, t) \Delta u_j(\tilde{X}_1 - \delta a, 0^{\pm}, t) d\tilde{X}_1, \tag{13.9}$$

where ${}_t\sigma_{2j}(\tilde{X}_1, 0, t)$ is the total traction and $\Delta u_j(\tilde{X}_1 - \delta a, 0^{\pm}, t) = \left[u_j(\tilde{X}_1 - \delta a, 0^{+}, t) - u_j(\tilde{X}_1 - \delta a, 0^{-}, t)\right]$ is the crack opening displacement.

It can be seen that the key difference lies in the replacement of the Cauchy stress by the total stress in the required calculations. Thus, the existing finite element codes with implementation of the virtual crack closure or crack extension technique can be readily extended to numerical evaluation of the generalized energy release rate. For example, the work required for crack closure for finite element representation of a crack modeled with two-dimensional four-node elements (Fig. 13.1) is obtained as

$$\tilde{G} = \frac{1}{2\delta a}[{}_tF_1^{(d)}(u_1^{(c)} - u_1^{(b)}) + {}_tF_2^{(d)}(u_2^{(c)} - u_2^{(b)})], \tag{13.10}$$

where ${}_tF_1^{(d)}$ and ${}_tF_2^{(d)}$ are the total shear and normal force components at node d, $u_1^{(b)}$ and $u_1^{(c)}$ are the shear displacement components at nodes b and c, $u_2^{(b)}$ and $u_2^{(c)}$ are the normal displacement components at nodes b and c.

In linearized magneto-electro-elasticity, it was shown in Section 6.3 that the dynamic energy release rate is related to the dynamic field intensity factors by

$$\tilde{G} = \frac{1}{4}(\tilde{K}_{II}, \tilde{K}_I, \tilde{K}_{III}, 0, 0) \cdot \tilde{\mathbf{H}}^i \cdot (\tilde{K}_{II}, \tilde{K}_I, \tilde{K}_{III}, \tilde{K}_D, \tilde{K}_B)^T, \tag{13.11}$$

where $\tilde{\mathbf{H}}^i$ is the dynamic counterpart of the Irwin matrix.

Equation (13.11) allows the evaluation of the dynamic energy release rate for magneto- or electro-sensitive materials via the dynamic field intensity factor method. It appears that the dynamic energy release rate is an odd function of the electric displacement intensity factor and the magnetic induction intensity factor, which is consistent with experimental evidence (e.g., Pak and Tobin, 1993; Tobin and Pak, 1993; Cao and Evans, 1994; Lynch *et al.*, 1995; Park and Sun, 1995a–b; Jiang and Sun, 1999, 2001; Zhang *et al.*, 2002; Chen and Lu, 2003; Soh *et al.*, 2003; Zhang and Gao, 2004; Zhang *et al.*, 2004; Chen and Hasebe, 2005; Schneider, 2007; Kuna, 2010) as reviewed in Chapter 4. In addition to small-scale yielding conditions, small-scale switching or small-scale saturation conditions should be satisfied in order to obtain a valid measurement of the critical values of the dynamic field intensity factors.

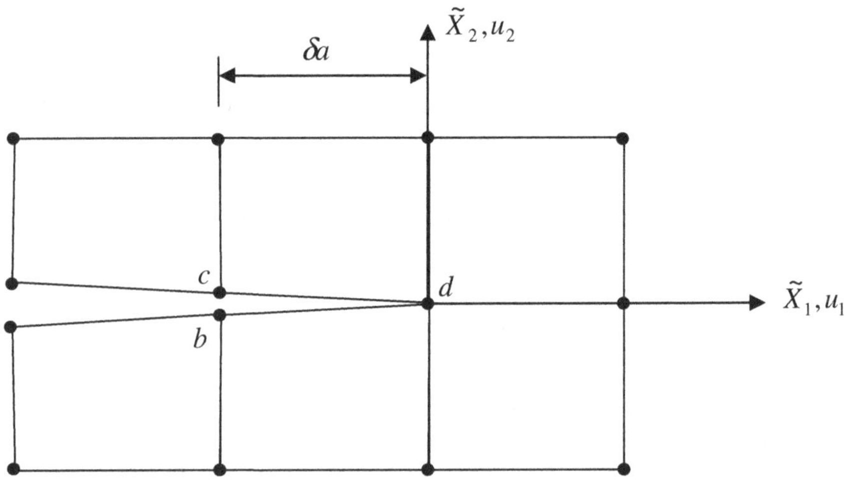

Fig. 13.1 Two-dimensional four-node finite element mesh for crack closure integral.

13.3 *J-R* Curve Method and its Generalization

Path-independent integrals have been widely used to study bodies with cracks and defects since the pioneering work of Eshelby (1951, 1956, 1970, 1975), Cherepanov (1967, 1968, 1979) and Rice (1968), as reviewed in Section 1.4. In particular, Rice (1968) applied the *J*-integral concept to nonlinear fracture mechanics problems. For a two-dimensional crack problem, the J-integral is given by

$$J = \int_{\Gamma}(w dx_2 - \sigma_{ij} n_j \frac{\partial u_i}{\partial x_1} ds), \tag{13.12}$$

where w is the strain energy density, σ_{ij} are the components of the Cauchy stress tensor, n_j are the components of the unit outer normal vector, u_i are the components of the displacement vector, ds is the length increment along the path Γ, and the x_2-direction is perpendicular to the crack line.

The J-integral criterion for crack initiation gives

$$J = J_c . \tag{13.13}$$

Stable crack growth may be maintained if the crack driving force and resistance curves satisfy the following conditions:

$$J = J_R , \tag{13.14}$$

$$T_{app} \le T_R , \tag{13.15}$$

where $T_{app} = (E/\sigma_0^2) dJ/da$ and $T_R = (E/\sigma_0^2) dJ_R/da$ are dimensionless tearing moduli (Paris *et al.*, 1979; Atkins and Mai, 1985; Anderson, 2005).

While the definition of the J-integral as a path-independent contour integral for linear or nonlinear elastic materials is very useful for fracture mechanics analysis, the energetic interpretation of the J-integral is most widely adopted for experimental characterization of fracture toughness for ductile materials. Since the J-integral is equivalent to the energy release rate for linear or nonlinear elastic materials, Begley and Landes (1972) developed the multi-specimen J-R curve method based on the interpretation of J as the energy release rate given by

$$J = -\frac{1}{B}\left(\frac{\partial U}{\partial a}\right)_{\Delta} . \tag{13.16}$$

Rice *et al.* (1973) proposed a method for estimating J from the load-displacement curve measured from a single specimen. ASTM Standard E813 describes the test procedure for determining J_{IC} as the critical plane-strain value at crack initiation for a wide range of ductile materials.

For a deeply cracked specimen, J is determined from

$$J = \frac{2U}{Bb}, \tag{13.17}$$

where U is computed from the total area under the load-deflection curve, $b = (W - a)$ is the specimen ligament length, and B is the specimen thickness.

The specimen dimension requirement for obtaining a valid J_{IC} value is given by

$$B, b > \frac{25 J_{IC}}{\sigma_y}. \tag{13.18}$$

ASTM Standard E1820 further covers the procedure for *J-R* curve testing. British Standard BS 7448: Part 1 is equivalent to ASTM Standard E1820, both of which combine *K*, *J*, and CTOD testing into a single standard.

In Chapters 5–12, the generalized $\tilde{J}$-integral method is proposed for quasi-static and dynamic fracture characterization of conservative and dissipative magneto- or electro-sensitive materials, with the generalized $\tilde{J}$-integral defined as

$$\tilde{J} = \frac{1}{\dot{A}} \int_{\tilde{\Gamma}} [\mathbf{n} \cdot {}_t\boldsymbol{\sigma} \cdot \mathbf{v} - \mathbf{n} \cdot \boldsymbol{S} + (\tilde{\rho}\tilde{h} + \tilde{\rho}\hat{k}) \mathbf{n} \cdot \mathbf{V}_C] d\tilde{\Gamma}, \tag{13.19}$$

where ${}_t\boldsymbol{\sigma}$ is the total stress, $\boldsymbol{S}$ is the Poynting vector in the co-moving frame, $\tilde{\rho}\hat{k}$ is the kinetic energy density, and $\tilde{\rho}\tilde{h}$ is the augmented Helmholtz free energy density, including the contribution of the energy of the free electromagnetic fields.

The physical interpretation of the crack-front generalized $\tilde{J}$-integral is the generalized energy release rate $\tilde{G}$ serving as the crack driving force, that is,

$$\tilde{J}_0 = \tilde{G}. \tag{13.20}$$

Crack initiation occurs under combined magnetic, electric, thermal, and mechanical loadings when $\tilde{J}_0$ reaches a critical value, $\tilde{J}_c$, i.e.,

$$\tilde{J}_0 = \tilde{J}_c. \tag{13.21}$$

Stable crack growth may be maintained if the following conditions are satisfied:

$$\tilde{J} = \tilde{J}_R, \tag{13.22}$$

$$\tilde{T}_{app} \le \tilde{T}_R, \tag{13.23}$$

where $\tilde{T}_{app} = (E/\sigma_0^2)d\tilde{J}/da$ and $\tilde{T}_R = (E/\sigma_0^2)d\tilde{J}_R/da$.

The invariant $\hat{J}$-integral defined by Eqs. (6.27), (8.149), (9.42), (10.48), (11.57), (11.58), (12.19), or (12.20), with addition of the domain integral terms to the generalized $\tilde{J}$-integral, is useful for numerical evaluation of the generalized energy release rate, since the generalized $\tilde{J}$- integral loses its path independence in the presence of unsteady state, mechanical body force, temperature change, electricity conduction, and time or loading path/history-dependent bulk dissipation rate.

13.4 Essential Work of Fracture Method and its Extension

The essential work of fracture (EWF) method is a simple yet elegant experimental technique developed by the Cotterell–Mai research group at the University of Sydney (Cotterell and Reddel, 1977; Mai and Cotterell, 1980, 1986) from the unified theory of fracture (Broberg, 1971, 1975), as reviewed in Section 1.7. The total work of fracture for any increment of crack growth includes both the essential work in the inner fracture process zone and the nonessential work in the outer plastic zone. It has become a widely accepted technique for fracture characterization of many ductile materials, including metallic alloys, polymeric films, toughened polymers, and their blends (Mai *et al.*, 2000; Clutton, 2001). Finite element simulation of the EWF method has also been attempted for the complete failure process of deep double-edge notched tension (DENT), deep center notched tension (DCNT), single-edge notched tension (SENT), and centre-lined ligament loading (CLLL) samples with different ligament lengths (Chen *et al.*, 2000).

Based on the ESIS-TC4 testing protocol for measurement of the EWF established by the European Structural Integrity Society-Technical Committee 4 (1997), the total work of fracture W_f can be separated into two components: the essential work W_e performed in the fracture process zone and the nonessential work W_p performed in the outer plastic zone, with the essential work of fracture W_e proportional to the ligament length l and the nonessential work of fracture W_p proportional to the square of the ligament length l^2, that is,

$$W_f = W_e + W_p, \tag{13.24}$$

$$W_e = w_e l B, \tag{13.25}$$

$$W_p = \beta w_p l^2 B, \tag{13.26}$$

where w_e is the specific essential work of fracture, w_p is the specific nonessential work of fracture, and β is a geometry-dependent plastic-zone shape factor.

As a result, the specific total work of fracture $w_f = W_f / Bl$ can be expressed as

$$w_f = w_e + \beta w_p l. \tag{13.27}$$

On the assumption that w_e is a material property and w_p and β are independent of l in all testing specimens, there should exist a linear relation when w_f is plotted against l according to Eq. (13.27). By extrapolation of this straight line to zero ligament length, the intercept at the y axis and the slope of the line gives w_e and βw_p, respectively.

In Chapters 11 and 12, the extended EWF method was proposed for quasi-static and impact fracture characterization of magnetosensitive elastomers, electroactive polymers, piezoelectric semiconductors, and switchable ferroelectrics or ferromagnetics, with the partition of the total work of fracture W_f into the essential work of fracture W_e and the nonessential work of fracture W_{ne} as

$$W_f = W_e + W_{ne}, \tag{13.28}$$

$$W_e = \int_{t_0}^{t_f} w_e \dot{A} dt = \int_{t_0}^{t_f} \tilde{R} \dot{A} dt, \tag{13.29}$$

$$\begin{aligned} W_{ne} = \Delta H + \Delta E^k &+ \int_{t_0}^{t_f} \int_{V_t} \rho \hat{s} \dot{T} dV dt + \int_{t_0}^{t_f} \int_{V_t} \mathbf{E} \cdot (\mathbf{j}_e - q_f \mathbf{v}) dV dt \\ &+ \int_{t_0}^{t_f} \int_{V_t} \rho \hat{\Lambda} dV dt, \end{aligned} \tag{13.30}$$

or

$$\begin{aligned} W_{ne} = \Delta H + \Delta E^k &+ \int_{t_0}^{t_f} \int_{V_t} \rho \hat{s} \dot{T} dV dt + \int_{t_0}^{t_f} \int_{V_t} \mathbf{E} \cdot (\mathbf{j}_e - q_f \mathbf{v}) dV dt \\ &+ \int_{t_0}^{t_f} \sum_m \int_{V_t} j^{-1} \tilde{g}^{(m)} \dot{\alpha}^{(m)} dV dt. \end{aligned} \tag{13.31}$$

The generalized energy release rate method, the generalized $\tilde{J}$-integral method, and the extended essential work of fracture method should give consistent results, independent of material systems, loading combinations, and environmental conditions.

13.5 Closure

The development of a nonlinear field theory of fracture mechanics for evaluating the crack driving force in the presence of magneto-electro-thermo-mechanical coupling and dissipative effects overcomes the limitations of classical fracture mechanics theories and sets up a bridge between damage mechanics and fracture mechanics. On the basis of the developed theory, the generalized energy release rate method, the generalized $\tilde{J}$-integral method, and the extended essential work of fracture method are proposed, which are generally applicable to quasi-static and dynamic fracture characterization of conservative and dissipative multifunctional smart material systems. In comparison with conventional fracture mechanics methodologies, the main advantages of this formulation are:

(i) It does not only afford a uniform treatment of complex nonlinear material and fracture behaviors involving multifield coupling and dissipated effects, but also enables damage and fracture processes at the micro-, meso- and macroscale levels to be managed in a unified way.

(ii) It provides a physically sound criterion for quasi-static and dynamic crack propagation in conservative or dissipative, homogeneous or nonhomogeneous media, including FGMs, subjected to combined magnetic, electric, thermal, and mechanical loadings.

(iii) It includes the strain energy release rate criterion, dynamic energy release rate criterion, conventional J-integral method, configuration force (material force) method, dynamic contour integral method, path-domain independent integral method, rate-dependent criterion for viscoelastic/viscoplastic crack growth, and Schapery's crack-tip model for viscoelastic facture as special cases.

(iv) It lays the theoretical foundation for the application of the generalized energy release rate method, the generalized $\tilde{J}$-integral method, and the extended essential work of fracture method to quasi-static and impact fracture characterization of electro- and magneto-sensitive materials. The equivalence of the critical generalized energy release rate, $\tilde{G}_c$, the critical crack-front generalized $\tilde{J}$-integral, $\tilde{J}_c$, and the specific essential work of fracture, w_e, as a measure of crack resistance warrants consistent results from application of the generalized energy release rate and $\tilde{J}$-integral methods as well as the extended essential work of fracture method for fracture toughness measurement.

(v) A key feature lies in the incorporation of the time or loading path/history-dependent bulk dissipation in the general formulation of the nonlinear field theory of fracture mechanics. This large difference is akin to the difference between elastic fracture mechanics and inelastic fracture mechanics.

(vi) This formulation can be readily extended to include gradient effects in thin films and microelectromechanical systems (MEMS).

Since the development of the nonlinear field theory of fracture mechanics for electromagnetic materials is still in its infancy, much remains to be done, especially on multiscale modeling of damage and fracture involving various failure mechanisms.

Bibliography

Abraham, F.F., Brodbeck, D., Rudge, W.E. (1997). A molecular-dynamics investigation of rapid fracture mechanics, *Journal of the Mechanics and Physics of Solids*, 45, pp. 1595–1619.

Achenbach, J.D. (1974). Dynamic effects in brittle fracture, *Mechanics Today*, 1, S. Nemat-Nasser, Ed., Pergamon Press, New York, pp. 1–57.

Alshits, V.I., Darinskii, A.N., Lothe, J. (1992). On the existence of surface waves in half-infinite anisotropic elastic media with piezoelectric and piezomagnetic properties, *Wave Motion*, 16, pp. 265–283.

Alshits, V.I., Kirchner H.O.K., Ting, T.C.T. (1995). Angularly inhomogeneous piezoelectric piezomagnetic magnetoelastic anisotropic media, *Philosophical Magazine Letters*, 71, pp. 285–288.

Alshits, V.I. (2002). On the role of anisotropy in crystalloacoustics, in *Surface Waves in Anistropic and Laminated Bodies and Defects Detection*, R.V. Goldstein and G. A. Maugin, Eds., Kluwer Academic Publishers, Dordrecht/Boston/London, pp. 3–68.

Anderson, T.L. (2005). *Fracture Mechanics: Fundamentals and Applications*, 3rd edition, CRC Press, Boca Raton.

Anlas, G., Santare, M.H., Lambros, J. (2000). Numerical calculation of stress intensity factors in functionally graded materials, *International Journal of Fracture*, 104, pp. 131–143.

Anstis, G.R., Chantikul, P., Lawn, B.R., *et al.* (1981). A critical evaluation of indentation techniques for measuring fracture toughness: I, Direct measurements, *Journal of the American Ceramics Society*, 64, pp. 533–538.

Atkins, A.G., Mai, Y.-W. (1985). *Elastic and Plastic Fracture: Metals, Polymers, Ceramics, Composites, Biological Materials*, Ellis Horwood Ltd., Chichester.

Atkinson, C., Eshelby, J.D. (1968). The flow of energy into the tip of a moving crack, *International Journal of Fracture*, 4, pp. 3–8.

Atkinson, C., Popelar, C.H. (1979). Antiplane dynamic crack propagation in a viscoelastic strip, *Journal of the Mechanics and Physics of Solids*, 27, pp. 431–439.

Atluri, S.N. (1982). Path-independent integrals in finite elasticity and inelasticity, with body forces, inertia and arbitrary crack-face conditions, *Engineering Fracture Mechanics*, 16, pp. 341–364.

Atluri, S.N., Nishioka, T., Nakagaki, M. (1984). Incremental path-independent integrals in inelasticity and dynamic fracture mechanics, *Engineering Fracture Mechanics*, 20, pp. 209–244.

Baker, B.R. (1962). Dynamic stresses created by a moving crack, *Journal of Applied Mechanics*, 29, pp. 449–458.

Bar-Cohen, Y. (2002). Electro-active polymers: current capabilities and challenges, in *Proceedings of the SPIE EAPAD conference*, San Diego, California, March 18-21, pp. 4695–4702.

Barnett, D.M., Lothe, J. (1975). Dislocations and line charges in anisotropic piezoelectric insulators, *Physica Status Solidi (b)*, 67, pp. 105–111.

Barenblatt, G.I. (1959a). The formation of equilibrium cracks during brittle fracture: General ideas and hypotheses, axially symmetric cracks, *Applied Mathematics and Mechanics* (PMM), 23, pp. 622–636.

Barenblatt, G.I. (1959b). Equilibrium cracks formed during brittle fracture: Rectilinear cracks in plane plates, *Applied Mathematics and Mechanics* (PMM), 23, pp. 1009–1029.

Barenblatt, G.I. (1959c). Concerning equilibrium cracks forming during brittle fracture: The stability of isolated cracks, relationship with energetic theories, *Applied Mathematics and Mechanics* (PMM), 23, pp. 1273–1282.

Barenblatt, G.I. (1962). The mathematical theory of equilibrium cracks in brittle fracture, *Advances in Applied Mechanics*, 7, Academic Press, New York, pp. 55–129.

Barenblatt, G.I., Salganik, R.L., Cherepanov, G.P. (1962). On the nonsteady motion of cracks, *Applied Mathematics and Mechanics* (PMM), 26, pp. 469–477.

Begley, J.A., Landes, J.D. (1972). The J-integral as a fracture criterion, in *Fracture Toughness*, Part II, ASTM STP 514, American Society for Testing and Materials, Philadelphia, pp. 1–20.

Beom, H.G. (2001). Effect of polarization saturation on the stress intensity factor for a crack in an electrostrictive ceramic, *Philosophical Magazine A*, 81, pp. 2303–2314.

Beom, H.G., Atluri, S.N. (2003). Effect of electric fields on fracture behavior of ferroelectric ceramics, *Journal of the Mechanics and Physics of Solids*, 51, pp. 1107–1125.

Blackburn, W.S. (1972). Path independent integrals to predict onset of crack instability in an elastic-plastic materials, *International Journal of Fracture Mechanics*, 8, pp. 343–346.

Bleustein, J.L. (1968). A new surface wave in piezoelectric materials, *Applied Physics Letters*, 13, pp. 412–413.

Borovik-Romanov, A.S. (1960). Piezomagnetism in the antiferromagnetic fluorides of cobalt and manganese, *Soviet Physics JETP*, 11, p. 786.

Broberg, K.B. (1960). The propagation of a brittle crack, *Arkiv för Fysik*, 18, pp. 159–192.

Broberg, K.B. (1968). Critical review of some theories in fracture mechanics, *International Journal of Fracture Mechanics*, 4, pp. 11–18.

Broberg, K.B. (1971). Crack growth criteria and non-linear fracture mechanics, *Journal of the Mechanics and Physics of Solids*, 19, pp. 407–418.

Broberg, K.B. (1975). On stable crack growth, *Journal of the Mechanics and Physics of Solids*, 23, pp. 215–237.

Broberg, K.B. (1999). *Cracks and Fracture*, Academic Press, Cambridge.

Buehler, M. J., Abraham, F. F., Gao, H. (2003). Hyperelasticity governs dynamic fracture at a critical length scale, *Nature*, 426, pp. 141–146.

Bui, H.D. (1974). Dual path independent integrals in the boundary-value problems of cracks, *Engineering Fracture Mechanics*, 6, pp. 287–296.

Bustamante, R., Dorfmann, A., Ogden, R.W. (2009). Nonlinear electroelastostatics: a variational framework, *Zeitschrift für angewandte Mathematik und Physik*, 60, pp. 154–177.

BS 5447 (1974). *Methods of Testing for Plane Strain Fracture Toughess (K_{IC}) of Metallic Materials*, British Standards Institution, London.

BS 7448 (1991). *Fracture Mechanics Toughness Tests, Part 1, Method for Determination of KIC, Critical CTOD and Critical J Values of Metallic Materials*, British Standards Institution, London.

Caruthers, J.M., Adolf, D.B., Chambers, R.S., *et al.* (2004). A thermodynamically consistent, nonlinear viscoelastic approach for modeling glassy polymers, *Polymer*, 45, pp. 4577–4597.

Cao, H.C., Evans, A.G. (1994). Electric-field-induced fatigue crack growth in piezoelectric ceramics, *Journal of American Ceramic Society*, 77, pp. 1783–1786.

Chakraborty, A., Gopalakrishnan, S., Kausel, E. (2005). Wave propagation analysis in inhomogeneous piezo-composite layer by the thin-layer method, *International Journal for Numerical Methods in Engineering*, 64, pp. 567–598.

Chantikul, P., Anstis, G.R., Lawn, B.R., *et al.* (1981). A critical evaluation of indentation techniques for measuring fracture toughness: II, Strength Method, *Journal of the American Ceramics Society*, 64, pp. 539–543.

Chen, X. (2007). Coupled hygro-thermo-viscoelastic fracture theory, *International Journal of Fracture*, 148, pp. 47–55.

Chen, X. (2009a). Crack driving force and energy-momentum tensor in electroelastodynamic fracture, *Journal of the Mechanics and Physics of Solids*, 57, pp. 1–9.

Chen, X. (2009b). Energy release rate and path-independent integral in dynamic fracture of magneto-electro-thermo-elastic solids, *International Journal of Solids and Structures*, 46, pp. 2706–2711.

Chen, X. (2009c). Dynamic crack propagation in a magneto-electro-elastic solid subjected to mixed loads: transient mode-III problem, *International Journal of Solids and Structures*, 46, pp. 4025–4037.

Chen, X. (2009d). On magneto-thermo-viscoelastic deformation and fracture, *International Journal of Non-Linear Mechanics*, 44, pp. 244–248.

Chen, X. (2009e). Nonlinear field theory of fracture mechanics for paramagnetic and ferromagnetic materials, *Journal of Applied Mechanics*, 76, 041016.

Chen, X. (2010). Nonlinear electro-thermo-viscoelasticity, *Acta Mechanica*, 211, pp. 49–59.

Chen, X., Mai, Y.-W., Tong, P., *et al.* (2000). Numerical simulation of the essential fracture work method, in *Fracture of Polymers, Composites and Adhesives, ESIS 27*, J.G. Williams and A. Pavan, Eds., Elsevier, Amsterdam, pp. 175–186.

Chen, X., Wang, S.S. (2006). A thermodynamic approach to long-term deformation and damage for polymeric materials in hygrothermal environment, in *Fracture of Materials: Moving Forwards*, H.-Y. Liu, X. Hu, and M. Hoffman, Eds., Trans Tech Publications, Dürnten, Switzerland, pp. 21–26.

Chen, X.-H., Ing, Y.-S., Ma, C.-C. (2007). Transient analysis of dynamic crack propagation in piezoelectric materials, *Journal of the Chinese Institute of Engineers*, 30, pp. 491–502.

Chen, X.-H., Ma, C.-C., Ing, Y.-S., *et al.* (2008). Dynamic interfacial crack propagation in elastic–piezoelectric bi-materials subjected to uniformly distributed loading, *International Journal of Solids and Structures*, 45, pp. 959–997.

Chen, Y.-H., Hasebe, N. (2005). Current understanding on fracture behaviors of ferroelectric/piezoelectric materials, *Journal of Intelligent Material Systems and Structures*, 16, pp. 673–687.

Chen, Y.-H., Lu, T.J. (2003). Recent developments and applications of invariant integrals, *Applied Mechanics Reviews*, 56, pp. 515–552.

Chen, Z.T., Yu, S.W. (1997). Anti-plane dynamic fracture mechanics in piezoelectric materials, *International Journal of Fracture*, 85, pp. L3–L12.

Chen, Z.T., Karihaloo, B.L. (1999). Dynamic response of a cracked piezoelectric ceramic under arbitrary electro-mechanical impact, *International Journal of Solids and Structures*, 36, pp. 5125–5133.

Cherepanov, G.P. (1967). Crack propagation in continuous media, *Applied Mathematics and Mechanics*, 31, pp. 467–488.

Cherepanov, G.P. (1968). Cracks in solids, *International Journal of Solids and Structures*, 6, pp. 811–831.

Cherepanov, G.P. (1979). *Mechanics of Brittle Fracture*, McGraw-Hill, New York.

Christensen, R.M. (1979). A rate-dependent criterion for crack growth, *International Journal of Fracture*, 15, pp. 3–21.

Christensen, R.M. (1980). Response: Discussion: A rate-dependent criterion for crack growth, *International Journal of Fracture*, 16, pp. R233–237.

Christensen, R.M. (1981). Viscoelastic crack growth – a response note, *International Journal of Fracture*, 17, pp.R169–176.

Christensen, R.M. (1982). *Theory of Viscoelasticity – An Introduction*, 2nd edition, Academic Press, New York.

Christensen, R.M., McCartney, L.N. (1983). Viscoelastic crack growth, *International Journal of Fracture*, 23, pp. R11–13.

Chue, C.H., Ou, Y.L. (2005). Mode III crack problems for two bonded functionally graded piezoelectric materials, *International Journal of Solids and Structures*, 42, pp. 3321–3337.

Chung, M.Y., Ting, T.C.T. (1995). The Green function for a piezoelectric piezomagnetic anisotropic elastic medium with an elliptic hole or rigid inclusion, *Philosophical Magazine Letters*, 72, pp. 405–410.

Clutton, E. (2001). Essential work of fracture, in *Fracture Mechanics Testing Methods for Polymers, Adhesives and Composites*, D.R. Moore, A. Pavan, and J.G. Williams, Eds., Elsevier Science, Oxford, pp. 177–195.

Coleman, B.D., Noll, W. (1960). An approximation theorem for functionals, with applications in continuum mechanics, *Archive for Rational Mechanics and Analysis*, 6, 355–370.

Coleman, B.D. (1964). Thermodynamics of materials with memory, *Archive for Rational Mechanics and Analysis*, 17, 1–46.

Coleman, B.D., Gurtin, M.E. (1967). Thermodynamics with internal state variables. *Journal of Chemical Physics*, 47, pp. 597–613.

Copson, E.T. (1961). On certain dual integral equations, *Proceedings of the Glasgow Mathematical Association*, 5, pp. 19–24.

Cost, T.L. (1973). A free energy functional for thermorheologically simple materials, *Acta Mechanica*, 17, pp. 153–167.

Cotterell, B., Mai, Y.-W. (1996). *Fracture Mechanics of Cementitious Materials*, Chapman & Hall, London, Chapter 1.

Cotterell, B., Reddel, J.K. (1977). The essential work of plane stress fracture, *International Journal of Fracture*, 13, pp. 267–277.

Cox, B.N., Gao, H., Cross, D., Rittel, D. (2005). Modern topics and challenges in dynamic fracture, *Journal of the Mechanics and Physics of Solids*, 53, pp. 565–596.

Craggs, J.W. (1960). On the propagation of a crack in an elastic-brittle material, *Journal of the Mechanics and Physics of Solids*, 8, pp. 66–75.

Crochet, M.J., Naghdi, P.M. (1969). A class of simple solids with fading memory, *International Journal of Engineering Science*, 7, pp. 1173–1198.

Curie, J., Curie, P. (1880). Development par compression de l'electricite polaire dans les cristaux hemiedres a faces inclines, *Comptes Rendus de l'Acadamie des Sciences Paris*, 91, pp. 294.

Curie, J., Curie, P. (1884). Contractions et dilations produces par des tensions electriques dans les cristaux hemiedres a faces inclines, *Comptes Rendus de l'Acadamie des Sciences Paris*, 93, pp. 1137.

Dascalu, C., Maugin, G.A. (1994). Energy-release rates and path-independent integrals in electroelastic crack propagation, *International Journal of Engineering Science*, 32, pp. 755–765.

Dascalu, C., Maugin, G.A. (1995). On the dynamic fracture of piezoelectric materials, *The Quarterly Journal of Mechanics and Applied Mathematics*, 48, pp. 237–254.

Deeg, W.F. (1980). *The analysis of dislocation, crack, and inclusion problems in piezoelectric solids*, Ph.D. thesis, Stanford University.

De Hoop, A.T. (1958). *Representation theorems for the displacement in an elastic solid and their application to elastodynamic diffraction theory*, Doctoral Dissertation, Technical University of Delft.

deLorenzi, H.G. (1982). On the energy release rate and the J-integral of 3-D crack configurations, *International Journal of Fracture*, 19, pp. 183–193.

deLorenzi, H.G. (1985). Energy release rate calculations by the finite element method, *Engineering Fracture Mechanics*, 21, pp. 129–143.

Dineva, P., Gross, D., Müller, R., Rangelov, T. (2010). BIEM analysis of dynamically loaded anti-plane cracks in graded piezoelectric finite solids, *International Journal of Solids and Structures*, 47, pp. 3150–3165.

Dolinek, J., Jaglii, Z. (2002). Slow relaxation of the thermoremanent magnetization and aging in icosahedral Tb–Mg–Zn quasicrystals, *Journal of Alloys and Compounds*, 342, pp. 377–380.

Dorfmann, A., Ogden, R.W. (2003). Nonlinear magnetoelastic deformations of elastomers, *Acta Mechanica*, 167, pp. 13–28.

Dorfmann, A., Ogden, R.W. (2004). Nonlinear magnetoelastic deformations, *The Quarterly Journal of Mechanics and Applied Mathematics*, 57, pp. 599–622.

Dorfmann, A., Ogden, R.W. (2005). Nonlinear electroelasticity, *Acta Mechanica*, 174, pp. 167–183.

Dorfmann, A., Ogden, R.W. (2006). Nonlinear electroelastic deformations, *Journal of Elasticity*, 82, pp. 99–127.

Du, J.K., Shen, Y.P., Ye, D.Y., Yue, F.R. (2004). Scattering of anti-plane shear waves by a partially debonded magneto-electro-elastic circular cylindrical inhomogeneity, *International Journal of Engineering Science*, 42, pp. 887–913.

Dugdale, D.C. (1960). Yielding of steel sheets containing slits, *Journal of the Mechanics and Physics of Solids*, 8, pp. 100–104.

Dunn, M.L. (1994). The effects of crack face boundary conditions on the fracture mechanics of piezoelectric solids, *Engineering Fracture Mechanics*, 48, pp. 25–39.

D5045 (1993). *Standard Test Methods for Plane Strain Fracture Toughness and Strain Energy Release Rate of Plastic materials*, American Society for Testing and Materials, Philadelphia.

E399 (1990). *Standard Test Method for Fracture Toughness of Metallic Materials*, American Society for Testing and Materials, Philadelphia.

E813 (1987). *Standard Test Method for J_{IC}, A Measure of Fracture Toughness*, American Society for Testing and Materials, Philadelphia.

E1820 (2001). *Standard Test Method for Measurement of Fracture Toughness*, American Society for Testing and Materials, Philadelphia.

ESIS (1997). *European Structural Integrity Society Test Protocol for "Essential Work of Fracture"*, Version 5, European Structural Integrity Society, October 1997.

Eischen, J.W. (1987a). Fracture of nonhomogeneous materials, *International Journal of Fracture*, 34, pp. 3–22.

Eischen, J.W. (1987b). An improved method for computing the J_2 integral, *Engineering Fracture Mechanics*, 26, pp. 691–700.

Erdogan, F. (1995). Fracture mechanics of functionally graded materials, *Composites Engineering*, 5, pp. 753–770.

Eringen, A.C. (1980). *Mechanics of Continua*, 2nd edition, Robert E. Krieger Publishing Company, Inc., Malabar, FL.

Eshelby, J.D. (1951). The force on an elastic singularity, *Philosophical Transactions of the Royal Society of London*, A244, pp. 87–112.

Eshelby, J.D. (1956). The continuum theory of lattice defects, in *Solid State Physics, Vol. 3*, F. Seitz and D. Turnbull, Eds., Academic Press, New York, pp. 79–144.

Eshelby, J.D. (1957). The determination of the elastic field of an ellipsoidal inclusion and related problems, *Proceedings of the Royal Society of London*, A241, pp. 376–396.

Eshelby, J.D. (1970). Energy relations and the energy-momentum tensor in continuum mechanics, in *Inelastic Behavior of Solids*, M.F. Kanninen, W.F. Adler, A.R. Rosenfield, and R.I. Jaffe, Eds., McGraw-Hill, New York, pp. 77–114.

Eshelby, J.D. (1975). The elastic energy-momentum tensor, *Journal of Elasticity*, 5, pp. 321–335.

Feng, W.J., Su, R.K.L. (2006). Dynamic internal crack problem of a functionally graded magneto-electro-elastic strip, *International Journal of Solids and Structures*, 43, pp. 5196–5216.

Feng, W., Su, R.K.L.(2007). Dynamic fracture behaviors of cracks in a functionally graded magneto-electro-elastic plate, *European Journal of Mechanics A: Solids*, 26, pp. 363–379.

Feng, W.J., Pan, E., Wang, X. (2007). Dynamic fracture analysis of a penny-shaped crack in a magnetoelectroelastic layer, *International Journal of Solids and Structures*, 44, pp. 7955–7974.

Ferry, J.D. (1980). *Viscoelastic Properties of Polymers*, 3rd edition, Wiley, New York.

Fomethe, A., Maugin, G.A. (1997). Propagation of phase-transition fronts and domain walls in thermoelastic ferromagnets, *International Journal of Applied Electromagnetics and Mechanics*, 8, pp. 143–165.

Frassine, R., Pavan, A. (1990). An application of viscoelastic fracture criteria to steady crack propagation in a polymeric material under fixed deformation, *International Journal of Fracture*, 43, pp. 303–317.

Freund, L.B. (1972a). Energy flux into the tip of an extending crack in an elastic solid, *Journal of Elasticity*, 2, pp. 341–349.

Freund, L.B. (1972b). Crack propagation in an elastic solid subjected to general loading. I: Constant rate of extension, *Journal of the Mechanics and Physics of Solids*, 20, pp. 120–140.

Freund, L.B. (1972c). Crack propagation in an elastic solid subjected to general loading. II: Nonuniform rate of extension, *Journal of the Mechanics and Physics of Solids*, 20, pp. 141–152.

Freund, L.B. (1973). Crack propagation in an elastic solid subjected to general loading. III: Stress wave loading, *Journal of the Mechanics and Physics of Solids*, 21, pp. 47–61.

Freund, L.B. (1974a). Crack propagation in an elastic solid subjected to general loading. IV: Obliquely incident stress pulse, *Journal of the Mechanics and Physics of Solids*, 22, pp. 137–146.

Freund, L.B. (1974b). The stress intensity factor due to normal impact loading of the faces of a crack, *International Journal of Engineering Science*, 12, pp. 179–189.

Freund, L.B. (1990). *Dynamic Fracture Mechanics*, Cambridge University Press, Cambridge.

Freund, L.B., Hutchinson, J.W. (1985). High strain-rate crack growth in rate-dependent plastic solids, *Journal of the Mechanics and Physics of Solids*, 33, pp. 169–191.

Fu, R., Zhang, T.-Y. (2000). Influence of temperature and electric field on the bending strength of lead zirconate titanate ceramics, *Acta Materialia.*, 48, pp. 1729–1740.

Fulton, C.C., Gao, H. (2001). Effect of local polarization switching on piezoelectric fracture, *Journal of the Mechanics and Physics of Solids*, 49, pp. 927–952.

Fung, Y.C., Tong, P. (2001). *Classical and Computational Solid Mechanics*, World Scientific, Singapore.

Gamby, D., Chaoufi J. (1999). Application of viscoelastic fracture criteria to progressive crack propagation in polymer matrix composites, *International Journal of Fracture*, 100, pp. 379–399.

Gamby, D., Delaumenie, V. (1993). On the application of viscoelastic fracture theories to the slow propagation of a crack in a carbon/epoxy composite, *International Journal of Fracture*, 64, R69–73.

Gao, C.-F., Mai, Y.-W., Wang, B.-L. (2008). Effects of magnetic fields on cracks in a soft ferromagnetic material, *Engineering Fracture Mechanics*, 75, pp. 4863–4875.

Gao, C.-F, Mai, Y.-W., Zhang, N. (2010a). Solution of a crack in an electrostrictive solid, *International Journal of Solids and Structures*, 47, pp. 444–453.

Gao, C.-F, Mai, Y.-W., Zhang, N. (2010b). Solution of collinear cracks in an electrostrictive solid, *Philosophical Magazine*, 90, pp. 1245–1262.

Gao, C.-F, Mai, Y.-W., Zhang, N. (2011). Corrigendum to "Solution of a crack in an electrostrictive solid" [*International Journal of Solids and Structures* 47 (2010) 444–453], *International Journal of Solids and Structures*, 48, pp. 1082–1083.

Gao, C.-F., Tong, P., Zhang, T.-Y. (2003). Interfacial crack problems in magneto-electroelastic solids, *International Journal of Engineering Science*, 41, pp. 2105–2121.

Gao, C.-F., Tong, P., Zhang, T.-Y. (2004). Fracture mechanics for a Mode III crack in a magnetoelectroelastic solid, *International Journal of Solids and Structures*, 41, pp. 6613–6629.

Gao, H. (1996). A theory of local limiting speed in dynamic fracture, *Journal of the Mechanics and Physics of Solids*, 44, pp. 1453–1474.

Gao, H., Ji, B. (2003). Modeling fracture in nanomaterials via a virtual internal bond method, *Engineering Fracture Mechanics*, 70, pp. 1777–1791.

Gao, H., Klein, P. (1998). Numerical simulation of crack growth in an isotropic solid with randomized internal cohesive bonds, *Journal of the Mechanics and Physics of Solids*, 46, pp. 187–218.

Gao, H., Zhang, T.-Y., Tong, P. (1997). Local and global energy release rate for an electrically yielded crack in a piezoelectric ceramic, *Journal of the Mechanics and Physics of Solids*, 45, pp. 491–510.

Geubelle, P.H. (1995). Finite deformation effects in homogeneous and interfacial fracture, *International Journal of Solids and Structures*, 32, pp. 1003–1016.

Geubelle, P.H., Knauss, W.G. (1995). Propagation of a crack in homogeneous and bimaterial sheets under general in-plane loading: nonlinear analysis, *Journal of Applied Mechanics*, 62, pp. 601–606.

Goodier, J.N., Field, F.A. (1963). Plastic energy dissipation in crack propagation, in *Fracture of Solids*, D.C. Drucker and J. J. Gliman, Eds., Wiley, New York, pp. 103–118.

Griffith, A.A. (1921). The phenomena of rupture and flow in solids, *Philosophical Transactions of the Royal Society of London*, A 221, pp. 163–198.

Griffith, A.A. (1924). The theory of rupture, in *Proceedings of the First International Conference of Applied Mechanics*, Delft, pp. 55–63.

Groot, S.R., Mazur, P. (1962). *Non-equilibrium Thermodynamics*, Interscience Publishers Inc., New York.

Gu, P., Dao, M., Asaro, R.J. (1999). A simplified method for calculating the crack-tip field of functionally graded materials using the domain integral, *Journal of Applied Mechanics*, 66, pp. 101–108.

Gulyaev, Y.V. (1969). Electroacoustic surface waves in solids, *Soviet Physics JETP Letters*, 9, pp. 37–38.

Gurney, C., Hunt, J. (1967). Quasi-static crack propagation, *Proceedings of the Royal Society of London*, A299, pp. 508–524.

Gurney, C. (1994). Continuum mechanics and thermodynamics in the theory of cracking, *Philosophic Magazine A*, 69, pp. 33–43.

Gurtin, M.E. (1979). On a path-independent integral for thermoelasticity, *International Journal of Fracture*, 15, pp. R169–170.

Gurtin, M.E. (2000). *Configurational Forces as Basic Concepts of Continuum Physics*, Springer-Verlag, New York.

Gurtin, M.E., Sternberg, E. (1962). On the linear theory of viscoelasticity, *Archive for Rational Mechanics and Analysis*, 11, pp. 291–356.

Hahn, T. (2005). *International Tables for Crystallography*, Volume A: Space Group Symmetry, 5th edition, Springer-Verlag, Berlin-New York.

Hao, T.H., Shen Z.Y. (1994). A new electric boundary condition of electric fracture mechanics and its applications, *Engineering Fracture Mechanics*, 47, pp. 793–802.

Harshe, G., Dougherty, J.P., Newnham, R.E. (1993). Theoretical modeling of 3–0/0–3 magnetoelectric composites, *International Journal of Applied Electromagnetics and Mechanics*, 4, pp. 161–171.

Haug, A., McMeeking, R. (2006). Cracks with surface charge in poled ferroelectrics, *European Journal of Mechanics A: Solids*, 25, pp. 24–41.

Hellen, T.K. (1975). On the method of virtual crack extensions, *International Journal for Numerical Methods in Engineering*, 9, pp. 187–207.

Hillerborg, A., Modeer, M., Peterson, P.E. (1976). Analysis of crack formation and crack growth in concrete by means of fracture mechanics and finite elements, *Cement and Concrete Research*, 6, pp. 773–782.

Hodge, I.M. (1995). Physical aging in polymer glasses, *Science*, 267, pp. 1945–1947.

Honein, T., Herrmann, G. (1997). Conservation laws in nonhomogeneous plane elastostatics, *Journal of the Mechanics and Physics of Solids*, 45, pp. 789–805.

Horstemeyer, M.F., Bammann, D.J. (2010). Historical review of internal state variable theory for inelasticity, *International Journal of Plasticity*, 26, pp. 1310–1334.

Hu, K., Li, G. (2005). Constant moving crack in a magnetoelectroelastic material under anti-plane shear loading, *International Journal of Solids and Structures*, 42, pp. 2823–2835.

Hu, K.Q., Kang, Y.L., Qin, Q.H. (2007). A moving crack in a rectangular magnetoelectroelastic body, *Engineering Fracture Mechanics*, 74, pp. 751–770.

Huang, J.H., Chiu, Y.H., Liu, H.K. (1998). Magneto-electro-elastic Eshelby tensors for a piezoelectric-piezomagnetic composite reinforced by ellipsoidal inclusions, *Journal of Applied Physics*, 83, pp. 5364–5370.

Hudnut, S., Almajid, A., Taya, M. (2000). Functionally gradient piezoelectric bimorph type actuator, in *Proceedings of SPIE*, 3992, pp. 376–386.

Hutchinson, J.W. (1968). Singular behavior at the end of a tensile crack tip in a hardening material, *Journal of the Mechanics and Physics of Solids*, 16, pp. 13–31.

Hwang, S.C., Lynch, C.S., McMeeking, R.M. (1995). Ferroelectric/ferroelastic interactions and a polarization switching model, *Acta Metallurgica et Materialia*, 43, pp. 2073–2084.

Ikeda, T (1990). *Fundamentals of Piezoelectricity*, Oxford Science Publications, New York.

Ing, Y.-S., Wang, M.-J. (2004a). Explicit transient solutions for a mode-III crack subjected to dynamic concentrated loading in a piezoelectric material, *International Journal of Solids and Structures*, 41, pp. 3849–3864.

Ing, Y.-S., Wang, M.-J. (2004b). Transient analysis of a mode-III crack propagating in a piezoelectric medium, *International Journal of Solids and Structures*, 41, pp. 6197–6214.

Inglis, C.E. (1913). Stresses in a plate due to the presence of cracks and sharp corners, *Transactions of the Institute of Naval Architects*, 55, pp. 219–241.

Irwin, G.R. (1948). Fracture dynamics, in *Fracturing of Metals*, American Society for Metals, Cleveland, OH, pp. 147–166.

Irwin, G.R. (1956). Onset of fast crack propagation in high strength steel and aluminum alloys, *Sagamore Research Conference Proceedings, Vol. 2*, pp. 289–305.

Irwin, G.R. (1957). Analysis of stresses and strains near the end of a crack traversing a plate, *Journal of Applied Mechanics* 24, pp. 361–364.

Irwin, G.R. (1958). Fracture, in *Handbuch de Physik, Vol. 6*, S. Flugge, Ed., Springer-Verlag, Berlin, pp. 551–590.

Irwin, G.R. (1961). Plastic zone near a crack and fracture toughness, *Sagamore Research Conference Proceedings*, 4, Syracuse University Research Institute, Syracuse, NY, pp. 63–78.

Ji, B., Gao, H. (2004). A study of fracture mechanisms in biological nano-composites via the virtual internal bond model, *Materials Science and Engineering*, A366, pp. 96–103.

Jiang, L.Z., Sun, C.T. (1999). Crack growth behavior in piezoceramics under cyclic loads, *Ferroelectrics*, 233, pp. 211–223.

Jiang, L.Z., Sun, C.T. (2001). Analysis of indentation cracking in piezoceramics, *International Journal of Solids and Structures*, 38, pp. 1903–1918.

Jih, C.J., Sun, C.T. (1990). Evaluation of a finite element based crack-closure method for calculating static and dynamic strain energy release rates, *Engineering Fracture Mechanics*, 37, pp. 313–322.

Jin, Z.-H. (2003). Fracture mechanics of functionally graded materials, in *Advances in Mechanics and Mathematics*, D.Y. Gao and R.W. Ogden, Eds., Kluwer Academic Publishers, Boston, pp. 1–108.

Jin, Z.-H., Sun, C.T. (2007). Integral representation of energy release rate in graded materials, *Journal of Applied Mechanics*, 74, pp. 1046–1048.

Kalyanam, S., Sun, C.T. (2009). Modeling the fracture behavior of piezoelectric materials using a gradual polarization switching model, *Mechanics of Materials*, 41, pp. 520–534.

Kanninen, M.F., Popelar, C.H. (1985). *Advanced Fracture Mechanics*, Oxford University Press, New York.

Kirchner, H.O.K., Alshits, V.I. (1996). Elastically anisotropic angularly inhomogeneous media II. The Green's function for piezoelectric, piezomagnetic and magnetoelectric media, *Philosophical Magazine*, 74, pp. 861–885.

Kishimoto, K., Aoki, S., Sakata, M. (1980). On the path independent integral- $\hat{J}$, *Engineering Fracture Mechanics,* 13, pp. 841–850.

Klein, P., Gao, H. (1998). Crack nucleation and growth as strain localization in a virtual-bond continuum, *Engineering Fracture Mechanics*, 61, pp. 21–48.

Knauss, W.G. (1970). Delayed failure – The Griffith problem for linearly viscoelastic materials, *International Journal of Fracture Mechanics*, 6, pp. 7–20.

Knauss, W.G. (1973). The mechanics of polymer fracture, *Applied Mechanics Review*, 26, pp. 1–17.

Knauss, W.G. (1974). On the steady propagation of a crack in a viscoelastic sheet: experiments and analysis, in *Deformation and Fracture of High Polymers*, H. H. Kausch, J. A. Hassell, and R. Jaffee, Eds., Plenum Press, New York, pp. 501–541.

Knowles, J.K., Sternberg, E. (1972). On a class of conservation laws in linearized and finite elastostatics, *Archive for Rational Mechanics and Analysis*, 44, pp. 187–211.

Koh, J.-H., Jeong, S.-J., Ha, M.-S., *et al.* (2006). Dynamic observation in piezoelectric aging behavior of Pb(MgNb)O3-Pb(ZrTi)O3 multilayer ceramic actuators, *Ferroelectrics*, 332, pp. 117–122.

Kotousov, A.G. (2002). On a thermo-mechanical effect and criterion of crack propagation, *International Journal of Fracture*, 114, pp. 349–358.

Krueger, R. (2004). Virtual crack closure technique: History, approach, and applications, *Applied Mechanics Reviews*, 57, pp. 109–143.

Kuang, Z.-B. (2008). Some variational principles in elastic dielectric and elastic magnetic materials, *European Journal of Mechanics A: Solids*, 27, pp. 504–514.

Kuna, M. (2010). Fracture mechanics of piezoelectric materials – Where are we right now?, *Engineering Fracture Mechanics*, 77, pp. 309–326.

Kwon, S.M. (2004). On the dynamic propagation of an anti-plane shear crack in a functionally graded piezoelectric strip, *Acta Mechanica*, 167, pp. 73–89.

Landis, C.M. (2004). Energetically consistent boundary conditions for electromechanical fracture, *International Journal of Solids and Structures*, 41, pp. 6291–6315.

Landau, L.D., Lifshitz, E.M. (1960). *Electrodynamics of Continuous Media*, Pergamon Press, Oxford.

Lax, M., Nelson, D. F. (1976). Maxwell equations in material form, *Physics Review B*, 13, pp. 1777–1784.

Lekhnitskii, S.G. (1950). *Theory of Elasticity of an Anisotropic Elastic Body*, Gostekhizdat, Moscow (in Russian).

Lemaitre, J. (1996). *A Course On Damage Mechanics, Second Revised and Enlarged Edition*, Springer-Verlag, Berlin/Heidelberg, New York.

Li, C., Weng, G.J. (2002a). Antiplane crack problem in functionally graded piezoelectric materials, *Journal of Applied Mechanics*, 69, pp. 481–488.

Li, C., Weng, G.J. (2002b). Yoffe-type moving crack in a functionally graded piezoelectric material, *Proceedings of the Royal Society of London*, A458, pp. 381–399.

Li, F.Z., Shih, C.F., Needleman, A. (1985). A comparison of methods for calculating energy release rates, *Engineering Fracture Mechanics*, 21, pp. 405–421.

Li, J.Y. (2000). Magnetoelectroelastic multi-inclusion and inhomogeneity problems and their applications in composite materials, *International Journal of Engineering Science*, 38, pp. 1993–2011.

Li, S., Mataga, P.A. (1996a). Dynamic crack propagation in piezoelectric materials-Part I. Electrode solution, *Journal of the Mechanics and Physics of Solids*, 44, pp. 1799–1830.

Li, S., Mataga, P.A. (1996b). Dynamic crack propagation in piezoelectric materials: Part II. Vacuum solution, *Journal of the Mechanics and Physics of Solids*, 44, pp. 1831–1866.

Li, S. (2003a). On global energy release rate of a permeable crack in a piezoelectric ceramic, *Journal of Applied Mechanics*, 70, pp. 246–252.

Li., S. (2003b). On saturation-strip model of a permeable crack in a piezoelectric ceramic, *Acta Mechanica*, 165, pp. 47–71.

Li, X.-F., Yang J. (2005). Electromagnetoelastic behavior induced by a crack under antiplane mechanical and inplane electric impacts, *International Journal of Fracture*, 132, pp. 49–64.

Lippman, H.G. (1881). Sur le principe de la conservation de l'electricite ou second principe de la theories des phenomenes electriques, *Comptes Rendus de l'Acadamie des Sciences Paris*, 29, pp. 1049.

Liu, J.X., Liu, X.L., Zhao, Y.B. (2001). Green's functions for anisotropic magnetoelectro-elastic solids with an elliptical cavity or a crack, *International Journal of Engineering Science*, 39, pp. 1405–1418.

Lothe, J., Barnett, D.M. (1976). Integral formalism for surface waves in piezoelectric crystals, *Journal of Applied Physics*, 47, pp. 1799–1807.

Lothe, J., Barnett, D.M. (1977). Further development of the theory for surface waves in piezoelectric crystals, *Physica Norvegica*, 8, pp. 239–254.

Lustig, S.R., Shay, R.M., Caruthers, J.M. (1996). Thermodynamic constitutive equations for materials with memory on a material time scale, *Journal of Rheology*, 40, pp. 69–106.

Lynch, C.S., Yang, W., Collier, L., Suo, Z., McMeeking, R.M. (1995). Electric field induced cracking in ferroelectric ceramics, *Ferroelectrics*, 166, pp. 11–30.

Ma, C.-C., Chen, S.-K. (1992). Investigation on the stress intensity factor field for unstable dynamic crack growth, *International Journal of Fracture*, 58, pp. 345–359.

Ma, L., Li, J., Abdelmoula, R., *et al.* (2007). Mode III crack problem in a functionally graded magneto-electro-elastic strip, *International Journal of Solids and Structures*, 44, pp. 5518–5537.

Mai, Y.-W., Cotterell, B. (1980). Effects of pre-strain on plane stress ductile fracture in α-brass, *Journal of Materials Science*, 15, pp. 2296–2306.

Mai, Y.-W., Cotterell, B. (1986). On the essential work of ductile fracture in polymers, *International Journal of Fracture*, 32, pp. 105–125.

Mai, Y.-W., Wong, S.-C., Chen, X.-H. (2000). Application of fracture mechanics for characterization of toughness of polymer blends, in *Polymer Blends*, D.R. Paul and C.B. Bucknall, Eds., Vol. 2, Chapter 20, John Wiley & Sons, New York, pp. 17–58.

Maignan, A., Sundaresan, A., Varadaraju, U. V., *et al.* (1998). Magnetization relaxation and aging in spin-glass $(La,Y)_{1-x}Ca_xMnO_3$ (x=0.25, 0.3 and 0.5) perovskite, *Journal of Magnetism and Magnetic Materials*, 184, pp. 83–88.

Makowski, J., Stumpf, H. (2001). Thermodynamically based concept for the modeling of continua with microstructure and evolving defects, *International Journal of Solids and Structures*, 38, pp. 1943–1961.

Maugin, G.A. (1988). *Continuum Mechanics of Electromagnetic Solids*, North-Holland, Amsterdam.

Maugin, G.A. (1992). *The Thermomechanics of Plasticity and Fracture*. Cambridge University Press, Cambridge.

Maugin, G.A. (1994). On the J-integral and energy-release rates in dynamic fracture, *Acta Mechanica*, 105, pp. 33–47.

Maugin, G.A., Berezovski, A. (1999). Material formulation of finite-strain thermoelasticity and applications, *Journal of Thermal Stresses*, 22, pp. 421–449.

Maugin, G.A., Epstein, M. (1991). The electroelastic energy-momentum tensor, *Proceedings of the Royal Society of London*, A433, pp. 299–312.

Maugin, G.A., Epstein, M., Trimarco, C. (1992). Pseudomomentum and material forces in inhomogeneous materials (Application to the fracture of electromagnetic materials in electromagnetoelastic fields), *International Journal of Solids and Structures*, 29, pp. 1889–1900.

Maugin, G.A., Trimarco, C. (1992). Pseudomomentum and material forces in nonlinear elasticity: variational formulations and application to brittle fracture, *Acta Mechanica*, 94, pp. 1–28.

Mazur, P., Nijboer, B.R.A. (1953). On the statistical mechanics of matter in electromagnetic fields, *Physica*, 19, pp. 971–986.

McCartney, L.N. (1977). Crack growth laws for a variety of viscoelastic solids using energy and COD fracture criteria, *International Journal of Fracture*, 15, pp. 31–40.

McCartney, L.N. (1980). Discussion: A rate-dependent criterion for crack growth, *International Journal of Fracture*, 16, R229–232.

McCartney, L.N. (1981). Response to discussion concerning kinetic criteria for crack in viscoelastic materials, *International Journal of Fracture*, 17, R161–168.

McMeeking, R.M. (1989). Electrostrictive stresses near crack like flaws, *Zeitschrift für angewandte Mathematik und Physik*, 40, pp. 615–627.

McMeeking, R.M. (1990). A J-integral for the analysis of electrically induced mechanical stress at cracks in elastic dielectrics, *International Journal of Engineering Science*, 28, pp. 605–613.

McMeeking, R.M. (1999). Crack tip energy release rate for a piezoelectric compact tension specimen, *Engineering Fracture Mechanics*, 64, pp. 217–244.

McMeeking, R.M. (2001). Towards a fracture mechanics for brittle piezoelectric and dielectric materials, *International Journal of Fracture*, 108, pp. 25–41.

McMeeking, R.M. (2004). The energy release rate for a Griffith crack in a piezoelectric material, *Engineering Fracture Mechanics*, 71, pp. 1149–1163.

McMeeking, R.M., Evans, A.G. (1982). Mechanics of transformation toughening in brittle materials, *Journal of American Ceramic Society*, 65, pp. 242–246.

McMeeking, R.M., Landis, C.M. (2005). Electrostatic forces and stored energy for deformable dielectric materials, *Journal of Applied Mechanics*, 72, pp. 581–590.

McMeeking, R.M., Landis, C.M., Jimenez, S.M.A. (2007). A principle of virtual work for combined electrostatic and mechanical loading of materials, *International Journal of Non-Linear Mechanics*, 42, pp. 831–838.

Meguid, S.A., Wang, X.D. (1998). Dynamic anti-plane behavior of interacting cracks in a piezoelectric material, *International Journal of Fracture*, 91, pp. 391–403.

Melkumyan, A. (2005). Comments on ''Dynamic crack propagation in piezoelectric materials—Part I. Electrode solution'' by Shaofan Li, Peter A. Mataga [*Journal of Mechanics and Physics of Solids* 44 (1996) 1799–1830]. *Journal of the Mechanics and Physics of Solids*, 53, pp. 1918–1925.

Moran, B., Shih, C.F. (1987a). Crack tip and associated domain integrals from momentum and energy balance, *Engineering Fracture Mechanics*, 27, pp. 615–642.

Moran, B., Shih, C.F. (1987b). A general treatment of crack tip contour integrals, *International Journal of Fracture*, 35, pp. 295–310.

Mott, N.F. (1947). Brittle fracture in mild steel plates, *Engineering*, 165, pp. 16–18.

Muller, I., Ruggeri, T. (1993). *Extended Thermodynamics*, Springer-Verlag, New York.

Nan, C.W. (1994). Magnetoelectric effect in composites of piezoelectric and piezomagnetic phases, *Physical Review B*, 50, pp. 6082–6088.

Narita, F., Shindo, Y. (1998). Dynamic anti-plane shear of a cracked piezoelectric ceramic, *Theoretical and Applied Fracture Mechanics*, 29, pp. 169–180.

Neelakanta, P.S. (1995). *Handbook of Electromagnetic Materials: Monolithic and Composite Versions and their Applications*, CRC Press, Boca Raton.

Nguyen, T.D., Govindjeeb, S., Kleinc, P.A., Gao, H. (2005). A material force method for inelastic fracture mechanics, *Journal of the Mechanics and Physics of Solids*, 53, pp. 91–121.

Nikitin, L.V. (1984). Application of the Griffith's approach to analysis of rupture in viscoelastic bodies, *International Journal of Fracture*, 24, pp. 149–157.

Niraula, O.P., Wang, B.L. (2006). A magneto-electro-elastic material with a penny-shaped crack subjected to temperature loading, *Acta Mechanica*, 187, pp. 151–168.

Noble, B. (1958). *Methods based on the Wiener–Hopf Technique*, Pergamon, New York.

Onat, E.T., Breuer, S. (1963). On uniqueness in linear viscoelasticity, in *Progress in Applied Mechanics*, The Prager Anniversary Volume, D.C. Drucker, Ed., Macmillan, New York, pp. 349–353.

Orowan, E., 1948. Fracture and strength of solids, *Reports on Progress in Physics, XII*, pp. 185–232.

Pak, Y.E., Hermann, G. (1986a). Conservation laws and the material momentum tensor for the elastic dielectric, *International Journal of Engineering Science*, 24, pp. 1365–1374.

Pak, Y.E., Hermann, G. (1986b). Crack extension force in a dielectric medium, *International Journal of Engineering Science*, 24, pp. 1375–1388.

Pak, Y.E. (1990). Crack extension force in a piezoelectric material, *Journal of Applied Mechanics*, 57, pp. 647–653.

Pak, Y.E. (1992). Linear electro-elastic fracture mechanics of piezoelectric materials, *International Journal of Fracture*, 54, pp. 79–100.

Pak, Y.E., Tobin, A. (1993). On electric field effects in fracture of piezoelectric materials, *Mechanics of Electromagnetic Materials and Structures AMD-*Vol.161/MD-Vol.42, ASME, pp. 51–62.

Pao, Y.H., Yeh, C.S. (1973). A linear theory for soft ferromagnetic elastic solids, *International Journal of Engineering Science*, 11, pp. 415–436.

Paris, P.C., Tada, H., Zahoor, A., Ernst, H. (1979). The theory of instability of the tearing mode of elastic-plastic crack growth, in *Elastic-Plastic Fracture, ASTM STP 668*, American Society for Testing and Materials, Philadelphia, pp. 5–36.

Park, S.B., Sun, C.T. (1995a). Effect of electric field on fracture of piezoelectric ceramics, *International Journal of Fracture*, 70, pp. 203–216.

Park, S.B., Sun, C.T. (1995b). Fracture criteria for piezoelectric ceramics, *Journal of American Ceramic Society*, 78, pp. 1475–1480.

Parks, D.M. (1974). A stiffness derivative finite element technique for determination of crack tip stress intensity factors, *International Journal of Fracture*, 10, pp. 487–502.

Parks, D.M. (1977). The virtual crack extension method for nonlinear material behavior, *Computer Methods in Applied Mechanics and Engineering*, 12, pp. 353–364.

Parton, V.Z. (1976). Fracture mechanics of piezoelectric materials, *Acta Astronautica*, 3, pp. 671–683.

Popelar, C.H., Atkinson, C. (1980). Dynamic crack propagation in a viscoelastic strip, *Journal of the Mechanics and Physics of Solids*, 28, pp. 79–93.

Qin, Q.-H. (2001). *Fracture Mechanics of Piezoelectric Materials*, WIT Press, Boston.

Qin, Q.-H., Mai, Y.-W. (1998). Thermoelectroelastic Green's function and its application for bimaterial of piezoelectric materials, *Archive of Applied Mechanics*, 68, pp. 433–444.

Rao, B.N., Kuna, M. (2008). Interaction integrals for fracture analysis of functionally graded piezoelectric materials, *International Journal of Solids and Structures*, 45, pp. 5237–5257.

Rao, B.N., Kuna, M. (2010). Interaction integrals for thermal fracture of functionally graded piezoelectric materials, *Engineering Fracture Mechanics*, 77, pp. 37–50.

Ravi-Chandar, K. (2004). *Dynamic Fracture*, Elsevier Ltd, Oxford.

Rice, J.R. (1968). A path independent integral and the approximate analysis of strain concentration by notches and cracks, *Journal of Applied Mechanics*, 35, pp. 379–386.

Rice, J.R. (1971). Inelastic constitutive relations for solids: an internal-variable theory and its application to metal plasticity, *Journal of the Mechanics and Physics of Solids*, 19, pp. 433–455.

Rice, J.R. (1978). Thermodynamics of quasi-static growth of Griffith cracks, *Journal of the Mechanics and Physics of Solids*, 26, pp. 61–78.

Rice, J.R., Paris, P.C., Merkle, J.G. (1973). Some further results of J-integral analysis and estimates, *Progress in Flaw Growth and Fracture Toughness Testing, ASTM STP 536*, American Society for Testing and Materials, Philadelphia, pp. 231–245.

Rice, J.R., Rosengren, G.F. (1968). Plane strain deformation near a crack tip in a power-law hardening material, *Journal of the Mechanics and Physics of Solids*, 16, pp. 1–12.

Ru, C.Q. (1999). Effect of electrical polarization saturation on stress intensity factors in a piezoelectric ceramic, *International Journal of Solids and Structures*, 36, pp. 869–883.

Ru, C.Q., Mao, X. (1999). Conducting cracks in a piezoelectric ceramic of limited electrical polarization, *Journal of the Mechanics and Physics of Solids*, 47, pp. 2125–2146.

Rybicki, E.F., Kanninen, M.F. (1977). A finite element calculation of stress intensity factors by a modified crack closure integral, *Engineering Fracture Mechanics*, 9, pp. 931–938.

Sabir, M., Maugin, G.A. (1996). On the fracture of paramagnets and soft ferromagnets, *International Journal of Non-Linear Mechanics*, 31, pp. 425–440.

Schapery, R.A. (1964). Application of thermodynamics to thermomechanical, fracture, and birefringent phenomena in viscoelastic media, *Journal of Applied Physics*, 35, pp. 1451–1465.

Schapery, R.A. (1969). On the characterization of nonlinear viscoelastic materials, *Polymer Engineering and Science*, 9, pp. 295–310.

Schapery, R.A. (1975a). A theory of crack-initiation and growth in viscoelastic media, Part I. Theoretical development, *International Journal of Fracture*, 11, pp. 141–159.

Schapery, R.A. (1975b). A theory of crack-initiation and growth in viscoelastic media, Part II. Approximate methods of analysis, *International Journal of Fracture*, 11, pp. 369–388.

Schapery, R.A. (1975c). A theory of crack-initiation and growth in viscoelastic media, Part III. Analysis of continuous growth, *International Journal of Fracture*, 11, pp. 549–562.

Schapery, R.A. (1978). A method for predicting crack growth in nonhomogeneous viscoelastic media, *International Journal of Fracture*, 14, pp. 293–309.

Schapery, R.A. (1984). Correspondence principles and a generalized J integral for large deformation and fracture analysis of viscoelastic media, *International Journal of Fracture*, 25, pp. 195–223.

Schapery, R.A. (1994). Nonlinear viscoelastic constitutive equations for composites based on work potentials, *Applied Mechanics Review*, 47, S269–S275.

Schapery, R.A. (1997). Nonlinear viscoelastic and viscoplastic constitutive equations based on thermodynamics, *Mechanics of Time-Dependent Materials*, 1, pp. 209–240.

Schapery, R.A. (1999). Nonlinear viscoelastic and viscoplastic constitutive equations with growing damage, *International Journal of Fracture*, 97, pp. 33–66.

Schapery, R.A. (2000). Nonlinear viscoelastic solids, *International Journal of Solids and Structures*, 37, pp. 359–366.

Schneider, G.A. (2007). Influence of electric field and mechanical stress on the fracture of ferroelectrics, *Annual Review of Materials Research*, 37, pp. 491–538.

Schuller, T., Lauke, B. (2006). Finite-element-simulation of interfacial crack propagation: cross-check of simulation results with the essential work of interfacial fracture method, in *Fracture of Materials: Moving Forwards*, H.-Y. Liu, X. Hu, and M. Hoffman, Eds., Trans Tech Publications, Dürnten, Switzerland, pp. 9–14.

Shih, C.F., Moran, B., Nakamura, T. (1986). Energy release rate along a three-dimensional crack front in a thermally stressed body, *International Journal of Fracture*, 30, pp. 79–102.

Shilo, D., Burcsu, E., Ravichandran, G., Bhattacharya, K. (2007). A model for large electrostrictive actuation in ferroelectric single crystals, *International Journal of Solids and Structures*, 44, pp. 2053–2065.

Shin, J.W., Kim, T.U., Kim, S.J., *et al.* (2009). Dynamic propagation of a finite eccentric crack in a functionally graded piezoelectric ceramic strip, *Journal of Mechanical Science and Technology*, 23, pp. 1–7.

Shin, J.W., Lee Y.-S. (2010). A moving interface crack between two dissimilar functionally graded piezoelectric layers under electromechanical loading, *International Journal of Solids and Structures*, 47, pp. 2706–2713.

Shindo, Y. (1977). The linear magnetoelastic problem for a soft ferromagnetic elastic solid with a finite crack, *Journal of Applied Mechanics*, 44, pp. 47–50.

Shindo, Y. (1978). Magnetoelastic interaction of a soft ferromagnetic elastic solid with a penny-shaped crack in a constant axial magnetic field, *Journal of Applied Mechanics*, 45, pp. 291–296.

Shindo, Y, Katsura, H., Yan, W. (1996). Dynamic stress intensity factor of a cracked dielectric medium in a uniform electric field, *Acta Mechanica*, 117, pp. 1–10.

Sih, G.C., Chen, E.P. (1977). Cracks moving at constant velocity and acceleration, in *Mechanics of Fracture* 4: *Elastodynamic Crack Problems*, G.C. Sih, Ed., Noordhoff, Amsterdam, pp. 59–117.

Sih, G.C., Zuo, J.Z. (2000). Multiscale behavior of crack initiation and growth in piezoelectric ceramics, *Theoretical and Applied Fracture Mechanics*, 34, pp. 123–141.

Sih, G.C., Jones, R., Song, Z.F. (2003). Piezomagnetic and piezoelectric poling effects on mode I and II crack initiation behavior of magnetoelectroelastic materials, *Theoretical and Applied Fracture Mechanics*, 40, pp. 161–186.

Simo, J.C., Honein, T. (1990). Variational formulation, discrete conservation laws, and path-domain independent integrals for elasto-viscoplasticity, *Journal of Applied Mechanics*, 57, pp. 488–497.

Singh, B.M., Rokne, J., Dhaliwal, R.S., *et al.* (2009). Scattering of anti-plane shear wave by an interface crack between two bonded dissimilar functionally graded piezoelectric materials, *Proceedings of the Royal Society of London*, A465, pp. 1249–1269.

Sladek, J., Sladek, V., Zhang, C., *et al.* (2007a). Evaluation of fracture parameters in continuously nonhomogeneous piezoelectric solids, *International Journal of Fracture*, 145, pp. 313–326.

Sladek, J., Sladek, V., Zhang, C. (2007b). A local integral equation method for dynamic analysis in functionally graded piezoelectric materials, in *Advances in Boundary Element Technique VIII*, V. Minutoto and M.H. Aliabadi, Eds., EC Ltd., United Kingdom, pp. 141–148.

Smith, T.E., Warren, W.E. (1966). Some problems in two-dimensional electrostriction, *Journal of Mathematical Physics*, 45, pp. 45–51.

Smith, T.E., Warren, W.E. (1968). Corrigenda to "Some problems in two-dimensional electrostriction", [*Journal of Mathematical Physics* 45 (1966) 45–51] *Journal of Mathematical Physics*, 47, pp. 109–110.

Soh, A.K., Lee, K.L., Fang, D.-N. (2003). On the effects of an electric field on the fracture toughness of poled piezoelectric ceramics, *Materials Science and Engineering A*, 360, pp. 306–314.

Song, Z.F., Sih, G.C. (2003). Crack initiation behavior in magnetoelectroelastic composite under in-plane deformation, *Theoretical and Applied Fracture Mechanics*, 39, pp. 189–207.

Sosa, H. (1991). Plane problems in piezoelectric media with defects, *International Journal of Solids and Structures*, 28, pp. 491–505.

Sosa, H. A., Pak, Y.E. (1990). Three-dimensional eigenfunction analysis of a crack in a piezoelectric material, *International Journal of Solids and Structures*, 26, pp. 1–15.

Steinmann, P. (2000). Application of material forces to hyperelastic fracture mechanics I. Continuum mechanics setting, *International Journal of Solids and Structures*, 37, pp. 7371–7391.

Steinmann, P., Ackermann, D., Barth, F. J. (2001). Application of material forces to hyperelastic fracture mechanics. II. Computational setting, *International Journal of Solids and Structures*, 38, pp. 5509–5526.

Stroh, A.N. (1958). Dislocations and cracks in anisotropic elasticity, *Philosophical Magazine*, 3, pp. 625–646.

Struik, L.C.E. (1978). *Physical Aging in Amorphous Polymers and Other Materials*, Elsevier Scientific Publishing Company, Amsterdam/Oxford/New York.

Stumpf, H., Le, K.C. (1990). Variational principles of nonlinear fracture mechanics, *Acta Mechanica*, 83, pp. 25–37.

Sun, C.T., Park, S.B. (1995). Determination of fracture toughness of piezoceramics under the influence of electric field using Vickers indentation, in *Proceedings of the 1995 North American Conference on Smart Structures and Materials*, 26 February–3 March, San Diego.

Suo, Z., Kuo, C.M., Barnett, D.M., Willis, J.R. (1992). Fracture mechanics for piezoelectric ceramics, *Journal of the Mechanics and Physics of Solid*, 40, pp. 739–765.

Suo, Z., Zhao, X., Greene, W. (2008). A nonlinear field theory of deformable dielectrics, *Journal of the Mechanics and Physics of Solids*, 56, pp. 467–486.

Suresh, S., Mortensen, A. (1998). *Fundamentals of Functionally Graded Materials*, Cambridge University Press, Cambridge.

Takagi, K., Li, J.F., Yokoyama, S., *et al.* (2002). Design and fabrication of functionally graded PZT/Pt piezoelectric bimorph actuator, *Science and Technology of Advanced Materials*, 3, pp. 217–224.

Takagi, K., Li, J.F., Yokoyama, S., *et al.* (2003). Fabrication and evaluation of PZT/Pt piezoelectric composites and functionally graded actuators, *Journal of the European Ceramic Society*, 10, pp. 1577–1583.

Tarantino, A.M. (2005). Crack propagation in finite elstodynamics, *Mathematics and Mechanics of Solids*, 10, pp. 577–601.

Ting, T.C.T. (1996). *Anisotropic Elasticity*, Oxford University Press, Oxford.

Ting, T.C.T. (2000). Recent development in anisotropic elasticity, *International Journal of Solids and Structures*, 37, pp. 401–409.

Tobin, A., Pak, Y.E. (1993). Effects of electric fields on fracture behavior of PZT ceramics, in *Smart Materials, SPIE 1916*, V.K. Varadan, Ed., pp. 78–86.

Trimarco, C. (2002). A Lagrangian approach to electromagnetic bodies, *Technische Mechanik*, 22, pp. 175–180.

Trimarco, C. (2009). On the Lagrangian electrostatics of elastic solids, *Acta Mechanica*, 204, pp. 193–201.

Trimarco, C., Maugin, G.A. (2001). Material mechanics of electromagnetic bodies, in *Configurational Mechanics of Materials, CISM courses and lectures*, 427, Springer-Verlag, New York, pp. 129–172.

Truesdell, C. (1984). *Rational Thermodynamics*, 2nd edition, Springer-Verlag, New York.

Truesdell, C., Noll, W. (2004). *The Non-Linear Field Theories of Mechanics*, 3rd edition, S.S. Antman, Ed., Springer-Verlag, Berlin/Heidelberg.

Tvergaard, V., Hutchinson, J.W. (1992). The relation between crack growth resistance and fracture process parameters in elastic-plastic solids, *Journal of the Mechanics and Physics of Solids*, 40, pp. 1377–1397.

Ueda, S. (2005). Impact response of functionally graded piezoelectric plate with a vertical crack, *Theoretical and Applied Fracture Mechanics*, 44, pp. 329–342.

Ueda, S. (2006). A finite crack in a semi-infinite strip of a graded piezoelectric material under electric loading, *European Journal of Mechanics A: Solids*, 25, pp. 250–259.

Ueda, S. (2007). Electromechanical impact of an impermeable parallel crack in a FGPM strip, *European Journal of Mechanics A: Solids*, 26, pp. 123–136.

Ueda, S. (2008). Transient thermoelectroelastic response of a functionally graded piezoelectric strip with a penny-shaped crack, *Engineering Fracture Mechanics*, 75, pp. 1204–1222.

Van Suchtelen, J. (1972). Product properties: a new application of composite materials, *Philips Research Reports*, 27, pp. 28–37.

Volokh, K.Y., Gao, H. (2005). On the modified virtual internal bond method, *Journal of Applied Mechanics*, 72, pp. 969–971.

Voigt, W. (1910). *Lehrbuch der Kristallphysik*, Teubner, Leipzig.

Vu, D.K., Steinmann, P. (2007). Nonlinear electro- and magneto-elastostatics: material and spatial settings, *International Journal of Solids and Structures*, 44, pp. 7891–7905.

Walsh, P., Li, W., Kalia, R.K., *et al.* (2001). Structural transformation, amorphization, and fracture in nanowires: a multimillion-atom molecular dynamics study, *Applied Physics Letters,* 78, pp. 3328–3330.

Wang, B.-L., Han, J.-C., Du, S.-Y. (2009). *Coupled Thermo-Electro-Magneto-Mechanical Cracking of Non-Homogeneous Media*, Nova Science Publishers, New York.

Wang, B.-L., Mai, Y.-W. (2003). Crack tip field in piezoelectric/piezomagnetic media, *European Journal of Mechanics A: Solids*, 22, pp. 591–602.

Wang, B.-L., Mai, Y.-W. (2005). A periodic array of cracks in functional graded materials subjected to thermo-mechanical loading, *International Journal of Engineering Science,* 43, pp. 432–446.

Wang, B.-L., Mai, Y.-W. (2006). A periodic array of cracks in functionally graded materials subjected to transient loading, *International Journal of Engineering Science*, 44, pp. 351–364.

Wang, B.-L., Mai, Y.-W. (2007a). Applicability of the crack-face electromagnetic boundary conditions for fracture of magnetoelectroelastic materials, *International Journal of Solids and Structures*, 44, pp. 387–398.

Wang, B.-L., Mai, Y.-W. (2007b). Investigation of a mode III crack in functionally graded magnetoelectroelastic materials, *Key Engineering Materials*, 348/349, pp. 713–716.

Wang, B.-L., Mai, Y.-W., Niraula, O.P. (2007c). Horizontal shear surface wave in magnetoelectroelastic materials, *Philosophical Magazine Letters*, 87, pp. 53–58.

Wang, J., Zhang, T.-Y. (2007). Phase field simulations of polarization switching-induced toughening in ferroelectric ceramics, *Acta Materialia*, 55, pp. 2465–2477.

Wang, X., Yu, S. (2000). Transient response of a crack in piezoelectric strip subjected to the mechanical and electrical impacts: mode-III problem, *International Journal of Solids and Structures*, 37, pp. 5795–5808.

Wang, X.D., Meguid, S.A. (2000). Modelling and analysis of the dynamic behaviour of piezoelectric materials containing interacting cracks, *Mechanics of Materials*, 32, pp. 723–737.

Wang, X.D., Jiang, L.Y. (2004). The nonlinear fracture behavior of an arbitrarily oriented dielectric crack in piezoelectric material, *Acta Mechanica*, 172, pp. 195–210.

Ward, I.M. (1983). *Mechanical Properties of Solid Polymers*, 2nd edition, Wiley, New York.

Weichert, R., Schonert, K. (1978). Heat generation at the tip of a moving crack, *Journal of the Mechanics and Physics of Solids*, 26, pp. 151–161.

Wells, A.A. (1961). Unstable crack propagation in metals: cleavage and fast fracture, *Proceedings of the crack propagation symposium*, 1, Paper 84, Cranfield.

Wells, A.A. (1963). Application of fracture mechanics at and beyond general yielding, *British Welding Journal*, 10, pp. 563–570.

Williams, J.G. (1973). *Stress Analysis of Polymers*, Halsted Press, New York.

Williams, J.G. (1987). *Fracture Mechanics of Polymers,* Ellis Horwood, Chichester.

Williams, M.L. (1957). On the stress distribution at the base of a stationary crack, *Journal of Applied Mechanics*, 24, pp. 109–114.

Willis, J.R. (1967). Crack propagation in viscoelastic media, *Journal of the Mechanics and Physics of Solids*, 15, pp. 229–240.

Wilson, W.K., Yu, L.-W. (1979). The use of the J-integral in thermal stress crack problems, *International Journal of Fracture*, 15, pp. 377–387.

Wu, C.M., Kahn, M., Moy, W. (1996). Piezoelectric ceramics with functional gradients: a new application in material design, *Journal of the American Ceramic Society*, 79, pp. 809–812.

Xu, X.L., Rajapakse, R.K.N.D (2001). On a plane crack in piezoelectric solids, *International Journal of Solids and Structures*, 38, pp. 7643–7658.

Xu, X.-P., Needleman, A. (1994). Numerical simulations of fast crack growth in brittle solids, *Journal of the Mechanics and Physics of Solids*, 42, pp. 1397–1434.

Yang, J. (2004). Effects of electromagnetic coupling on a moving crack in polarized ceramics, *International Journal of Fracture*, 126, pp. L83–L88.

Yang, W., Suo, Z. (1994). Cracking in ceramics actuators caused by electrostriction, *Journal of the Mechanics and Physics of Solids*, 42, pp. 649–663.

Yang, W., Zhu, T. (1998). Switch-toughening of ferroelectrics subjected to electric fields, *Journal of the Mechanics and Physics of Solids*, 46, pp. 291–311.

Yoffe, E.H. (1951). The moving Griffith crack, *Philosophical Magazine*, 42, pp. 739–750.

Zhang, Q.M., Zhao, J., Cross, L.E. (1996). Aging of the dielectric and piezoelectric properties of relaxor ferroelectric lead magnesium niobate-lead titanate in the electric field biased state, *Journal of Applied Physics*, 79, pp. 3181–3187.

Zhang, Q.M., Zhao, J., Shrout, T.R., *et al.* (1997). The effect of ferroelectric coupling in controlling the abnormal aging behavior in lead magnesium niobate-lead titanate relaxor ferroelectrics, *Journal of Materials Research*, 12, pp. 1777–1784.

Zhang, T.-Y., Tong, P. (1996). Fracture mechanics for a mode III crack in a piezoelectric material, *International Journal of Solids and Structures*, 33, pp. 343–359.

Zhang, T.-Y., Qian, C.F., Tong, P. (1998). Linear electro-elastic analysis of a cavity or a crack in a piezoelectric material, *International Journal of Solids and Structures*, 35, pp. 2121–2149.

Zhang, T.-Y., Zhao, M.H., Tong, P. (2002). Fracture of piezoelectric ceramics, in *Advances in Applied Mechanics*, 38, Academic Press Inc., San Diego, pp. 147–289.

Zhang, T.-Y., Gao, C.F. (2004). Fracture behaviors of piezoelectric materials, *Theoretical and Applied Fracture Mechanics*, 41, pp. 339–379.

Zhang, T.-Y., Zhao, M.H., Gao, C.F. (2005). The strip dielectric breakdown model, *International Journal of Fracture*, 132, pp. 311–327.

Zhang, X.P., Galea, S., Ye, L., Mai Y.-W. (2004). Characterisation of the effects of applied electric fields on fracture toughness and cyclic electric field induced fatigue crack growth for piezoceramic PIC 151, *Smart Materials and Structures*, 13, pp. N9–N16.

Zhao, M.-H., Fan, C.-Y. (2008). Strip electric-magnetic breakdown model in magnetoelectroelastic medium, *Journal of the Mechanics and Physics of Solids*, 56, pp. 3441–3458.

Zhong, X.-C., Li, X.-F. (2006). A finite length crack propagating along the interface of two dissimilar magnetoelectroelastic materials, *International Journal of Engineering Science*, 44, pp. 1394–1407.

Zhong, X.-C., Liu, F., Li, X.-F. (2009). Transient response of a magnetoelectroelastic solid with two collinear dielectric cracks under impacts, *International Journal of Solids and Structures*, 46, pp. 2950–2958.

Zhou, Z.-G., Wu, L.-Z., Wang, B. (2005). The behavior of a crack in functionally graded piezoelectric/piezomagnetic materials under anti-plane shear loading, *Archive of Applied Mechanics*, 74, pp. 526–535.

Zhu, T., Yang, W. (1997). Toughness variation of ferroelectrics by polarization switch under non-uniform electric field, *Acta Materialia*, 45, pp. 4695–4702.

Zhu, T., Yang, W. (1999). Fatigue crack growth in ferroelectrics driven by cyclic electric loading, *Journal of the Mechanics and Physics of Solids*, 47, pp. 81–97.

Zhu, X., Xu, J., Meng, Z., *et al.* (2000). Microdisplacement characteristics and microstructures of functionally gradient piezoelectric ceramic actuator, *Materials and Design*, 21, pp. 561–566.

Index